Wendell T. Hill and Chi H. Lee

Light-Matter Interaction

1807–2007 Knowledge for Generations

Each generation has its unique needs and aspirations. When Charles Wiley first opened his small printing shop in lower Manhattan in 1807, it was a generation of boundless potential searching for an identity. And we were there, helping to define a new American literary tradition. Over half a century later, in the midst of the Second Industrial Revolution, it was a generation focused on building the future. Once again, we were there, supplying the critical scientific, technical, and engineering knowledge that helped frame the world. Throughout the 20th Century, and into the new millennium, nations began to reach out beyond their own borders and a new international community was born. Wiley was there, expanding its operations around the world to enable a global exchange of ideas, opinions, and know-how.

For 200 years, Wiley has been an integral part of each generation's journey, enabling the flow of information and understanding necessary to meet their needs and fulfill their aspirations. Today, bold new technologies are changing the way we live and learn. Wiley will be there, providing you the must-have knowledge you need to imagine new worlds, new possibilities, and new opportunities.

Generations come and go, but you can always count on Wiley to provide you the knowledge you need, when and where you need it!

William J. Pesce
President and Chief Executive Officer

Peter Booth Wiley
Chairman of the Board

Wendell T. Hill and Chi H. Lee

Light-Matter Interaction

Atoms and Molecules in External Fields
and Nonlinear Optics

BICENTENNIAL 1807 WILEY 2007 BICENTENNIAL

WILEY-VCH Verlag GmbH & Co. KGaA

The Authors

Prof. Wendell T. Hill
University of Maryland
Institute for Physical Science and Technology
USA - College Park MD 20742
wth@glue.umd.edu

Prof. Chi H. Lee
University of Maryland
Electrical and Computer Engineering
USA - College Park Md 20742
chlee@glue.umd.edu

Cover
False color representation of the probability density (square of the wavefunction) for an electron in the $3d_{m_l=0}$ state of the hydrogen atom.

Library of Congress Card No.:
applied for

British Library Cataloguing-in-Publication Data
A catalogue record for this book is available from the British Library.

Bibliographic information published by the Deutsche Nationalbibliothek
Die Deutsche Nationalbibliothek lists this publication in the Deutsche Nationalbibliografie; detailed bibliographic data are available in the Internet at <http://dnb.d-nb.de>.

Typesetting Uwe Krieg, Berlin
Printing Strauss GmbH, Mörlenbach
Binding Litges & Dopf GmbH, Heppenheim

Printed in the Federal Republic of Germany
Printed on acid-free paper

ISBN: 978-3-527-40661-6

Contents

Light-Matter Interaction: Atoms and Molecules in External Fields and Nonlinear Optics.
W. T. Hill and C. H. Lee

ISBN: 978-3-527-40661-6

Preface

With several Nobel Prizes to its credit over the past decade, research in atomic, molecular and optical (AMO) science and engineering finds itself at the apex of investigations ranging from fundamental studies in quantum mechanics to the development of new nano-technologies. Whereas in years past, most researchers working in AMO fields had strong backgrounds in physics, optics and electromagnetic theory, this is not always the case today as the field and its participants become more and more diverse. Consequently, there is a need for a text designed to provide an overview of the field to get researchers new to AMO areas up and running as quickly as possible. This two-volume series entitled *Light-Matter Interactions* is designed to meet this need and to be user-friendly to investigators from physics, chemistry and the engineering disciplines. The purpose of Volume II is to provide a firm understanding of atoms and molecules, their interaction with external fields and how to exploit nonlinear optical processes to manipulate light. The text is geared toward the advanced undergraduate and the first- or second-year graduate student. While some familiarity with quantum mechanics and electromagnetic theory is assumed, readers with limited or no knowledge of atoms, molecules and nonlinear optics will find this text a good introduction. The more advanced readers should appreciate having available in one reference a vast array of ideas.

The number of subjects included in this volume is far too great for all to be covered in detail. As the details of each subject can be found readily in the literature, much of which we cite, it is not necessary or even prudent to write such a text. Our approach has been, rather, to give key results, for the most part. The derivations we have included were chosen to help illuminate the physics. To encourage the reader to delve deeper, a list of references and suggested extra reading is given at the end of each chapter. Again, our goal is to present enough information for readers from a variety of backgrounds to understand the important concepts.

The book is organized into three parts. Atoms and molecules are the subject of the first six chapters. The focus will be on the valence electrons as they

Light-Matter Interaction: Atoms and Molecules in External Fields and Nonlinear Optics.
W. T. Hill and C. H. Lee

ISBN: 978-3-527-40661-6

are primarily responsible for how the system interacts with its environment – chemical reactions, cooling and trapping, etc. Nonlinear optics will occupy the balance of the main portion of the text. Finally, to aid the reader's effort to become familiar with the numerous energy units, atomic parameters and special functions used in AMO research, we have compiled a set of appendices summarizing those most frequently used.

Our discussion of atoms and molecules will be confined to three specific types of systems – one-electron atoms, multi-electron atoms and diatomic molecules. In Chapter 1 we begin with hydrogen-like ions and look in some detail at their structure both nonrelativistically and relativistically; we introduce spin, spin-orbit interaction and the radiative hamiltonian all via the Dirac equation. We present the one-electron wavefunction, parity, the Rydberg constant and series as well as bound and continuum states in this chapter. Chapter 2 is devoted to multi-electron atoms where we classify states, discuss angular momentum coupling schemes and fine structure. Static field effects, Stark and Zeeman, are discussed in Chapter 3. Hyperfine structure as well as field ionization are also subjects covered. Radiative transitions in the presence of electromagnetic fields dominate the discussion in Chapter 4. Oscillator strengths, cross sections, line strengths, multiphoton transitions, selection rules (for electric and magnetic dipole along with electric quadrupole transitions) and strong-field ionization processes round out our discussion in this chapter.

We turn our attention to diatomic molecules in Chapters 5 and 6. Chapter 5 begins by presenting solutions to the separable nuclear and electronic equations. The consequences of this separation are discussed as well. Electronic states along with Hund's four primary cases are discussed in this chapter. In addition, molecular symmetries, molecular labeling conventions and molecular orbital theory are reviewed. Radiative transitions in diatomic systems is the subject of discussion in Chapter 6, which includes a summary of selection rules for electric dipole transitions for each Hund's case.

The second part of this volume deals with nonlinear optics. Ultrashort optical pulses and extremely high field strengths (intensities) are now routinely generated in contemporary laboratories. When exposed to such fields, condensed and gas phase systems respond nonlinearly. In particular, these fields induce a nonlinear polarization leading to a variety of phenomena, the most basic of which is optical second harmonic generation. From its inception in the early 1960s, nonlinear optics has blossomed into a research field of its own. As is the case with atoms and molecules, there are many excellent texts on nonlinear optics. We have made no attempt to duplicate these presentations. Rather, we seek to provide the reader with a straightforward and practical view of nonlinear optics. Although nonlinear interactions include higher order effects, involving interaction with lattice vibration, etc., the basic principle can be understood in terms of optical second harmonic generation. Consequently, the main focus of this section will be harmonic generation and parametric interaction.

Chapter 7 provides a phenomenological description of nonlinear optics. Nonlinear phenomena, including the electro-optic effect, sum- and difference-frequency generation, optical rectification, terahertz generation and parametric generation, are described in terms of the nonlinear susceptibility, with various combinations of frequencies. More complicated nonlinear effects can be described by a nonlinear polarization expressed as a product of higher order nonlinear susceptibilities and higher order products of optical fields. These give rise to the phenomena of four wave mixing, intensity dependent index changes. Temporal effects give rise to self-phase modulation, while the spatial effects cause self-focusing or spatial solitons. All of this is discussed in Chapter 7. Space requirements dictate omitting some subjects, however. We have chosen not to include important phenomena requiring more of a background in condensed matter than this volume provides.

Once the nonlinear polarization is obtained, it can be regarded as the driving source term for the nonlinear wave equation. This leads us to a general, albeit somewhat detailed, treatment of wave propagation in a nonlinear medium, the subject of Chapter 8. Since the source wave and the generated harmonic wave propagate at different speeds through the medium, concepts and practical methods for achieving phase-matching are discussed. To introduce parametric processes (oscillation and amplification), we extend our discussion of degenerate 3-wave interaction to three waves with different frequencies.

For those interested in learning how to calculate the nonlinear susceptibilities from a quantum picture, Chapter 9 provides an overview of the quantum theory of nonlinear processes. The density matrix formulism is introduced and combined with the perturbation method. This is followed by a treatment of multiphoton absorption. Calculations of linear and nonlinear susceptibilities, up to the third order, are then presented. Doppler free two-photon absorption is presented in this chapter as an example of using nonlinear effects to realize high resolution measurement in the presence of Doppler broadening.

Finally, a few additional applications of nonlinear optical effects are given in Chapter 10. Since many well-known applications are covered extensively in standard texts on nonlinear optics, we decided to cite a few applications that have received less attention. These include optical harmonics generation at oblique incidence angles where a nonlinear Brewster's angle is introduced, harmonic generation from surfaces, nonlinear electro-reflectance, near field second harmonic microscopy and terahertz pulse generation by optical rectification.

Five appendices appear at the end of the book covering atomic physics definitions, mathematics related to AMO calculations, atomic and molecular data, coupling angular momenta and tensor algebra. These appendices are intended as a reference source and include links to several web-based databases where the latest atomic and molecular information can be downloaded.

We wish to thank Professor Marshall Ginter for valuable discussions and helpful advice and Professor William DeGraffenreid for generating many of the figures. We acknowledge collaborations with our graduate students who helped generate some of the data presented in this text and the technical assistance of Mrs. Patsy Keehn and Mr. Alen Shu-Zee Lo who typed portions of the manuscript. Much of the material came from teaching graduate courses at the University of Maryland over the years. We thank the many graduate students for feedback that helped focus our thoughts. We trust the readers will find this volume useful and we welcome suggestions for improvement.

College Park, Maryland, summer 2006 *Wendell T. Hill, III and Chi Lee*

Part 1
Light-Matter Interaction: Atoms, Molecules and External Fields

Light-Matter Interaction: Atoms and Molecules in External Fields and Nonlinear Optics.
W. T. Hill and C. H. Lee

ISBN: 978-3-527-40661-6

1
Hydrogen-Like Ion: An Atom (Ion) With One Electron

The atom can be viewed in a variety of ways. From a macroscopic point of view, the atom is the smallest entity of significance and the building blocks of matter. From a microscopic point of view, the atom is a perplexing composite of more basic particles of which only the electron appears to be fundamental or elementary; the proton and neutron are composed of more bizarre constituents. Our focus will not be at the subatomic level where the strong and weak forces must be considered along with the electromagnetic force nor the macroscopic and mesoscopic levels involving clusters of particles. Rather, we will concentrate on the atomic level where the Coulomb interaction between the electrons and the nucleus as well as between electrons (and between different nuclear centers in molecules) is the dominant interparticle force of concern, at least in the absence of an applied strong field. We will assume the nucleus (composed of protons and neutrons) and the electron to be the basic constituents of concern. Our primary goal is to model atoms and molecules both in the absence and presence of weak and strong external electromagnetic fields. We will treat the motion of the electrons and nuclei quantum mechanically, for the most part, and the external field classically. Consequently, we will seek solutions to Schrödinger's equation,

$$i\hbar\frac{\partial}{\partial t}\Psi\left(\vec{r},t\right) = \widehat{H}\Psi\left(\vec{r},t\right), \tag{1.1}$$

for specific Hamiltonians, $\widehat{H}$.

We begin our discussion by considering the hydrogen-like, one-electron atom, in the absence of an external field. This atom will be described by a family of stationary states of well-defined energies, that can be ascertained from the time-independent Schrödinger equation

$$\widehat{H}\psi\left(\vec{r}\right) = E\psi\left(\vec{r}\right). \tag{1.2}$$

Equation (1.2) is obtained from Eq. (1.1) by letting $\Psi\left(\vec{r},t\right) \rightarrow e^{-iEt/\hbar}\psi\left(\vec{r}\right)$. The Hamiltonian is taken as the quantum mechanical analog of the classical Hamiltonian given by $T + V$, where $T = p^2/2m_e$ and the potential, $V\left(r\right)$, is just the Coulomb potential, $-Ze^2/4\pi\varepsilon_0 r$, with Z being the charge on the

Light-Matter Interaction: Atoms and Molecules in External Fields and Nonlinear Optics.
W. T. Hill and C. H. Lee

ISBN: 978-3-527-40661-6

nucleus ($Z = 1$ for neutral atoms). Since $p \rightarrow -i\hbar\nabla$ (see Appendix B.8), the quantum mechanical Hamiltonian for this single-particle nonrelativistic case takes the simple form

$$\widehat{H} = -\frac{\hbar^2}{2m_e}\nabla^2 - \frac{Ze^2}{4\pi\varepsilon_0 r}, \tag{1.3}$$

where m_e is the mass of the electron, e is the elementary charge (which we take to be positive in this book), $\hbar$ is Planck's constant divided by 2π and ε_0 is the vacuum permittivity. We will refer to this as our one-electron, field-free Hamiltonian.

1.1
Bohr Model of the Atom

Before solving Eq. (1.2) with the Hamiltonian given in Eq. (1.3), it is interesting to see that it is possible to determine the energy spectrum classically by assuming the existence of stationary states – Bohr's hypothesis that the electron does not radiate when circling the nucleus. For stationary states to exist, the Coulomb force ($-Ze^2/4\pi\varepsilon_0 r^2$) must equal the centripetal force ($-m_e v^2/r$, where v is the tangential velocity). This leads to a kinetic energy of

$$T = \frac{1}{2}m_e v^2 = \frac{1}{2}\left(\frac{Ze^2}{4\pi\varepsilon_0 r}\right) \tag{1.4a}$$

and a total energy of

$$E = T + V = -\frac{1}{2}\left(\frac{Ze^2}{4\pi\varepsilon_0 r}\right). \tag{1.4b}$$

Here, we adopt the usual convention that negative energy means that the system is bound. Prior to the development of quantum mechanics, it was known that atoms emitted light at specific wavelengths. In the late 19th century it was determined that the energy of the light, in wavenumbers (see Eq. (A.4) of Appendix A), associated with a transition between energy levels n and n' was found to obey[1]

$$\tilde{v}_{n'n} = C\left(\frac{1}{n^2} - \frac{1}{n'^2}\right), \tag{1.5}$$

where C is a constant that is related to the Rydberg constant. If one further postulates that the energy levels are quantized such that $r \rightarrow n^2 r_1$, where r_1 is

1) This general formula was proposed by Rydberg in 1889. Prior to this date, Balmer used the empirical formula $(4/B)\left(1/2^2 - 1/n^2\right)$ to fit the visible hydrogen spectral lines he observed. With $n = 3, 4, 5, 6, \ldots$, he extracted a value of 3645.6 Å for B.

the minimum stable radius, Eq. (1.4b) becomes

$$E_n = -\frac{1}{2}\left(\frac{Ze^2}{4\pi\varepsilon_o n^2 r_1}\right). \tag{1.6}$$

It follows that

$$hc\tilde{\nu}_{n'n} = E_{n'} - E_n = \frac{1}{2}\left(\frac{Ze^2}{4\pi\varepsilon_o r_1}\right)\left(\frac{1}{n^2} - \frac{1}{n'^2}\right). \tag{1.7}$$

If we identify the constant as

$$C = \frac{1}{2hc}\left(\frac{Ze^2}{4\pi\varepsilon_o r_1}\right), \tag{1.8}$$

the energy spectrum for hydrogen and hydrogen-like ions [2] takes this simple form:

$$E_n = -hcC\frac{1}{n^2}. \tag{1.9}$$

Since the potential is not harmonic, the energy levels are not equally spaced. Figure 1.1 shows that the second level, the first excited state, is three-quarters of the way to the ionization limit while subsequent levels become more closely spaced.

It is possible to determine C by fitting the spectrum to Eq. (1.5), from which we can estimate r_1. Furthermore, it is possible to predict C and r_1 theoretically from the correspondence principle.[3] When the radius of the electron orbit becomes large, n becomes large and the circulating electron will radiate, as any charged particle must classically, with a frequency given by

$$\nu = \frac{v}{2\pi n^2 r_1}. \tag{1.10}$$

When a classical particle radiates, it loses energy. Quantum mechanically, radiation is accompanied by transitions between quantum levels, as implied by Eq. (1.5). To remain classical, radiation must be between levels with large n. Consider a transition between two adjacent levels where $n' = n + 1$. Equation (1.5) then becomes

$$\tilde{\nu}_{n'n} = C\left(\frac{1}{n^2} - \frac{1}{(n+1)^2}\right) \simeq \frac{2C}{n^3}. \tag{1.11}$$

2) Hydrogen-like ions are ions with a single electron plus the nucleus. These include H, He^+, as well as all multiply-charged ions that have all but one electron removed; Ca^{19+} would be an example.

3) The correspondence principle states that in the limit of large sizes the quantum description must approach the classical description.

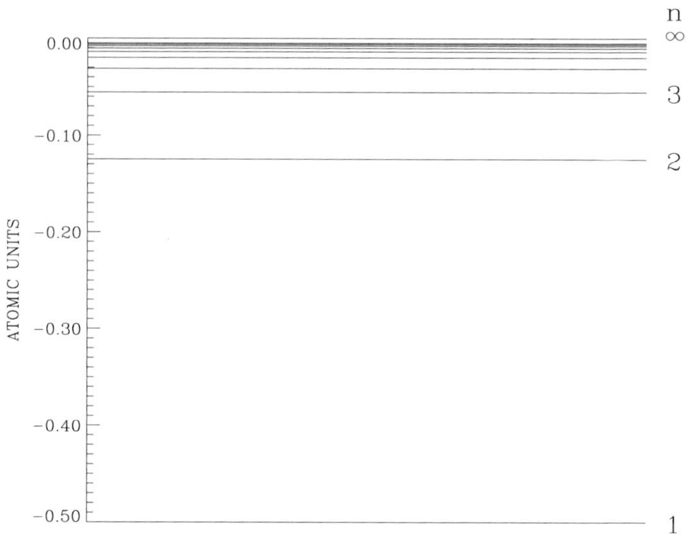

Fig. 1.1 Bound energy levels of the traditional Bohr hydrogen atom.

Multiplying Eq. (1.11) by c, the speed of light, and equating it with Eq. (1.10), using Eq. (1.4a) to express v and Eq. (1.8) to express $\mathcal{C}$, we can show that

$$r_1 \simeq \left(\frac{4\pi\varepsilon_0}{e^2}\right)\frac{\hbar^2}{Zm_e}, \tag{1.12}$$

from which we identify the minimum radius for $Z = 1$,

$$a_o \equiv \left(\frac{4\pi\varepsilon_0}{e^2}\right)\frac{\hbar^2}{m_e}, \tag{1.13}$$

called the Bohr radius.[4] The Rydberg constant, for an infinitely massive nucleus,[5] is then defined as

$$\mathcal{R}_\infty \equiv \frac{1}{4\pi}\left(\frac{e^2}{4\pi\varepsilon_0}\right)\frac{1}{\hbar c a_o} = \frac{\mathcal{C}}{Z^2}. \tag{1.14}$$

The energy spectrum can then be expressed as

$$E_n = -\frac{1}{2}\left(\frac{e^2}{4\pi\varepsilon_0}\right)\frac{Z^2}{n^2 a_o} = -hc\mathcal{R}_\infty\frac{Z^2}{n^2} = -\frac{1}{2}\left(\frac{e^2}{4\pi\varepsilon_0}\right)^2\frac{Z^2 m_e}{n^2\hbar^2}, \tag{1.15}$$

which depends only on the integers, Z and n, and fundamental constants.

4) We note that $r_1 = a_o/Z$.

5) An atom with an infinite nuclear mass is sometimes called a *fixed nucleus* or *stationary nucleus* atom.

For an infinite nuclear mass, we identify $n^2 a_0$ as the radius of the orbit of the nth state for hydrogen. This is not exact because the nucleus has finite mass m_N. To account for this, we can let $m_e \to \mu$, the reduced mass. The nth radius then becomes

$$r_n = \frac{m_e}{\mu}\frac{n^2 a_0}{Z} \simeq \frac{n^2 a_0}{Z}, \tag{1.16}$$

assuming that the nucleus has charge Z, where the reduced mass is given by

$$\mu \equiv \left(\frac{1}{m_e} + \frac{1}{m_N}\right)^{-1}. \tag{1.17}$$

If we let $m_e \to \mu$ in Eq. (1.15), however, we shift the energy levels, requiring the *finite-mass* Rydberg constant to be defined as

$$\mathcal{R}_M = \mathcal{R}_\infty \left(1 + \frac{m_e}{M_N}\right)^{-1}. \tag{1.18}$$

1.2 Hydrogen-Like Ions, Quantum Approach: Bound States

Although we were able to determine the spectrum for the simple hydrogen-like ions from our classical approach, to study any of the details of the structure (energy levels) requires the time-independent Schrödinger equation (Eq. (1.2)) to be solved. In this section we will look at exact solutions in the nonrelativistic approximation for hydrogen-like ions.

1.2.1 Angular Wavefunctions

When we generalize the field-free Hamiltonian (Eq. (1.3)) by letting $m_e \to \mu$, the reduced mass (Eq. (1.17)), the Schrödinger equation takes the form

$$\nabla^2 \psi(\vec{r}) + \frac{2\mu}{\hbar^2}\left(E + \frac{Ze^2}{4\pi\varepsilon_0 r}\right)\psi(\vec{r}) = 0. \tag{1.19}$$

Again, Z is the residual charge on the nucleus. That is, $Z = 1$ for H, $Z = 2$ for He^+, etc. The Coulomb potential only depends on $|\vec{r}|$ and, thus, falls into the class of centrally symmetric potentials. It is natural to seek solutions to Eq. (1.19) in spherical coordinates in this case. If we write the Laplacian in spherical coordinates,

$$\nabla^2 \to \frac{1}{r^2}\frac{\partial}{\partial r}\left(r^2 \frac{\partial}{\partial r}\right) + \frac{1}{r^2 \sin\vartheta}\frac{\partial}{\partial \vartheta}\left(\sin\vartheta \frac{\partial}{\partial \vartheta}\right) + \frac{1}{r^2 \sin^2\vartheta}\frac{\partial^2}{\partial \varphi^2}, \tag{1.20}$$

the radial and angular variables separate and the solutions to Eq. (1.19) take the form

$$\psi(\vec{r}) = R_{nl}(r) Y_{lm}(\vartheta, \varphi), \tag{1.21}$$

where $R_{nl}(r)$ is known as the radial wavefunction and $Y_{lm}(\vartheta, \varphi)$ the spherical harmonics (see Appendix B.7). Such wavefunctions describe stationary states with definite energy E and angular momentum l. For centrally symmetric potentials, the angular momentum is conserved.[6] The $Y_{lm}(\vartheta, \varphi)$s describe the angular part of the electron distribution and are given by

$$Y_{lm}(\vartheta, \varphi) = (-1)^m \sqrt{\frac{(2l+1)}{4\pi} \frac{(l-m)!}{(l+m)!}} P_l^m(\cos\vartheta)\, \mathrm{e}^{im\varphi}, \tag{1.22}$$

with $P_l^m(\cos\theta)$ being the associated Legendre polynomial (see Appendix B.6) given by

$$P_l^m(x) = \frac{1}{2^l l!} \left(1 - x^2\right)^{m/2} \frac{d^{l+m}}{dx^{l+m}} \left(x^2 - 1\right)^l, \tag{1.23}$$

where $m = -l, -(l-1), \ldots, 0, \ldots, l-1, l$.[7] The spherical harmonics are orthogonal, normalized and obey

$$\int_0^{2\pi} d\varphi \int_0^{\pi} Y^*_{l'm'}(\vartheta, \varphi) Y_{lm}(\vartheta, \varphi) \sin\vartheta d\vartheta = \delta_{l'l} \delta_{m'm}, \tag{1.24}$$

where δ_{ij} is the Kronecker delta (see Appendix B.1). The first six spherical harmonics are

$$\left.\begin{aligned}
Y_{00}(\vartheta, \varphi) &= \sqrt{1/4\pi} \\
Y_{10}(\vartheta, \varphi) &= \sqrt{3/4\pi} \cos\vartheta \\
Y_{1\pm1}(\vartheta, \varphi) &= \mp\sqrt{3/8\pi} \sin\vartheta \mathrm{e}^{\pm i\varphi} \\
Y_{20}(\vartheta, \varphi) &= \sqrt{5/4\pi} \left(\tfrac{3}{2}\cos^2\vartheta - \tfrac{1}{2}\right) \\
Y_{2\pm1}(\vartheta, \varphi) &= \mp\sqrt{15/8\pi} \sin\vartheta \cos\vartheta \mathrm{e}^{\pm i\varphi} \\
Y_{2\pm2}(\vartheta, \varphi) &= \sqrt{15/32\pi} \sin^2\vartheta \mathrm{e}^{\pm 2i\varphi}
\end{aligned}\right\} . \tag{1.25}$$

Figure 1.2 shows spherical polar representations of the first few Y_{lm}s.

6) We will find that when the spherical symmetry is broken, there will be a preferred direction in space and l will not be conserved in gen-

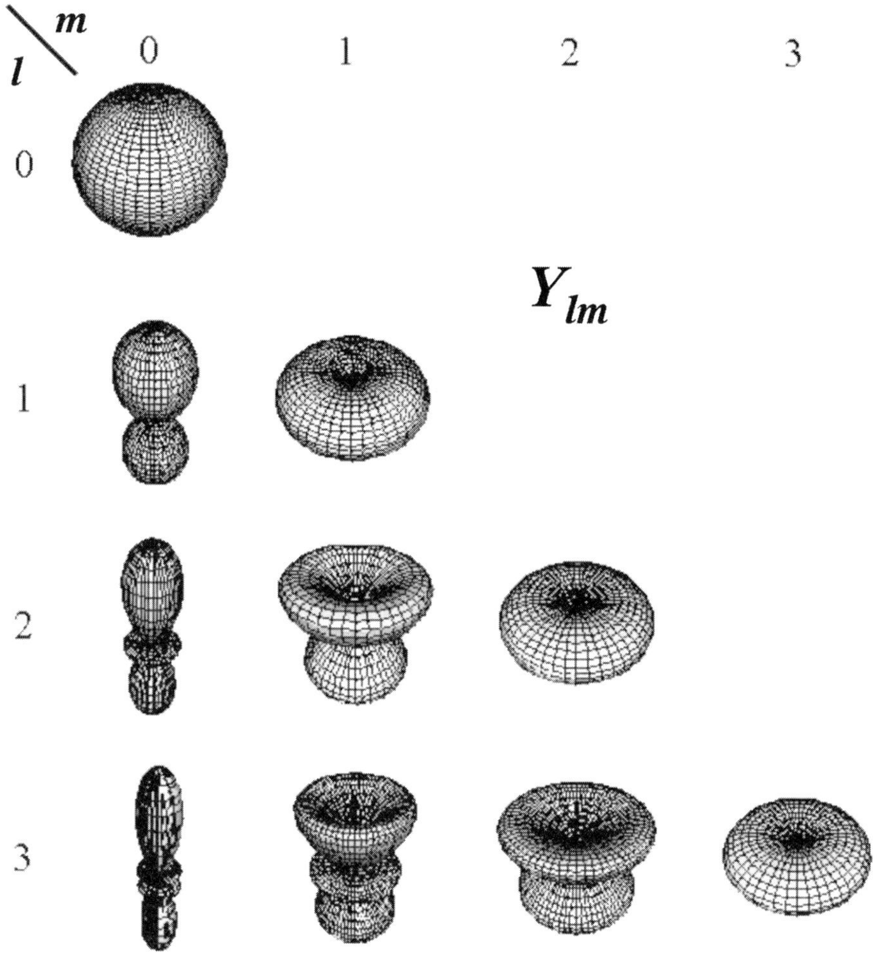

Fig. 1.2 Spherical polar plots of the spherical harmonics, $|Y_{lm}(\vartheta, \varphi)|^2$, for $l = 0, \ldots, 3$ and $m = -l, \ldots, l$. The z-axis is vertical in these 3D images. Note that only Y_{00} has a nonzero contribution at the origin. The Y_{l0} and Y_{ll} harmonics are oriented along and perpendicular to the z-axis, respectively.

The radial wave function is a solution to the equation

$$\left\{\frac{d^2}{dr^2} + \frac{2}{r}\frac{d}{dr} + \frac{2\mu}{\hbar^2}\left[E - V_{eff}(r;l)\right]\right\} R_{nl}(r) = 0, \tag{1.26}$$

eral. This occurs in diatomic molecules, for example, or when an atom is placed in an external field. However, in both cases the projection of l along the axis of symmetry often will remain conserved.

7) The phases for the spherical harmonics are consistent with *Mathematical Methods for Physicists* by Arfken and Weber [1], *The Theory of Atomic Structure and Spectra* by Cowan [2] and MATHEMATICA®.

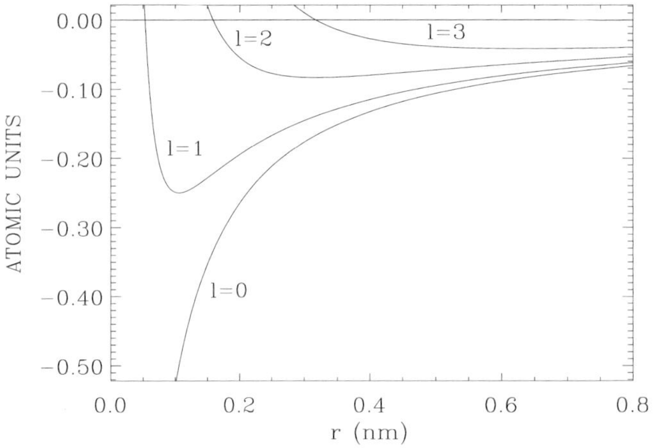

Fig. 1.3 Effective single-electron potential for $l = 0, 1, 2, 3$.

where

$$V_{eff}(r;l) = -\frac{Ze^2}{4\pi\varepsilon_0 r} + \frac{l(l+1)}{r^2}\frac{\hbar^2}{2\mu} \tag{1.27}$$

is the effective potential. Figure 1.3 shows that the potentials for $l \neq 0$ become repulsive at the origin. As we will see below, this causes the wavefunction for states with $l \neq 0$ to vanish at the origin. For $l = 0$ the wavefunctions do not have to vanish at the origin.

1.2.2
Radial Wavefunction and Energy States

It is straightforward to show that the radial wavefunction leads to the same energy spectrum predicted by the Bohr model.[8] To see this, consider the asymptotic behavior of Eq. (1.26). As $r \to \infty$, $V_{eff} \to 0$ and Eq. (1.26) reduces to

$$\left[\frac{d^2}{dr^2} + \frac{2\mu}{\hbar^2}E\right] R(r) = 0 \tag{1.28}$$

and has solutions of the form

$$R(r) = e^{\pm ir\sqrt{2\mu E/\hbar^2}}. \tag{1.29}$$

For positive energy, $E > 0$, the exponent is imaginary and the electron is described by unbounded plane waves. These correspond to continuum states.

8) Our treatment follows that found in Bethe and Salpeter, *Quantum Mechanics of One- and Two-Electron Atoms* [3] and Messiah, *Quantum Mechanics* [4].

For bound states we want $R \to 0$ as $r \to \infty$. This requires $E < 0$ and the minus sign in the exponent. The solution to Eq. (1.26) must then take the form

$$R(r) = f(r)\, \mathrm{e}^{-r\sqrt{-2\mu E/\hbar^2}}, \tag{1.30}$$

where $f(r)$ is a function that must be finite at the origin and blows up slower than $\exp\left[r\sqrt{-2\mu E/\hbar^2}\right]$ as $r \to \infty$. The minus sign was inserted under the radical to take care of the fact that $E < 0$. Substituting this expression for $R(r)$ into Eq. (1.26) produces

$$f''(r) + 2\left[\frac{1}{r} - \sqrt{-\frac{2\mu}{\hbar^2}E}\right] f'(r) + \left[\frac{2}{r}\left(\frac{Ze^2}{4\pi\varepsilon_0}\frac{\mu}{\hbar^2} - \sqrt{-\frac{2\mu}{\hbar^2}E}\right) - \frac{l(l+1)}{r^2}\right] f(r) = 0. \tag{1.31}$$

The general solution for $f(r)$ will be polynomials,

$$f(r) = r^{\beta} \sum_{\nu=0}^{\infty} b_{\nu} r^{\nu}. \tag{1.32}$$

Equation (1.32) can be written in the form of Laplace's equation with a regular solution of confluent hypergeometric series that will be introduced in the next section. Substituting this back into Eq. (1.31) gives

$$\sum_{\nu=0}^{\infty} b_{\nu}\left\{\left[(\beta+\nu)(\beta+\nu+1) - l(l+1)\right] r^{\beta+\nu-2} - \frac{2}{\hbar^2}\left[\hbar\sqrt{-2\mu E}\,(\beta+\nu+1) - \frac{Ze^2}{4\pi\varepsilon_0}\mu\right] r^{\beta+\nu-1}\right\} = 0. \tag{1.33}$$

In order for Eq. (1.33) to hold for all r, the coefficient of each term must vanish. For the lowest-order term, the $r^{\beta-2}$ term, we must demand that

$$\beta(\beta+1) = l(l+1) \to \beta = \begin{cases} l \\ -(l+1) \end{cases} \tag{1.34}$$

if b_0 is nonzero. In order that $R(r)$ be finite as $r \to 0$, we must choose $\beta = l$. Setting the coefficient of the second term in Eq. (1.33), the $r^{\beta-1}$ term, to zero with $\beta = l$ yields

$$b_1 = 2b_0 \frac{\sqrt{-2\mu E/\hbar^2}\,(l+1) - \left(Ze^2/4\pi\varepsilon_0\right)\mu/\hbar^2}{(l+1)(l+2) - l(l+1)},$$

from which the recursion relation,

$$b_\nu = 2b_{\nu-1}\frac{\sqrt{-2\mu E/\hbar^2}\,(l+\nu) - \left(Ze^2/4\pi\varepsilon_0\right)\mu/\hbar^2}{(l+\nu)\,(l+\nu+1) - l\,(l+1)}, \tag{1.35}$$

follows. It is convenient to write b_ν as

$$b_\nu = b_{\nu-1}\frac{2\zeta}{\nu}\frac{\nu+l-\lambda}{\nu+2l+1}, \tag{1.36}$$

where

$$\begin{aligned} \zeta &= \sqrt{\frac{-2\mu E}{\hbar^2}} \text{ and} \\ \lambda &= \frac{Ze^2/4\pi\varepsilon_0}{\sqrt{-2\mu E/\hbar^2}}\frac{\mu}{\hbar^2}. \end{aligned} \tag{1.37}$$

Using Eq. (1.36), we can write b_ν in terms of b_0 as

$$b_\nu = \left(\frac{2\zeta}{\nu}\frac{\nu+l-\lambda}{\nu+2l+1}\right)\cdot\left(b_{\nu-2}\frac{2\zeta}{\nu-1}\frac{\nu-1+l-\lambda}{\nu-1+2l+1}\right) \tag{1.38}$$

$$= \left(\frac{2\zeta}{\nu}\frac{\nu+l-\lambda}{\nu+2l+1}\right)\cdot\left(\frac{2\zeta}{\nu-1}\frac{\nu-1+l-\lambda}{\nu-1+2l+1}\right)\cdots\left(b_0\frac{2\zeta}{1}\frac{1+l-\lambda}{1+2l+1}\right) \tag{1.39}$$

$$= b_0\frac{(2\zeta)^\nu}{\nu!}\frac{(\nu+l-\lambda)!(2l+1)!}{(l-\lambda)!(\nu+2l+1)!}, \tag{1.40}$$

and $f(r)$ becomes

$$f(r) = r^l b_0 \sum_{\nu=0}^{\infty}\frac{(2r\zeta)^\nu}{\nu!}\frac{(\nu+l-\lambda)!(2l+1)!}{(l-\lambda)!(\nu+2l+1)!}. \tag{1.41}$$

We need to consider again the behavior as $r \to \infty$. This can be done by looking at the asymptotic form of $f(r)$ as ν becomes large. From Eq. (1.41) we write

$$f(r) \overset{\nu\to\infty}{\to} r^l b_0 \sum_{\nu=\text{large}}^{\infty}\frac{(2r\zeta)^\nu}{\nu!} = b_0 r^l \mathrm{e}^{2r\sqrt{-2\mu E/\hbar^2}},$$

from which it follows that

$$R\,(r) = \mathrm{e}^{-r\sqrt{-2\mu E/\hbar^2}} f\,(r) \to r^l \mathrm{e}^{r\sqrt{-2\mu E/\hbar^2}}.$$

This exponential divergence at large r can be avoided if we truncate the power series with a judicious choice for the value of λ, namely an integer. The integer must be chosen to make $b_{\nu=n-l} = 0$. Thus, $\lambda = l + 1 + k$ where $k = 0, 1, 2, 3, 4$ and we find that

$$E_{lk} = -\frac{1}{2}\left(\frac{e^2}{4\pi\varepsilon_0}\right)^2 \frac{Z^2\mu}{(l+1+k)^2\hbar^2}, \tag{1.42}$$

from Eq. (1.35), which we see is the energy spectrum for bound states. This energy spectrum is identical to Eq. (1.15) when we recognize $n \equiv \lambda = l+1+k$, the principal quantum number, and we let $m_e \to \mu$.

1.2.3
Exact Radial Solution, Hydrogen-Like Ions

The exact solutions to the radial equation, Eq. (1.26), can be determined from Eq. (1.41) by recognizing that

$$\sum_{\nu=0}^{\infty} \frac{(2r\zeta)^\nu}{\nu!} \frac{(\nu+l-\lambda)!(2l+1)!}{(l-\lambda)!(\nu+2l+1)!} = F(l+1-\lambda; 2l+2; 2r\zeta), \tag{1.43}$$

where $F(l+1-\lambda; 2l+2; 2r\zeta)$ is the confluent hypergeometric function (see Appendix B.4). Truncating the series at $n-l-1$ as discussed in the previous section leads to the normalized wavefunction expressed as

$$R_{nl}(r) = \frac{1}{(2l+1)!}\sqrt{\frac{(n+l)!}{2n(n-l-1)!}}\left(\frac{2Z}{na_o}\right)^{3/2} e^{-\frac{Zr}{na_o}} \left(\frac{2Zr}{na_o}\right)^l \times F\left(l+1-n, 2l+2; \frac{2Zr}{na_o}\right), \tag{1.44a}$$

which can also be expressed as an associated Laguerre polynomial (see Appendix B.5)

$$R_{nl}(r) = -\sqrt{\frac{(n-l-1)!}{2n[(n+l)!]^3}}\left(\frac{2Z}{na_o}\right)^{3/2} e^{-\frac{Zr}{na_o}} \left(\frac{2Zr}{na_o}\right)^l \times L_{n+l}^{2l+1}\left(\frac{2Zr}{na_o}\right). \tag{1.44b}$$

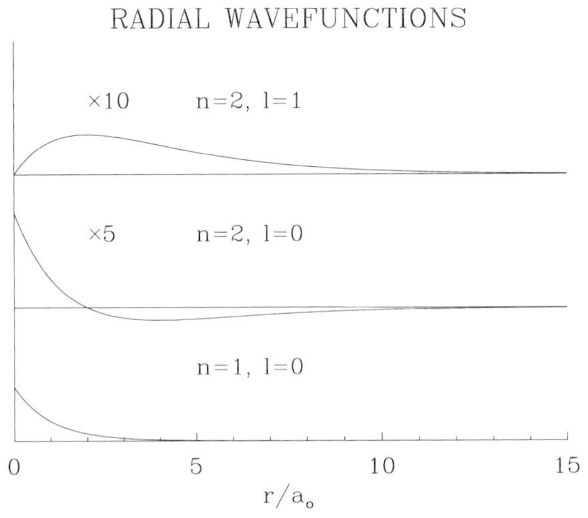

Fig. 1.4 First three radial wavefunctions, R_{nl}, for hydrogen-like ions.

Explicit expressions for the first four functions are

$$\left.\begin{aligned} R_{10}(r) &= 2\left(\frac{Z}{a_o}\right)^{3/2} \mathrm{e}^{-Zr/a_o} \\ R_{20}(r) &= 2\left(\frac{Z}{2a_o}\right)^{3/2}\left(1-\frac{Zr}{2a_o}\right) \mathrm{e}^{-Zr/2a_o} \\ R_{21}(r) &= \frac{2}{\sqrt{3}}\left(\frac{Z}{2a_o}\right)^{3/2}\left(\frac{Zr}{2a_o}\right) \mathrm{e}^{-Zr/2a_o} \\ R_{30}(r) &= 2\left(\frac{Z}{3a_o}\right)^{3/2}\left[1-2\left(\frac{Zr}{3a_o}\right)+\frac{2}{3}\left(\frac{Zr}{3a_o}\right)^2\right] \mathrm{e}^{-Zr/3a_o} \end{aligned}\right\} . \tag{1.45}$$

Additional radial functions can be generated from one of the two expressions using a symbolic manipulation program.

The radial wavefunctions are normalized and orthogonal and thus satisfy

$$\int R_{nl}(r)\, R_{n'l}(r)\, r^2 dr = \delta_{n'n}, \tag{1.46}$$

where $\delta_{n'n}$ is the Kronecker delta (see Appendix B). The first three solutions are shown graphically in Fig. 1.4. We note that (1) R_{nl} is finite at $r = 0$ for $l = 0$, while $R_{nl} \to 0$ as $r \to 0$ for $l \neq 0$, and (2) the number of times the wavefunction crosses the axis (number of zeros) is given by $n - l - 1$.

The probability density, $r^2 R_{nl}^2$,[9] *for finding the electron at radius r* in an effective potential (given by Eq. (1.27)) is displayed in Fig. 1.5 for a few wavefunctions. It is clear from this figure that we are more likely to find the electron

9) We will discuss this probability density at the end of this chapter.

at larger r as the principal quantum number increases. Specifically, the mean value of r is given by $\langle r\rangle = (a_o/2Z)\left[3n^2 - l\,(l+1)\right]$. This and other moments of r can be determined from

$$\langle r^\gamma\rangle = \int r^{2+\gamma} R_{nl}^2\,(r)\,dr. \tag{1.47}$$

A few common moments are[10]

$$\left.\begin{aligned}
\langle r^2\rangle &= \tfrac{a_o^2 n^2}{2Z^2}\left[5n^2 + 1 - 3l\,(l+1)\right]\\
\langle r^3\rangle &= \tfrac{a_o^3 n^2}{8Z^3}\left[35n^2\left(n^2-1\right) - 30n^2\,(l+2)\,(l-1)\right.\\
&\qquad \left. + 3\,(l+2)\,(l+1)\,(l-1)\right]\\
\langle r^4\rangle &= \tfrac{a_o^4 n^4}{8Z^4}\left[63n^4 - 35n^2\left(2l^2+2l-3\right) + 5l\,(l+1)\right.\\
&\qquad \left. \times\left(3l^2+3l-10\right) + 12\right]\\
\langle r^{-1}\rangle &= \tfrac{Z}{a_o n^2}\\
\langle r^{-2}\rangle &= \tfrac{Z^2}{a_o^2 n^3\left(l+\frac{1}{2}\right)}\\
\langle r^{-3}\rangle &= \tfrac{Z^3}{a_o^3 n^3 (l+1)\left(l+\frac{1}{2}\right)l}\\
\langle r^{-4}\rangle &= \tfrac{Z^4\left[3n^2 - l(l+1)\right]}{a_o^4 2n^5 (l+3/2)(l+1)\left(l+\frac{1}{2}\right)l\left(l-\frac{1}{2}\right)}
\end{aligned}\right\}. \tag{1.48}$$

Combining our radial and angular solutions, the three-dimensional nonrelativistic wavefunction for a one-electron atom can be written as

$$\Psi_{nlm}(r,\vartheta,\varphi) = R_{nl}(r)Y_{lm}(\vartheta,\varphi), \tag{1.49}$$

which is the same as Eq. (1.21). With this definition Ψ is normalized according to[11]

$$\int |\Psi_{nlm}(r,\vartheta,\varphi)|^2 r^2 \sin\theta dr d\vartheta d\varphi = 1. \tag{1.50}$$

We must remember, however, that even though Ψ depends on n, l and m, the energy,

$$E_n = -\frac{1}{2}\left(\frac{e^2}{4\pi\varepsilon_o}\right)^2 \frac{Z^2\mu}{n^2\hbar^2} = -hc\frac{Z^2}{n^2}\mathcal{R}_M, \tag{1.51}$$

10) These expressions are taken from Bockasten [5].

11) *Caution!* Some authors define the radial wavefunction as $rR_{nl}(r)$, leading to a different normalization from that in Eq. (1.50).

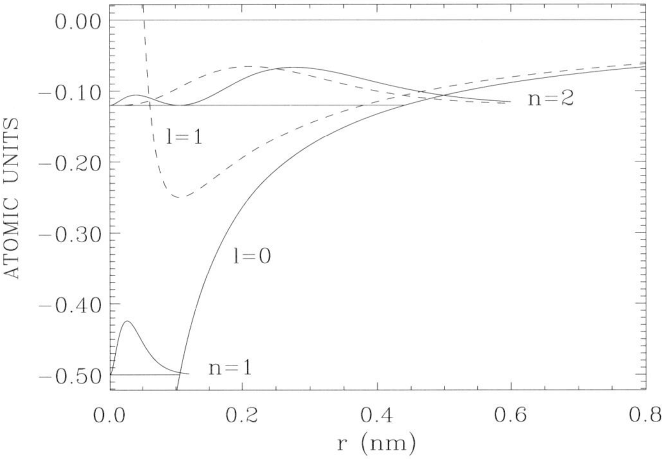

Fig. 1.5 Radial probability densities, $r^2 R_{nl}^2$, for the first three radial wavefunctions, $n, l = 1, 0; 2, 0$ and $2, 1$, respectively, for hydrogen-like ions displayed in conjunction with the associated effective potentials of Fig. 1.3. The solid curve corresponds to $l = 0$ and the dashed curve to $l = 1$.

only depends on n, nonrelativistically. The reader will note that the difference between Eqs. (1.15) and (1.51) is that the former describes the energy spectrum for an infinitely massive nucleus while the latter accounts for the finite mass of a particular nucleus; $\mathcal{R}_M$ and $\mathcal{R}_\infty$ are given respectively by Eqs. (1.18) and (1.14). This one-electron wavefunction is often called an *atomic orbital*. We will use the atomic orbitals to create states for more complex multielectron atoms and molecules in Chapters 2 and 5, respectively.

Probability Density. We are now in a position to put everything together and talk about how to visualize the electron in an atom. To that end, we recall that one of the main tenets of quantum mechanics, due to Max Born, is that $|\Psi_{nlm}(r, \vartheta, \varphi)|^2$ represents the probability per unit volume or the probability density *for measuring the electron* with a specific set of physical coordinates. One should not think of the electron, however, as a corpuscular particle actually whizzing about the nucleus as planets do the Sun so that at time t_1 it is at position $\vec{r}_1$ and at time t_2 it is at position $\vec{r}_2$ as one would classically. Rather, Ψ describes the electron as a wave – the electron's wavefunction – that exists in an abstract space. We can picture this density as a cloud with the electron everywhere within the cloud simultaneously. Figure 1.6 shows a few representations of the density for the first few states for the case of $Z = 1$. While Ψ contains all the information we can know about the electron, according to quantum mechanics, what we measure in the laboratory is related to $|\Psi|^2$. For physical coordinates where $|\Psi|^2$ is large, there is a large probability for finding the electron. The probability density should be understood in a statistical sense. When we make a measurement, we project the electron into a physical space with classical parameters, position and momentum, but subject to the uncertainty principle.[12,13]

Several comments can be made concerning Fig. 1.6, which can be deduced from Figs. 1.2 and 1.5 as well. In fact, the images in Fig. 1.6 were generated by mapping R_{nl}^2 onto the distributions in Fig. 1.2. The first observation we can make is that the atom increases in size as n increases ($\langle r \rangle$ scales as n^2) and the electron cloud becomes more diffuse. Only $l = 0$, s states that are spherically symmetric contribute to the electron density near the origin. Finally, we notice that the contribution to the density from the $np, m = \pm 1$ state is about half that of the $np, m = 0$ state. This is because each $m = \pm 1$ state is actually spread between two orientations. This is more easily seen in rectangular coordinates. To show this, we first recognize that $\Psi_{n,l=1,m}$ represents a spherical vector as discussed in Appendix E. From the general conversion between spherical and Cartesian vectors, Eq. (E.8), it is straightforward to show that

$$\psi_{np_z} = \Psi_{n10}, \tag{1.52}$$

$$\psi_{np_x} = \frac{1}{\sqrt{2}} \left(\Psi_{n11} + \Psi_{n1,-1} \right), \tag{1.53}$$

$$\psi_{np_y} = \frac{1}{i\sqrt{2}} \left(\Psi_{n11} - \Psi_{n1,-1} \right), \tag{1.54}$$

where $\psi_{np_{x,y,z}}$ are the Cartesian vectors – the wavefunctions in rectangular coordinates. Whereas $\psi_{np_z} = \Psi_{n10}$, ψ_{np_x} and ψ_{np_y} are reduced by $1/\sqrt{2}$ while being linear combinations of Ψ_{n11} and $\Psi_{n1,-1}$ and vice versa. In Cartesian coordinates, we would view the electron distribution as being double-lobed clouds along the x-, y- and z-axes, each similar to that of the $2p_{m=0}$ distribution in Fig. 1.6 (see also Y_{10} in Fig. 1.2). The Cartesian representation is the one usually presented in chemistry text books.

1.2.4
Energy Units and Atomic States

Generally, there are two conventions for energy levels and several energy units (see Appendix A) in use. Theorists tend to prefer atomic units (Hartree, see Appendix A) and to set the zero of the energy at the ionization threshold.

12) The wave nature of quantum mechanics leads quite naturally to an uncertainty in our knowledge of the position, Δx, and momentum, Δp, of the electron or any particle and a limit on the determination of both simultaneously, $\Delta x \Delta p \geq \hbar/2$.

13) It is interesting to note that Schrödinger actually misinterpreted Ψ as the density distribution of matter, with some regions of space richer in matter than others.

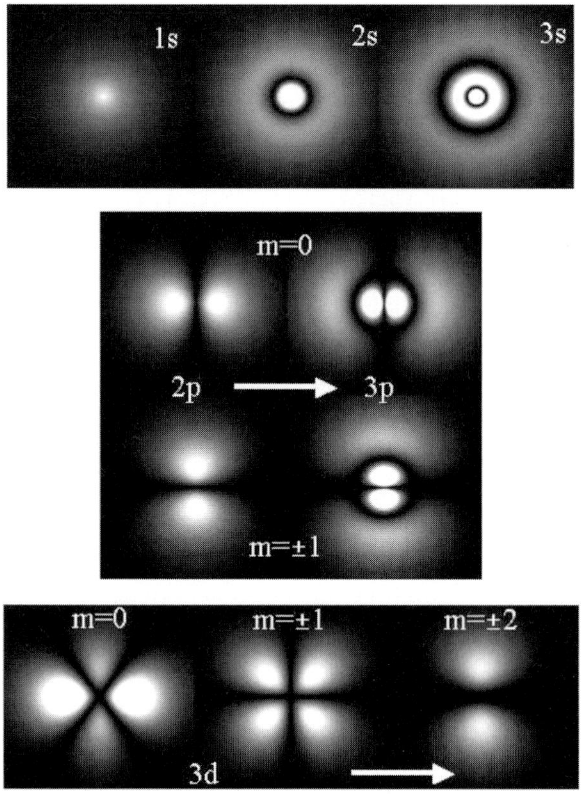

Fig. 1.6 The probability per unit volume, the probability density $|\Psi_{nlm}(r, \vartheta, \varphi)|^2$, for several states of the hydrogen atom. These are cross-sectional images of the densities with the symmetry axis – the z-axis – in the plane of the paper as indicated by the arrows. The top row represents the probability densities for the first three s states, the middle block all m values for the $2p$ and $3p$ states and the bottom row all m values for the $3d$ state. The spatial and intensity scales vary from image to image. Relative to the $1s$ image, the ratios between the spatial scales are 1:5:10:5:5:10:10:10:10:10, respectively, for $1s$:$2s$:$3s$:$2p$, m=0:$2p$, m=$\pm$1:$3p$, m=0:$3p$, m=$\pm$1:$3d$, m=0:$3d$, m=$\pm$1:$3d$, m=$\pm$2. That is, the $1s$ distribution extends over about 0.1 nm while the $2s$ distribution extends over about 0.5 nm. To make it visible, each image is multiplied by 1:200:2000:67:133:1000:2000:1000:1333:1000 respectively from top left to bottom right. We note that to make the weaker components visible, the stronger contributions are saturated in several of the images.

Bound states then will have negative values and continuum states will have positive values. When energy levels are determined experimentally, they are often expressed as wavenumbers,[14] as in the energy-level tables provided by

14) It is interesting to note that wavenumbers originated at a time before computers. While we think nothing of multiplying by hc today, it was very time consuming in the past. Wavenumbers were invented to aid calculations.

the National Institute for Standards and Technology (NIST).[15] Typically, the energies of both bound and continuum states have positive values when expressed as wavenumbers. This is due simply to shifting the zero of the scale down from the ionization threshold to the ground state. To see this, we write the energy of level n relative to level n' as

$$E_n = E_{n'} - hcZ^2\mathcal{R}_M\left(\frac{1}{n^2} - \frac{1}{n'^2}\right). \tag{1.55}$$

If we let $n' \to \infty$, then we have

$$E_n = E_\infty - hc\frac{Z^2}{n^2}\mathcal{R}_M, \tag{1.56}$$

the energy of level n relative to the ionization energy I_P. That is, we let $E_\infty = I_P$ instead of 0. Thus, in wavenumbers the energy of level n becomes

$$\overline{\nu}_n = I_P - \frac{Z^2}{n^2}\mathcal{R}_M. \tag{1.57}$$

We point out that the fact that such a shift is possible illustrates the important point that only the relative energy is important. Furthermore, it would be possible to use atomic units with the zero set to the ground state or wavenumbers with the zero set at the ionization threshold.

The energies of the first three states of hydrogen and the ionization threshold in wavenumbers and atomic units are

		n			
		1	2	3	∞
Energy	cm^{-1}	0	82 259.07	97 492.23	109 678.8
	a.u.	−0.50	−0.12	−0.06	0

1.3 Classification of Nonrelativistic States

The full nonrelativistic hydrogen (one-electron atom or ion) wavefunction, $\psi_{nlm}(r, \vartheta, \varphi)$, has three quantum numbers:

$n =$ the principal quantum number, integers ranging from 1 to ∞ in steps of 1

$l =$ the (orbital) angular momentum, integers ranging from 0 to $n - 1$ in steps of 1

15) Other possible units in use are eV (electron volt), Hz (s^{-1}) and kcal/mol (kilocalorie per mole, a unit often used by chemists). The conversion between these units is given in Appendix A.

$m =$ the magnetic quantum number, integers ranging from $-l$ to l in steps of 1 .

The states are named according to their angular momentum,

$$\begin{array}{lcccccc} l = & 0 & 1 & 2 & 3 & 4 & 5 \\ & s & p & d & f & g & h \end{array}$$

The state with $n = 3$ and $l = 2$ is called the $3d$ state. An electron in the $3d$ state is also said to be in the $3d$ subshell, where a shell is designated by nl.[16]

1.3.1 Parity

The wavefunction is said to have positive parity if $\psi(-\vec{r}) = \psi(\vec{r})$ (i.e., inversion $x \to -x$, $y \to -y$, $z \to -z$) and negative parity if $\psi(-\vec{r}) = -\psi(\vec{r})$. In spherical coordinates this translates into $r \to r$, $\vartheta \to \pi - \vartheta$ and $\varphi \to \varphi + \pi$. The parity is thus determined by the angular part of the wavefunction, $Y_{lm}(\vartheta, \varphi) \propto P_l^m(\cos\vartheta) e^{im\varphi}$. Since

$$\left.\begin{array}{lcr} \vartheta \to \pi - \vartheta & \to & P_l^m(\cos[\pi - \vartheta]) = (-1)^{l-m} P_l^m(\cos\vartheta) \\ \varphi \to \varphi + \pi & \to & e^{im(\varphi+\pi)} = (-1)^m e^{im\varphi} \end{array}\right\}, \tag{1.58}$$

$$Y_{lm}(\pi - \vartheta, \varphi + \pi) = (-1)^l Y_{lm}(\vartheta, \varphi).$$

Consequently, $\psi(-\vec{r}) = (-1)^l \psi(\vec{r})$ implies that

$s, d, \ldots$ *even parity,*
$p, f, \ldots$ *odd parity.*

Neither n nor m participate in the determination of parity.

1.3.2 Degeneracy

Each energy state in Eq. (1.42) is n^2 degenerate, with n^2 unique combinations of values for l and m. The energy, however, does not depend on l nor m. This is because the potential is spherically symmetric so that all the forces are isotropic. As we mentioned before, if an external field were applied to the atom the symmetry could be broken and part of the degeneracy would be lifted. It should be understood that the l-independent Hamiltonian of Eq. (1.3) is a result of ignoring the spin of the electron. When spin is included, we will see that part of the l degeneracy will be lifted.

16) Shell and subshell are often used interchangeably. We will reserve shell for states with the same n regardless of l.. As we will discuss in the next chapter, the $3d$ designation also refers to an atomic orbital.

1.4 Corrections to the Energy Levels

Thus far, we have examined idealized hydrogen-like ions in the absence of external fields. These ions were composed of a nonrelativistic spinless electron and a nucleus that we have treated as a spinless point charge. In reality, of course, the nucleus has a finite size (see Problem 1.3), the electron and nucleus have spin and the electron can move with relativistic velocities when near the nucleus. These lead to shifts of the energy levels (Eq. (1.42)) and symmetry breaking. In this section we will examine three corrections to the energy levels for hydrogen-like ions: (1) relativistic motion, (2) the spin of the electron and (3) the spin of the nucleus. The first two will give rise to what is called the fine structure and partially splits the l degeneracy, while the latter gives rise to one form of the hyperfine structure. In addition, we will set up the framework for looking at atoms in applied electric and magnetic fields.

1.4.1 Relativistic Motion

The relativistic energy of the electron is given by

$$\begin{aligned} E_R &= V + \sqrt{p^2c^2 + m_e^2c^4} \\ &\backsimeq V + m_ec^2 + \underbrace{\left[\frac{p^2}{2m_e} - \frac{p^4}{8m_e^3c^2} + \cdots\right]}_{\text{kinetic energy}}. \end{aligned} \tag{1.59}$$

The first term is the usual Coulomb potential, the second is the rest mass, the third is the nonrelativistic kinetic energy while the fourth is the first-order relativistic correction term to the kinetic energy. If we define

$$\widehat{H}_o = \frac{p^2}{2m_e} + V \tag{1.60}$$

$$\widehat{\zeta}_R = -\frac{p^4}{8m_e^3c^2} \tag{1.61}$$

$$= -\frac{1}{2m_ec^2}\left[\widehat{H}_o - V\right]^2, \tag{1.62}$$

we can apply first-order perturbation theory to $\widehat{\zeta}_R$ (i.e., find the expectation value of $\widehat{\zeta}_R$ with the state vectors of $\widehat{H}_o$) to determine the energy shift, ΔE_R. Doing this gives

$$\begin{aligned} \Delta E_R &= \left\langle \psi_n \middle| \widehat{\zeta}_R \middle| \psi_n \right\rangle \\ &= -\frac{1}{2m_ec^2}\left\langle \psi_n \middle| \widehat{H}_o^2 - 2\widehat{H}_oV + V^2 \middle| \psi_n \right\rangle. \end{aligned} \tag{1.63a}$$

But,[17]

$$\widehat{H}_o\left|\psi_n\right\rangle = E_n\left|\psi_n\right\rangle \rightarrow \left\langle\left|\widehat{H}_o^2\right|\right\rangle = E_n^2, \tag{1.63b}$$

while

$$\left\langle\left|\widehat{H}_o V\right|\right\rangle = -E_n\frac{Ze^2}{4\pi\varepsilon_0}\left\langle\frac{1}{r}\right\rangle = -\left(\frac{Z^2e^2}{4\pi\varepsilon_0}\right)\frac{E_n}{n^2a_0} \tag{1.63c}$$

and

$$\left\langle\left|V^2\right|\right\rangle = \left(\frac{Ze^2}{4\pi\varepsilon_0}\right)^2\left\langle\frac{1}{r^2}\right\rangle = \left(\frac{Z^2e^2}{4\pi\varepsilon_0}\right)^2\frac{1}{n^3a_0^2(l+\frac{1}{2})}. \tag{1.63d}$$

The expressions in Eq. (1.48) were used to evaluate the expectation values for $1/r$ and $1/r^2$. Collecting the terms leads to

$$\Delta E_R = -\frac{E_n^2}{2m_ec^2}\left[1+2\frac{Z^2e^2}{4\pi\varepsilon_0 n^2a_0}\frac{1}{E_n}+\left(\frac{Z^2e^2}{4\pi\varepsilon_0 n^2a_0}\right)^2\frac{n}{E_n^2\left(l+\frac{1}{2}\right)}\right]. \tag{1.64}$$

Finally, with the help of Eq. (1.42) the shift can be written as

$$\begin{aligned}\Delta E_R &= -\frac{E_n^2}{2m_ec^2}\left(\frac{4n}{l+\frac{1}{2}}-3\right) && (1.65a)\\ &= -\frac{\alpha^2E_n}{4}\left(\frac{4n}{l+\frac{1}{2}}-3\right)\frac{Z^2}{n^2} && (1.65b)\\ &= -\frac{\alpha^2\mathcal{R}_M}{4}\left(\frac{4n}{l+\frac{1}{2}}-3\right)\frac{Z^4}{n^4}, && (1.65c)\end{aligned}$$

where α is the fine-structure constant (see Appendix A). We see that the shift depends on both n and l; thus, the degeneracy between states with the same n but different l is broken.

1.4.1.1 **Electron Spin and the Dirac Equation**

The electron, a fermion, has half-integer spin and thus obeys Fermi–Dirac statistics. Spin, which plays a major role in atomic and molecular physics, is not included in the nonrelativistic Hamiltonian or its correction term (Eq. (1.61)). Instead of inserting a series of terms to account for spin and its interaction with other angular momenta, we will endeavor to understand the origin of

17) Here, $|\psi_n\rangle$ is an eigenstate of $\widehat{H}_o$ with eigenvalue E_n. See Appendix B.8 for further discussion of the mathematical formalism of $\widehat{H}_o|\psi_n\rangle = E_n|\psi_n\rangle$.

the spin terms in this section. Spin, as it turns out, is relativistic in origin. To derive the terms we need, we must start with an equation that is relativistically invariant. We will use the free-particle Dirac equation,

$$\left[c\underset{\sim}{\alpha} \cdot \vec{p} + \underset{\sim}{\beta} m_e c^2\right] \psi = E\psi, \tag{1.66}$$

where $\underset{\sim}{\alpha}$ and $\underset{\sim}{\beta}$ are 4×4 matrices given by[18]

$$\underset{\sim}{\alpha} = \begin{pmatrix} 0 & \underset{\sim}{\sigma} \\ \underset{\sim}{\sigma} & 0 \end{pmatrix} \qquad \underset{\sim}{\beta} = \begin{pmatrix} \underset{\sim}{I} & 0 \\ 0 & -\underset{\sim}{I} \end{pmatrix}, \tag{1.67}$$

$\underset{\sim}{\sigma}$ are the 2×2 Pauli spin matrices and $\underset{\sim}{I}$ is the identity matrix.[19]

As before, $\vec{p}$ is our momentum operator $(-i\hbar\vec{\nabla})$ and the wavefunction, ψ, is now a four-component column vector,

$$\psi = \begin{pmatrix} \psi_1 \\ \psi_2 \\ \psi_3 \\ \psi_4 \end{pmatrix}, \tag{1.68}$$

where we can define ψ_u and ψ_v as

$$\psi_u = \begin{pmatrix} \psi_1 \\ \psi_2 \end{pmatrix}, \qquad \psi_v = \begin{pmatrix} \psi_3 \\ \psi_4 \end{pmatrix}, \tag{1.69}$$

which correspond to the positive and negative energies, respectively. Inserting this wavefunction into the Dirac equation leads to two equations that must be solved simultaneously,

$$\begin{aligned} c\underset{\sim}{\sigma} \cdot \vec{p}\,\psi_v + \left(m_e c^2 - E\right)\psi_u &= 0, \\ c\underset{\sim}{\sigma} \cdot \vec{p}\,\psi_u - \left(m_e c^2 + E\right)\psi_v &= 0, \end{aligned} \tag{1.70}$$

In the nonrelativistic limit, $v \ll c$, these reduce to the Schrödinger equation. From the second equation in Eq. (1.70), we have

$$\psi_v = \frac{c\underset{\sim}{\sigma} \cdot \vec{p}}{m_e c^2 + E}\psi_u. \tag{1.71}$$

18) We will use a tilde under a symbol to indicate a matrix.

19) The Pauli spin matrices (see, for example, Eq. (4.73) of Ref. [6]) can be represented as

$$\underset{\sim}{\sigma}(x) = \begin{pmatrix} 0 & 1 \\ 1 & 0 \end{pmatrix}, \quad \underset{\sim}{\sigma}(y) = \begin{pmatrix} 0 & -i \\ i & 0 \end{pmatrix}, \quad \underset{\sim}{\sigma}(z) = \begin{pmatrix} 1 & 0 \\ 0 & -1 \end{pmatrix}, \quad \underset{\sim}{I} = \begin{pmatrix} 1 & 0 \\ 0 & 1 \end{pmatrix}.$$

When the velocities are small $\left|\overrightarrow{p}\right| \rightarrow m_e v$ and $E \rightarrow mc^2$, which implies

$$|\psi_v| \approx \frac{v}{2c} |\psi_u| \ll |\psi_u| \,. \tag{1.72}$$

Thus, most of the physics will be represented by the positive-energy wavefunctions at nonrelativistic velocities. Substituting Eq. (1.71) into the upper expression in Eq. (1.70) to eliminate ψ_v gives

$$c^2 \left(\underset{\sim}{\sigma} \cdot \overrightarrow{p}\right)^2 \psi_u = \left(m_e c^2 + E\right)\left(E - m_e c^2\right) \psi_u \tag{1.73a}$$

$$= E^2 - m_e^2 c^4 \psi_u \tag{1.73b}$$

$$\approx 2 m_e c^2 E^{nr} \psi_u, \tag{1.73c}$$

where $E^{nr} = E - m_e c^2$. Using the spinor identity

$$\left(\underset{\sim}{\sigma} \cdot \vec{A}\right)\left(\underset{\sim}{\sigma} \cdot \vec{B}\right) = \vec{A} \cdot \vec{B} + i \underset{\sim}{\sigma} \cdot \left(\vec{A} \times \vec{B}\right), \tag{1.74}$$

we can write

$$c^2 p^2 \psi_u = 2 m_e c^2 E^{nr} \psi_u, \tag{1.75}$$

which is just the free-particle Schrödinger equation.

To account for interactions with the Coulomb potential of the nucleus and external magnetic and time-varying electric fields, we introduce two additional terms into the free-particle equation (Eq. (1.70)). We incorporate the external fields with the vector potential, $\vec{A}$, by letting

$$\overrightarrow{p} \rightarrow \overrightarrow{p} - \frac{e}{c} \overrightarrow{A} \equiv \overrightarrow{\pi}, \tag{1.76}$$

where the electric and magnetic fields are given by

$$\vec{\mathbf{E}} = -\overrightarrow{\nabla} \phi - \frac{\partial \vec{\mathbf{A}}}{\partial t}, \tag{1.77a}$$

$$\vec{\mathbf{B}} = \overrightarrow{\nabla} \times \vec{\mathbf{A}}. \tag{1.77b}$$

A static potential, $-e\phi$, was added to Eq. (1.77a) to account for the nuclear charge. In Chapter 4, $\vec{\pi}$ will lead to the radiation Hamiltonian used in photoexcitation. The modified Dirac equation takes the form

$$\begin{aligned} c \underset{\sim}{\sigma} \cdot \left(\overrightarrow{p} - \tfrac{e}{c} \vec{A}\right) \psi_v &= \left(E - m_e c^2 - e\phi\right) \psi_u, \\ c \underset{\sim}{\sigma} \cdot \left(\overrightarrow{p} - \tfrac{e}{c} \vec{A}\right) \psi_u &= \left(E + m_e c^2 - e\phi\right) \psi_v. \end{aligned} \tag{1.78}$$

Following steps similar to those taken above, the lower expression in Eq. (1.71) becomes

$$\psi_v = \frac{c\underset{\sim}{\sigma} \cdot \left(\vec{p} - \frac{e}{c}\vec{A}\right)}{E + m_e c^2 - e\phi}\psi_u. \tag{1.79}$$

Substituting this into the upper expression and expanding the denominator,

$$\frac{1}{E + m_e c^2 - e\phi} \approx \frac{1}{2m_e c^2}\left[1 - \frac{E^{nr} - e\phi}{2m_e c^2} + \cdots\right] \tag{1.80}$$

(where terms of order v^2/c^2 have been kept), gives

$$\frac{1}{2m_e}\left[\left(\underset{\sim}{\sigma} \cdot \vec{\pi}\right)^2 - \left(\underset{\sim}{\sigma} \cdot \vec{\pi}\right)\frac{E^{nr} - e\phi}{2m_e c^2}\left(\underset{\sim}{\sigma} \cdot \vec{\pi}\right)\right]\psi_u = \left(E^{nr} - e\phi\right)\psi_u. \tag{1.81}$$

We will solve this equation by allowing $\psi = \xi\psi_u$, where

$$\xi = 1 + \frac{\left(\underset{\sim}{\sigma} \cdot \vec{\pi}\right)^2}{8m_e^2 c^2}. \tag{1.82}$$

To first order in v^2/c^2,

$$\xi^{-1} \approx 1 - \frac{\left(\underset{\sim}{\sigma} \cdot \vec{\pi}\right)^2}{8m_e^2 c^2}. \tag{1.83}$$

Making this substitution for ψ_u and multiplying the resulting equation on the left by ξ^{-1} gives

$$\begin{aligned}
&\left[\frac{1}{2m_e}\left(\underset{\sim}{\sigma} \cdot \vec{\pi}\right)^2 - \frac{\left(\underset{\sim}{\sigma} \cdot \vec{\pi}\right)^4}{8m_e^2 c^2} + \frac{\left(\underset{\sim}{\sigma} \cdot \vec{\pi}\right)^2}{8m_e^2 c^2}\left(E^{nr} - e\phi\right)\right.\\
&\left.- \frac{1}{4m_e^2 c^2}\left(\underset{\sim}{\sigma} \cdot \vec{\pi}\right)\left(E^{nr} - e\phi\right)\left(\underset{\sim}{\sigma} \cdot \vec{\pi}\right) + \left(E^{nr} - e\phi\right)\frac{\left(\underset{\sim}{\sigma} \cdot \vec{\pi}\right)^2}{8m_e^2 c^2}\right]\psi\\
&= \left(E^{nr} - e\phi\right)\psi.
\end{aligned} \tag{1.84}$$

Recognizing that

$$\begin{aligned}
\left(\underset{\sim}{\sigma} \cdot \vec{\pi}\right)^2 &= \left(\vec{p} - \frac{e}{c}\vec{A}\right)^2 - \frac{e\hbar}{c}\underset{\sim}{\sigma} \cdot \vec{\nabla} \times \vec{A},\\
\left(\underset{\sim}{\sigma} \cdot \vec{\pi}\right)\left(E^{nr} - e\phi\right) - \left(E^{nr} - e\phi\right)\left(\underset{\sim}{\sigma} \cdot \vec{\pi}\right) &= ie\hbar\left(\underset{\sim}{\sigma} \cdot \vec{\nabla}\phi\right),\\
\left(\underset{\sim}{\sigma} \cdot \vec{\pi}\right)\left(\underset{\sim}{\sigma} \cdot \vec{\nabla}\phi\right) - \left(\underset{\sim}{\sigma} \cdot \vec{\nabla}\phi\right)\left(\underset{\sim}{\sigma} \cdot \vec{\pi}\right) &= -i\hbar\vec{\nabla} \cdot \vec{\nabla}\phi + 2i\underset{\sim}{\sigma} \cdot \vec{\pi} \times \vec{\nabla}\phi,
\end{aligned} \tag{1.85}$$

we can rewrite Eq. (1.84) as

$$\left[\frac{1}{2m_e}\left(\vec{p}-\frac{e}{c}\vec{A}\right)^2-\frac{e\hbar}{2m_e c}\underset{\sim}{\sigma}\cdot\vec{\nabla}\times\vec{A}-\frac{\vec{p}^{\,4}}{8m_e^3c^3}\right.$$
$$\left.+\frac{e\hbar^2}{8m_e^2c^2}\vec{\nabla}\cdot\vec{\nabla}\phi-\frac{e\hbar}{4m_e^2c^2}\underset{\sim}{\sigma}\cdot\vec{\nabla}\phi\times\vec{p}\right]\psi=\left(E^{nr}-e\phi\right)\psi. \quad (1.86)$$

Again, only terms of order v^2/c^2 have been kept. The quantity in parentheses on the left- and the right-hand sides comprises the nonrelativistic Schrödinger equation with an interaction term represented by $-e\vec{A}/c$. The rest of the terms all have a relativistic origin and are responsible for spin-1/2 particle interaction with magnetic fields as well as spin–orbit interaction. A brief description of each term along with its approximate magnitude is given in Table 1.1.

The second perturbation in Table 1.1 allows us to define the Bohr magneton, $\mu_B = e\hbar/2m_e c$ while the fifth, the spin–orbit interaction, further splits the degeneracy between states of different l. Before we calculate the magnitude of this splitting, we will say a few words about spin.

1.4.1.2 Classification of Relativistic Hydrogen States

The spin quantum number, $\vec{s}$, assumes only one value, 1/2. Similar to the orbital angular momentum, the eigenvalues of $\vec{s}^{\,2}$ will be $s(s+1) = 3/4$. There is also a magnetic or z component of $\vec{s}$, $\vec{s}_z$, that will have eigenvalues $m_s = \pm 1/2$. We will define a new operator called the total angular momentum,

$$\vec{j} = \vec{l} + \vec{s}, \quad (1.87)$$

with eigenvalues (of $\vec{j}^{\,2}$) $j(j+1)$. The sum in Eq. (1.87) is a vector sum. The values that $\vec{j}$ can assume are

$$j = \begin{cases} l \pm \frac{1}{2}, & l \neq 0, \\ \frac{1}{2}, & l = 0. \end{cases} \quad (1.88)$$

The magnetic quantum number or the eigenvalue of the z component of $\vec{j}$ is m_j, with $2j+1$ values, $m_j = -j, -j+1, \ldots, j$.

In Section 1.3 we introduced the concept of a shell, designated by nl. Each l will have two values of j associated with it when $l > 0$. For example, an electron in a state of $n = 4$ and $l = 2$ will have $j = 3/2, 5/2$. Thus, we introduce the concept of a subshell designated by nl_j. This electron is said to be in the $4d$ shell and will be in either the $4d_{3/2}$ or $4d_{5/2}$ subshell. Note that now we have two angular momenta about which to be concerned, l and j. To

Tab. 1.1 Summary of relativistic and spin-dependent interaction terms.

Perturbation	Description	Magnitude
$\frac{1}{2m_e}\left(\vec{p} - \frac{e}{c}\vec{A}\right)^2$	The nonrelativistic motion, $p^2/2m_e$, and the interaction terms $\frac{e}{2m_e c}\vec{A}\cdot\vec{\nabla} + \frac{e^2}{2m_e c^2}\vec{A}\cdot\vec{A}$. The latter terms are responsible for absorption and emission; they can be written as $-e\vec{r}\cdot\vec{E}$ (see Section 4.1.1).	$\gtrsim 10^5$ cm^{-1}
$\frac{e\hbar}{2m_e c}\underset{\sim}{\sigma}\cdot\vec{\nabla}\times\vec{A}$	The spin-1/2 ($\vec{s} = \underset{\sim}{\sigma}/2$) interaction with a magnetic field ($\vec{B} \equiv \vec{\nabla}\times\vec{A}$). This term gives the correct g-factor, $g = 2$, and magnetic moment of the electron, $\mu = \frac{e\hbar}{2m_e c}g\vec{s}$.	≈ 1 cm^{-1}
$\frac{\vec{p}^{\,4}}{8m_e^3c^3}$	Relativistic mass correction term.	≈ 0.1 cm^{-1}
$\frac{e\hbar^2}{8m_e^2c^2}\vec{\nabla}\cdot\vec{\nabla}\phi$	The "Darwin" term responsible for s-state shifts. It represents the relativistic nonlocalizability of the electron and is related to both the negative-energy sea and its rapid motion.	< 0.1 cm^{-1}
$\frac{e\hbar}{4m_e^2c^2}\underset{\sim}{\sigma}\cdot\vec{\nabla}\phi\times\vec{p}$	Spin–orbit interaction. As shown in Problem 1.4, this term can be written as $\frac{e\hbar^2}{2m_e^2c^2}\frac{1}{r}\frac{d\phi}{dr}\left(\vec{l}\cdot\vec{s}\right)$, where $\vec{l}$ is the orbital angular momentum. In contrast to the previous term, this does not affect s states.	10–10^3 cm^{-1}

distinguish their magnetic quantum numbers, we will add subscripts l and j to the orbital, m_l, and total, m_j, quantum numbers.

1.4.1.3 Hydrogen-Like Ion Wavefunction Including Spin

We are now in a position to write down the three-dimensional wavefunction for a one-electron atom including the spin of the electron. This wavefunction takes the form

$$\Psi_{nlm_lm_s} = R_{nl}(r)Y_{lm_l}(\vartheta,\varphi)\phi_{m_s}(s_z), \tag{1.89}$$

where we have placed the subscripts l and s on the magnetic quantum numbers to distinguish orbital from spin. Since this wavefunction describes the state of a single electron, we only need to specify the direction of the spin.

1.4.2
Fine Structure and Spin–Orbit Interaction

In general, subshells have different energies. The splitting between the states is determined in part by the spin–orbit interaction

$$\widehat{H}_{so} = \frac{e\hbar^2}{2m_e^2c^2}\frac{1}{r}\frac{d\phi}{dr}\left(\vec{l}\cdot\vec{s}\right). \tag{1.90}$$

This interaction causes a level to shift by $\Delta E_{so} = \left\langle \widehat{H}_{so} \right\rangle$.[20] To calculate $\left\langle \widehat{H}_{so} \right\rangle$, it is convenient to write $\vec{l}\cdot\vec{s}$ as[21]

$$\vec{l}\cdot\vec{s} = \frac{\vec{j}^2 - \vec{l}^2 - \vec{s}^2}{2}. \tag{1.91}$$

The energy shift then becomes

$$\Delta E_{so} = \frac{e\hbar^2}{2m_e^2c^2}\left[\frac{j(j+1) - l(l+1) - s(s+1)}{2}\right]\left\langle \frac{1}{r}\frac{d\phi}{dr}\right\rangle. \tag{1.92}$$

But,

$$\frac{1}{r}\frac{d\phi}{dr} = \frac{Ze}{4\pi\varepsilon_0 r^3}. \tag{1.93}$$

Using the expressions in Eq. (1.48), it is possible to write

$$\Delta E_{so} = \frac{Z^4e^2\hbar^2}{4\pi\varepsilon_0}\frac{\hbar^2}{2m^2c^2a_0^3n^3}\frac{j(j+1) - l(l+1) - s(s+1)}{2l\left(l+\frac{1}{2}\right)(l+1)}. \tag{1.94}$$

Combining this with the relativistic energy correction (Eq. (1.65a)), we obtain

$$\Delta E_{nlj} = \Delta E_R + \Delta E_{so} = \alpha^2\mathcal{R}_M\left(\frac{3}{4n} - \frac{1}{j+\frac{1}{2}}\right)\frac{Z^4}{n^3}. \tag{1.95}$$

At the maximum value of $j + 1/2 = n$, ΔE_{nlj} is always < 0, which means that the combination of the relativistic and spin–orbit interactions binds the system more tightly as shown for a few levels in Fig. 1.7. The figure shows as well that the degeneracy has not been lifted completely. States of different l

20) We saw in Eq. (1.65a) that the relativistic correction shifted levels of different l by different amounts. Below, we will combine the spin–orbit with the relativistic correction to remove the degeneracy of a shell partially.

21) This can be shown using the algebra developed in Appendix E; see Eq. (E.29) and surrounding discussion.

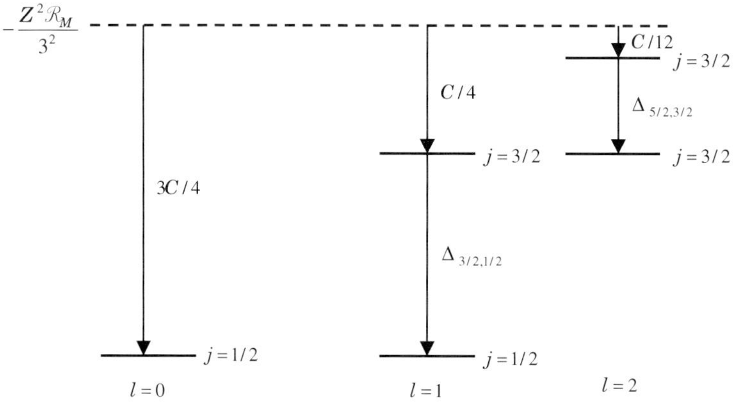

Fig. 1.7 Fine structure of hydrogen-like ions in wavenumbers for $n = 3$ where $C = \alpha^2 Z^4 \mathcal{R}_M / n^3$ and $\Delta_{j+1,j} = \Delta E_{nlj+1} - \Delta E_{nlj} = C/l\,(l+1)$. The levels are all shifted below $-Z^2\mathcal{R}_M/n^2$, the nonrelativistic energy.

but with the same j have the same energy. The splitting between $j = l \pm 1/2$ levels will be given by

$$\Delta_{j+1,j} = \frac{\alpha^2 Z^4 R_M}{n^3} \frac{1}{l(l+1)}. \tag{1.96}$$

Example. For $j = 1/2$ and $3/2$ in hydrogen, Eq. (1.95) gives $\Delta_{j+1,j} = 0.36$, $0.12, 0.044\ \text{cm}^{-1}$, respectively, for $n = 2, 3, 4$.

1.4.3
Rydberg Series

Figure 1.1 shows a progression of energy levels to the ionization limit as $n \to \infty$. To lowest order, the energy of each level in the series is given by Eq. (1.56). For each level n, there are n nearly degenerate l sublevels. For all but s ($l = 0$) states, there are two possible j ($= l \pm s$) values for each l and thus two series for each l. Consequently, there are an infinite number of series of levels marching their way toward the ionization limit. Each series, nl_j for fixed l and j, is called a *Rydberg series*. For hydrogen-like systems (ions with just one electron) these series have a simple behavior. Systems with two or more electrons behave very differently because the members of the series can interact in very profound ways leading to a variety of interesting phenomena such as autoionization. We will look at Rydberg series for multielectron atoms in Chapter 2.

1.5 Continuum States

Thus far, we have focused on bound states, which correspond to solutions to the radial equation (Eq. (1.26)) when $E < 0$. However, when $E > 0$ the electron energy lies above the $n = \infty$ level in Fig. 1.1. The asymptotic ($r \to \infty$) solutions for the positive-energy states take the form $R(r) = e^{\pm ir\sqrt{2\mu E/\hbar^2}}$ (Eq. (1.29)), which oscillates sinusoidally and does not tend to zero. The radial solution for these states can be obtained by following the steps we took for the negative-energy states up to the recursion relation in Eq. (1.35), which become complex when $E > 0$. Since the wavefunctions are finite at $r = \infty$ (Eq. (1.29) is bounded by 1), there is now no need to truncate the series. As a result, the state of the electron can assume all positive energies; hence, the name *continuum state*. Since Eq. (1.35) is still valid, the wavefunctions can be written in terms of hypergeometric functions; however, the principal quantum number, n, becomes imaginary in Eq. (1.37),

$$n \equiv \lambda = -i\frac{Ze^2/4\pi\varepsilon_0}{\sqrt{2\mu E/\hbar^2}}\frac{\mu}{\hbar^2}. \tag{1.97}$$

The solution, Eq. (4.23) of Bethe and Salpeter, can be written as

$$\begin{aligned} R_{\varepsilon l}(r) &= (-1)^{l+1}\frac{2\sqrt{\frac{Ze^2}{4\pi\varepsilon_0}}}{\sqrt{1-e^{-2\pi n'}}} \\ &\times \prod_{v=1}^{l}\sqrt{v^2-n'^2}\frac{(2kr)^l}{(2l+1)!}e^{-ikr}F(in'+l+1, 2l+2, 2ikr), \end{aligned} \tag{1.98}$$

where $n' = in$ is real, and $k = \sqrt{2\mu E/\hbar^2}$ also real with dimensions of $1/r$.

In general, each nl Rydberg series in the bound region is associated with a series of the same-parity energy states in the continuum region. The continuum states are designated by the label εl. An nl Rydberg series plus its εl continuum is called a *channel*. The channel is said to be *closed* when $E < 0$ (the electron is bound to the nucleus) and *open* when $E > 0$ (the electron is free).

Further Reading

1 H. A. Bethe and E. E. Salpeter, *Quantum Mechanics of One- and Two-Electron Atoms*, Plenum, New York, NY, 1977.

2 A. Messiah, *Quantum Mechanics*, Dover Publications, 2000. ISBN: 0486409244.

3 R. D. Cowan, *The Theory of Atomic Structure and Spectra*, University of California Press, Los Angeles, CA, 1981.

4 I. I. Sobelman, *Atomic Spectra and Radiative Transitions*, Springer Series in Chemical Physics, Springer, New York, NY, 1979.

5 J. Mathews and R. L. Walker, *Mathematical Methods of Physics*, Benjamin Cummings, San Francisco, CA, 1970. ISBN 0-8053-7002-1, Second edition.B

6 N. N. Lebedev, *Special Functions and Their Applications*, translated and edited by R. A. Silverman, Dover, New York, NY, 1972.

7 J. J. Sakurai, *Advanced Quantum Mechanics*, Addison-Wesley, Menlo Park, CA, 1973.

8 L. I. Schiff, *Quantum Mechanics*, McGraw-Hill, San Francisco, CA, 1968.

9 F. Yang and J. H. Hamilton, *Modern Atomic and Nuclear Physics*, McGraw-Hill, San Francisco, CA, 1996.

10 J. D. Bjorken and S. D. Drell, *Relativistic Quantum Mechanics*, McGraw-Hill, San Francisco, CA, 1964.

11 For additional problems with worked solutions see D. Budker, D. F. Kimball and D. P. DeMille, *Atomic Physics: An Exploration Through Problems and Solutions*, Oxford University Press, New York, NY, 2004.

Problems

1.1 Find the lowest-energy state of the one-dimensional hydrogen atom where the Schrödinger equation takes the form

$$-\frac{\hbar^2}{2\mu}\frac{d^2}{dx^2}\psi(x) - \left(\frac{1}{4\pi\varepsilon_o}\frac{Ze^2}{|x|} - E\right)\psi(x) = 0 \qquad (-\infty < x < \infty). \tag{1.99}$$

1.2 Calculate

$$\langle r^\gamma \rangle = \int r^{2+\gamma} R_{nl}^2(r)\, dr \tag{1.100}$$

for $\gamma = -3, -2, -1, 1, 2, 3$.

1.3 Since the wavefunctions of the s levels are finite at the origin, the potential for s electrons changes when the electron penetrates the nucleus and gives rise to a shift in the energy levels relative to Eq. (1.42). If the radius of the nucleus is r_o, the potential will be given by

$$\begin{aligned} &\frac{Q}{|r|} && \text{when} \quad r > r_o, \\ &\int_N \frac{\rho(\vec{r}')}{|\vec{r}-\vec{r}'|} d^3r' && \text{when} \quad r < r_o, \end{aligned} \tag{1.101}$$

where ρ is the charge density of the nucleus. Use first-order perturbation theory to show that the s states are shifted by

$$(\Delta E_s)_{\text{finite size}} = \frac{2}{3}\frac{e^2}{4\pi\varepsilon_0}\frac{\langle r^2\rangle_N}{a_0^3}, \tag{1.102}$$

where

$$\left\langle r^2 \right\rangle_N = \int_N r'^2 \rho\left(\vec{r}'\right) d^3r'. \tag{1.103}$$

If you take the size of the nucleus to be about 8 fm, estimate the magnitude of this shift for the ground state, i.e., calculate $(\Delta E_s)_{\text{finite size}} / \mathrm{E}_{1s}$.

1.4 Show that

$$\frac{e\hbar}{4m_e^2c^2}\underline{\sigma} \cdot \vec{\nabla}\phi \times \vec{p} \tag{1.104}$$

in Table 1.1 can be rewritten as

$$\frac{e\hbar^2}{2m_e^2c^2}\frac{1}{r}\frac{d\phi}{dr}\left(\vec{L} \cdot \vec{S}\right). \tag{1.105}$$

1.5 It is possible to derive the spin–orbit interaction by going to the rest frame of the electron and allowing the nucleus to orbit the electron. The electron then feels a magnetic field due to the circular current. Determine the expression for the Hamiltonian, $-\vec{\mu} \cdot \vec{B}$, that describes the interaction.

1.6 Sketch the probability densities similar to Fig. 1.6 for the 1s through 3d states of hydrogen in Cartesian coordinates. You might find it helpful first to express the spherical harmonics in terms of Cartesian coordinates. That is, relate the spherical components (l, m) to combinations of x, y and z.

2
The Structure of the Multielectron Atom

Thus far, we have restricted our investigation to atomic systems with a single electron. We will now turn our attention to atoms with more than one electron. We will focus primarily on one-electron and two-electron atoms – atoms with one and two electrons outside closed shells, respectively.

2.1
Overview

We begin by introducing the nonrelativistic multielectron Hamiltonian,

$$\widehat{H}_o = \sum_i \left[\frac{p_i^2}{2m_e} - \frac{Ze^2}{4\pi\varepsilon_0 r_i} \right] + \frac{1}{2} \sum_{i \neq j} \frac{Ze^2}{4\pi\varepsilon_0 r_{ij}} + \sum_i \zeta\left(r_i\right) \left(\vec{l}_i \cdot \vec{s}_i \right). \tag{2.1}$$

The first sum is the single-particle Hamiltonian, the second is the mutual interaction between electrons, also known as the electrostatic interaction, and the last is the spin–orbit interaction. It is the second sum that distinguishes the Hamiltonians for multielectron systems from those of hydrogen-like ions. Since the Hamiltonian involves more than two particles, we will not be able to generate an exact solution to Schrödinger's equation but must resort to approximations. The simplest approximation that captures much of the important physics is known as the *central field approximation*. We will first look at the problem without the last term. At the end we will put it back in as a perturbation.

This single-particle approximation assumes that the interaction between electrons can be represented as an average central potential that can be added to the Coulomb potential of the nucleus to give an effective potential $U(r)$. The Schrödinger equation,

$$\nabla^2 \psi\left(\vec{r}\right) + \frac{2\mu}{\hbar^2} \left[E - U\left(r\right)\right] \psi\left(\vec{r}\right) = 0, \tag{2.2}$$

takes a form similar to Eq. (1.19) with $-Ze^2/4\pi\varepsilon_0 r$ replaced by $U(r)$. Assuming that $\psi\left(\vec{r}\right)$ is again separable into radial and angular portions (Eq. (1.21)),

Light-Matter Interaction: Atoms and Molecules in External Fields and Nonlinear Optics.
W. T. Hill and C. H. Lee

ISBN: 978-3-527-40661-6

the radial wavefunction is then a solution to

$$\frac{1}{r^2}\frac{d}{dr}\left(r^2\frac{d}{dr}R_{nl}(r)\right)+\left[E-U_l(r)\right]R_{nl}(r)=0, \tag{2.3}$$

where

$$U_l(r)=U(r)+\frac{\hbar^2}{2\mu}\frac{l(l+1)}{r^2}. \tag{2.4}$$

The solutions $R_{nl}(r)$, from which we can describe the structure and dynamics of multielectron systems depend on the exact expression for $U(r)$. Before looking at details, we will state a few general features of multielectron systems.

1. Limiting values for $U(r)$ when $E<0$:

   $$U(r)\to V(r)=-Ze^2/4\pi\varepsilon_0 r \qquad \text{as} \qquad r\to 0$$

   $$U(r)\to 0 \qquad \text{as} \qquad r\to\infty$$

   The first is true because as $r\to 0$, all other electrons will have larger orbits and the electron will sense only the nucleus. The second must be true to have bound states.

2. In general, the system will be uniquely determined by E, l, m_l. For hydrogen-like ions we saw in the nonrelativistic approximation that there will be $2l+1$ degenerate states. In multielectron systems part of this degeneracy is lifted. We will discuss this in more detail in (4) below.

3. The states of atomic systems consist of *shells* and *subshells*, which are composed of collections of atomic orbitals (one-electron wavefunctions, Eq. (1.49)).

   - An *atomic orbital* is designated as nl, e.g., $1s$, $2p$, etc., where s and p indicate $l=1,2$, respectively.
   - A *subshell* is very similar to an orbital, the difference being that one designates the number of electrons contained within. We would call a $2s$ orbital with two electrons the $2s^2$ subshell. In general, a subshell with # electrons would be designated by $nl^{\#}$, where the maximum value # can assume is given by $(2s+1)(2l+1)=2(2l+1)$. The maximum number of electrons per subshell for

$l = 0$–3 is

l	Designation	Maximum number
0	s	2
1	p	6
2	d	10
3	f	14

- A *shell* is composed of all the states for a given n. For example, there are two subshells ($2s$ and $2p$) that form the $n = 2$ shell. One will often find shells referred to by following X-ray absorption names:

K shell	$n = 1$	$l = 0$
L shell	$n = 2$	$l = 0, 1$
M shell	$n = 3$	$l = 0, 1, 2$

etc.

- A shell or subshell is said to be *closed* when it contains its maximum number of electrons. A closed subshell bears the special property of being spherically symmetric, i.e., $\vec{\mathbf{J}} = 0$.[1] It should be understood that when *shell* is used in the literature, it can be in reference to what we have defined as a subshell or to what we have defined as a shell. Consequently, a *closed shell* can mean a shell with its maximum number of electrons, e.g., $2s^2 2p^6$, or a subshell with its maximum number of electrons, e.g., $2s^2$.
- A *configuration* describes the states of all the electrons in an atom. For example, the ground-state configuration for carbon, which has six electrons, is $1s^2 2s^2 2p^2$. In general, a configuration will lead to several different terms (which we define next). With the exception of a closed shell or subshell, however, one should view the placement of electrons into specific configurations as only an approximation to the truth. Configurations should be used primarily to aid bookkeeping. Although there are cases where a pure configuration can be a reasonable approximation for defining the state of an atom, in general, including ground states of atoms, admixtures

1) When an orbital contains its maximum number of electrons, the spherical symmetry is due to the fact that $\vec{\mathbf{J}} = \sum_i \vec{\mathbf{j}}_i = 0$ or $\vec{\mathbf{L}} = \sum_i \vec{\mathbf{l}}_i = 0$ and $\vec{\mathbf{S}} = \sum_i \vec{\mathbf{s}}_i = 0$ so that $\vec{\mathbf{L}} + \vec{\mathbf{S}} = \vec{\mathbf{J}} = 0$, for example, depending on the appropriate coupling scheme (discussed in Section 2.2).

of several configurations are required to determine the energy of a particular state of an atom adequately.

- A *term* designates a particular grouping or coupling of all the electrons in an atom. For example, the electrons in the lowest subshells of carbon ($1s^2 2s^2 2p^2$) couple together to give several LS terms, ${}^3P_{0,1,2}$, ${}^1D_2$ and ${}^1S_0$ (see Section 2.2 for these LS designations), each having a different energy. A complete designation of the term requires the configuration (nl) to be specified as well. Since closed subshells produce spherical charge distributions, it is customary to give only the highest or all open subshells. For our carbon example this would be $2p^2\,{}^3P$. In some cases, particularly for heavier atoms, other coupling schemes are more appropriate (see Section 2.2) leading to multiple jj terms, for example. Ground-state configurations are listed in Appendix C.4.

4. In many-electron systems, the electrons typically fall into one of two groups – outer or valence electrons and inner electrons, which form what we refer to as the core electrons or the core.

   - *Valence electrons.* The outer electrons determine the ground state and the bound excited electronic states of an atom. They are responsible for the chemical behavior of atoms in general. Valence electrons typically respond to low-frequency light, frequencies associated with energies of about 1 to 10 eV (near-infrared to ultraviolet wavelengths). The wavefunction of the valence electrons extends from the nucleus to the outer edges of the atom, in general. While outside the core, the valence electron in neutral atoms experiences a Coulomb attraction to the nucleus with an effective charge of one electric charge. Inside the core, the nuclear charge is only partially screened and the effective nuclear charge increases along with stronger attraction to the nucleus. At the same time, the valence electrons interact with the core electrons more strongly when inside the core. The net result is that the valence electrons are more tightly bound than an equivalent electron (one with the same n) in the hydrogen atom. Consider Na, for example. The binding energy of a $3s$ electron in hydrogen, according to Eq. (1.15), is about 1.5 eV, whereas a bit more than 5.1 eV is required to remove the ground-state valence electron from Na. Figure 2.1 compares the hydrogen energy levels with those of several alkali atoms.

     In Section 1.2.3 we learned that the wavefunctions for s states penetrate the nucleus while those for states with $l \neq 0$ are excluded, with those of higher l being pushed further away. This causes low-l states to spend more time near the nucleus than high-l states and,

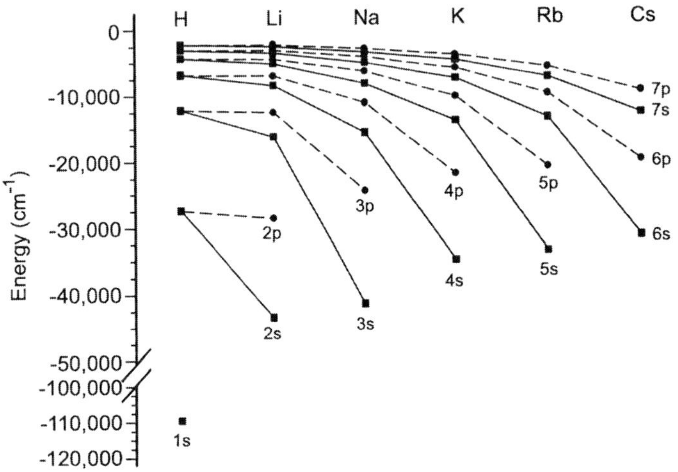

Fig. 2.1 A few observed energy levels of alkali atoms (Li–Cs) vs those of hydrogen.

thus, experience more interactions with the core electrons and a stronger attraction to the nucleus. At the same time, the average radius for the electron for hydrogen-like ions is given by

$$\langle r \rangle = \frac{3}{2}\frac{n^2}{Z}a_0. \tag{2.5}$$

Thus, we expect the shift from the hydrogen position to decrease as n and l increase as a consequence of the electron spending less time in the core. The fact that the shift is larger for s states (see Fig. 2.1) is just a reflection of the simple fact that s states penetrate the core most deeply.

- *Effective Quantum Number.* Even though many-electron atoms are inherently more complicated than hydrogen-like ions, it turns out that we can specify the location of bound levels in Fig. 2.1, at least to lowest order, with a simple equation similar to Eq. (1.51) or (1.57). This is possible because we can account for the average effect of the core and other valence electrons with one parameter we call the *quantum defect*, μ. If we define an *effective quantum number*,[2]

$$n^* = n - \mu, \tag{2.6}$$

2) The effective quantum number is sometimes given the label ν instead of n^*, which is not to be confused with the energy of the state in Eq. (2.7), $\overline{\nu}_n$, or the frequency of the light associated with transitions between levels.

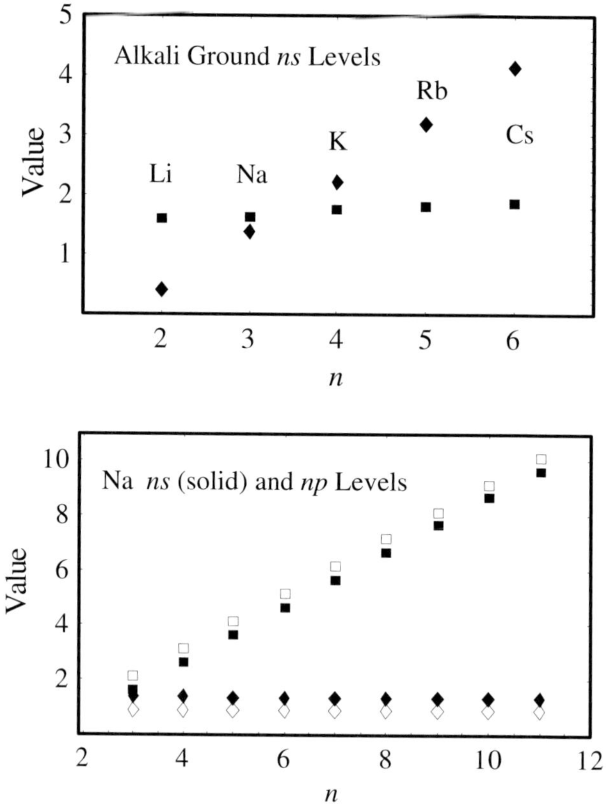

Fig. 2.2 Effective quantum numbers (squares) and the associated quantum defects (diamonds) for neutral alkali atoms (upper) and along the ns (filled) and np Rydberg sequences for neutral Na. Data for these plots were taken from the NIST Atomic Spectra Database [7, 8].

the energy of a particular level can be determined from Eq. (1.57) when we replace the principal quantum number with n^*,

$$\overline{\nu}_n = I_P - \frac{Z^2}{n^{*2}}\mathcal{R}_M. \tag{2.7}$$

Equation (2.7) leads to Rydberg series for multiple electrons as well. Figure 2.2 shows how n^* and μ depend on n and l. In the upper plot, one sees that n^* is nearly constant (changes by less than 20%) for the ground states of Li through Cs. At the same time, μ changes by one unit for each step down the column. This change, however, is due to the fact that n changes by one unit along the sequence. In the lower plot we see that n^* increases with n for both s and p states, while μ is nearly constant in both cases. In addition, we see that μ is different for s and p states, as we surmised above. These observations are associated with the fact that the physics is embodied in the noninteger portion of n^* or μ.

5. What we learned in the preceding paragraphs under (4) can be used to explain the electronic structure of the ground states of the atoms in the periodic table. In general, the energy increases with the sum $n + l$. A state of $n, l + 2$ can be above a state of $n + 1, l$, however. For example, the $4p$ state sits at a higher energy than the $3d$ state, which sits higher than the $4s$ state. As a consequence, in going from K to Kr along the fourth row of the periodic table, the $4s$ subshell fills first followed by the $3d$ and finally the $4p$ subshell. A similar behavior is seen in the fifth, sixth and seventh rows of the periodic table as shown in tables in Appendix C.

6. The degeneracy of the energy within a shell is partially lifted not only by the spin–orbit interaction, the last sum in the many-electron Hamiltonian, Eq. (2.1), but also by the electrostatic interaction, the second sum in Eq. (2.1). This interaction is also responsible for purely quantum effects associated with correlated motion between the electrons as well as the fact that the electrons obey Fermi–Dirac statistics (Pauli exclusion principle).

7. The electronic state of the atom is defined not in terms of the individual orbital and spin angular momenta but by what is called the total angular momentum, $\vec{J}$. The total angular momentum defines the collective angular momentum of the system and is the result of the coupling of the individual orbital and spin angular momenta in one of a variety of possible ways. At one extreme, $\vec{J}$ is the vector sum of the total orbital and spin angular momenta of the system, $\vec{L}$ and $\vec{S}$, respectively, where $\vec{L} = \sum_i \vec{l}_i$ and $\vec{S} = \sum_i \vec{s}_i$, the vector sums of individual orbital and spin angular momenta. The individual total angular momenta $\vec{j}_i (= \vec{l}_i + \vec{s}_i$ for the ith electron in the system) are not good quantum numbers in this case. At the other extreme, the total angular momentum is given by $\vec{J} = \sum_i \vec{j}_i$. In this case, $\vec{L}$ and $\vec{S}$ are not good quantum numbers. In these as well as other cases discussed in Section 2.2, $\vec{J}$ is a good quantum number. In the absence of an applied external field, there will be $2J + 1$ degenerate sublevels (M_J) for each J.

8. When the spin of the nucleus, I, is included, the grand total angular momentum for the atom will be $F = I + J$. The interaction between I and J tends to be weak, so that J remains a good quantum number. For the hyperfine case, there are $2F + 1$ degenerate sublevels (M_F) for each F.

2.2
Angular Momentum Coupling Schemes

The presence of more than one electron necessitates understanding how the various angular momenta (orbital and spin) couple to give a total angular momentum for the atom. In general, the coupling scheme depends on several features. However, the relative magnitudes of the last two sums in the many-electron Hamiltonian in Eq. (2.1) set two important limits. These two terms describe two distinct interactions between the electrons: (i) electrostatic and (ii) spin–orbit. When either (i) or (ii) clearly dominates, the electrons couple according to the *LS*-coupling scheme, (i) > (ii), or the *jj*-coupling scheme, (ii) > (i). The ground state of an atom is usually well represented by *LS* coupling while *jj* coupling describes systems where one electron is highly excited, for example in a Rydberg state. When neither dominates, an intermediate coupling scheme must be employed, which can occur for low-lying excited states of heavy systems.

2.2.1
LS or Russell–Saunders Coupling

When the second sum in Eq. (2.1) (the electrostatic interaction) is larger than the last (the spin–orbit interaction of either electron), *LS* coupling is a good approximation. While the name might imply that there is a strong coupling between $\vec{l}$ and $\vec{s}$, this is not the case! Quite the contrary, in this scheme the orbital angular momenta of the individual electrons couple to give a total $\vec{L}$ and the individual spins couple to give a total $\vec{S}$. The name originates from the coupling of the total orbital angular momentum, $\vec{L}$, to the total spin, $\vec{S}$, to give $\vec{J}$. The maximum and minimum values of J are $\left|\vec{L}\right| + \left|\vec{S}\right|$ and $\left|\vec{L} - \vec{S}\right|$, respectively. The electrostatic interaction usually dominates in low-lying states of light atoms where the electron(s) in the outer shell interact strongly with electrons in the inner shells known as the core. This coupling scheme is often appropriate for the ground states of atoms as well.

As an example of *LS* coupling, consider a closed-shell system such as He, with a ground-state configuration of $1s^2$. The values of L, S and J are all zero. This is true for L because both electrons have $l = 0$. This is true for S because there are only two possible values for m_s, $\pm 1/2$, commonly referred to as *spin up* ($+1/2$ or $\uparrow$) and *spin down* ($-1/2$ or $\downarrow$). Since no two fermions can be in the same state at the same time (Pauli's exclusion principle), one electron must be in the $m_s = +1/2$ state while the other will be in the $m_s = -1/2$ state. The only value M_S can then have is zero, which means that $S = 0$. Since L and S are zero J must be zero as well.

To understand the general case, consider two electrons with quantum numbers ζ_1 and ζ_2. We can form a joint state of the system by multiplying the wavefunctions of the two electrons:

$$\Psi = \psi_1(\zeta_1)\psi(\zeta_2),$$

where 1 and 2 refer to the electrons 1 and 2. However, since the electrons are identical particles, we have to take into account the fact that if we exchange the two, we will not be able to tell. For example, the wavefunction will be symmetric upon exchange of electrons if both electrons have spin up ($\uparrow\uparrow$) or spin down ($\downarrow\downarrow$). However, the wavefunction will be antisymmetric upon exchange of electrons if one electron has spin up while the other has spin down. Thus, we end up with two types of states:

$$\begin{aligned}\Psi_s &= \psi_1(\zeta_1)\psi(\zeta_2) + \psi_1(\zeta_2)\psi(\zeta_1) \quad \text{(triplet state)}\\ \Psi_a &= \psi_1(\zeta_1)\psi(\zeta_2) - \psi_1(\zeta_2)\psi(\zeta_1) \quad \text{(singlet state)}\end{aligned}$$

In general, these will be two different states. Typically, LS states with different L, S and J will have different energies. We label the states in the LS-coupling scheme as

$$nl^{\#\ 2S+1}L_J \text{ or } nl^{\#\ 2S+1}L^{\mathrm{o}}_J,$$

where $nl^{\#}$ is the last subshell to be filled and # is the number of electrons in the subshell. If there are two open subshells, the second is often indicated by adding $n'l'^{\#'}$ to the expression. In the labeling scheme $2S+1$ is called the multiplicity of the state (see Section 2.3) and the superscript ^{o} designates an odd-parity state (see Section 1.3.1) determined by

$$\psi\left(-\vec{r}\right) = (-1)^{\sum_i \left|\vec{l}_i\right|}\psi\left(\vec{r}\right), \tag{2.8}$$

where the sum is over all the electrons in open subshells. In analogy to the names of the subshells, a state of total L is indicated by

$$\begin{aligned} L = {} & 0 \quad 1 \quad 2 \quad 3 \quad 4\\ & S \quad P \quad D \quad F \quad G\end{aligned}$$

but with upper case letters. As mentioned earlier, these are usually referred to as LS terms.

Returning to our He example with a $1s^2$ configuration, since an s shell has even parity, the LS-coupled state formed by this configuration also has even parity. The term representing the state (the ground state in this case) is $1s^{2\,1}S_0$. The superscript 1 preceding S indicates that this is a singlet. In general, closed shells or subshells are spherically symmetric (S states), have even parity (an even number of electrons) and zero spin. These states are then "singlet S zero states."

Within each term there will be one or more states of specific J and its projection along the z-axis, M_J. We will represent these states by a *state vector*,[3]

$$\left|\gamma;\left(l_i m_{l_i},\ldots\right) LM_L \left(s_i m_{s_i},\ldots\right) SM_S; JM_J\right\rangle,$$

where γ represents the other parameters necessary to define the state such as position, n, etc. Within the state vector it is understood that the items inside the parentheses are coupled first to give L and S before L and S are coupled to give J. There is a lot of freedom in writing this vector. It can be written more compactly as

$$\left|\gamma;\left(l_i,\ldots\right) L \left(s_i,\ldots\right) S; JM_J\right\rangle \qquad \text{or} \qquad \left|\gamma; LM_L, SM_S; JM_J\right\rangle. \tag{2.9}$$

Often, the right-hand vector in Eq. (2.9) is written without M_L and M_S, while the left-hand vector is written without the subscript on M_J.

As another example, we will consider the open-shell system of sodium. Its ground-state configuration is $1s^2 2s^2 2p^6 3s$. The first 10 electrons are in closed shells and thus can be designated by ${}^1S_0$. The last electron is in the $3s$ subshell with $s = 1/2$ and $l = 0$. Thus, $L = 0$, $S = 1/2$ (there is only one electron) and $J = 1/2$. We can designate the ground state of Na as

$$1s^2\left({}^1S_0\right) 2s^2\left({}^1S_0\right) 2p^6\left({}^1S_0\right) 3s\,{}^2S_{1/2}, \tag{2.10}$$

or more compactly as $3s^2\ S_{1/2}$. This term is called a "doublet S one-half state." The state vector takes the form $|00, 1/2 \pm 1/2; 1/2 \pm 1/2\rangle$. The ${}^1S_0$ term in Eq. (2.10) contributes nothing to the final designation because $S = J = 0$, i.e., the total spin and total angular momentum are zero.

If the $3s$ electron were promoted to the $3p$ subshell, the total angular momentum could have two values, $L + S = 3/2$ and $L - S = 1/2$. As a result, there are two final states with different J values, $3p^2\ P_{3/2}$ and $3p^2\ P_{1/2}$. The state vectors for these two cases could be written as $|1M_L, 1/2 \pm 1/2; 3/2M\rangle$ and $|1M_L, 1/2 \pm 1/2; 1/2 \pm 1/2\rangle$. These states are called "doublet P three-halves" and "doublet P one-half" states, respectively.

The meaning of the multiplicity is now clear. It is the number of different J states possible for a specific value of L and not the number of possible M_J states as Eq. (2.10) might imply. For S states ($L = 0$) there is only one J value possible regardless of the value of the total spin. Although misleading, the multiplicity, $2S + 1$, is traditionally given for S states ($L = 0$) as well, as was done for the $3s$ configuration. To circumvent this anomaly with S terms, one

3) See Appendix B.8 for a discussion of the mathematical formalism of state vectors (the properties of bras, $|\rangle$ and kets, $\langle|$) and the complex algebra they obey.

must remember that the number of J levels is given by $2S + 1$ (the multiplicity) when $L \geqslant S$ and $2L + 1$ when $L < S$.

As a last example, we will look at the two-electron system of carbon with a ground-state configuration of $1s^2 2s^2 2p^2$. As in the previous examples, the $1s^2 2s^2$ subshells are closed so that we only need to look at the two $2p$ electrons. Each electron can be in one of six possible states.

$$s = \tfrac{1}{2} \quad m_s = -\tfrac{1}{2} \quad \tfrac{1}{2}$$

$$l = 1 \quad m_l = -1 \quad 0 \quad 1$$

If we couple two orbital angular momenta with $l = 1$ indiscriminately, we would find that there are three possible values for L: 0, 1 and 2. The possible values for S are 0 and 1. Consequently, there would be $36 = 6^2$ possible states with J ranging in value from 0 to 3. However, all of these states will not be allowed in every situation because of Pauli's exclusion principle. Some combinations are not allowed when so-called equivalent electrons, that is, those that have the same n and l, are coupled. The only allowed states will be those where m_{l_1} and m_{l_2} are different and/or m_{s_1} and m_{s_2} are different. The number of states meeting this criterion is given by the binomial coefficient $\binom{6}{2} = 15$. That is, of the 36 possible combinations, only 15 exist where their quantum numbers are not identical; these will be the allowed states. To determine which 15 do not violate Pauli's exclusion principle, we will use Table 2.1 to help us keep track of the quantum numbers for each electron.

In the first two columns, we give the magnetic quantum numbers for l and s of the first and second electrons as $m_l^{m_s}$. In the third through fifth columns we give $M_L = m_{l_1} + m_{l_2}$, $M_S = m_{s_1} + m_{s_2}$ and $M_J = M_L + M_S$ values, respectively. To determine the terms to which the entries in Table 2.1 are to be associated, we will group the occurrences of M_L and M_S. We recall that the maximum values that L and S can assume are 2 and 1, respectively. When $L = 0$, M_L takes on one value, 0, while when $L = 1$ and 2, $M_L = 0, \pm1$ and 0, ±1 and ±2, respectively. We next must count the occurrences of the various values for M_L and M_S, which we tabulate in Table 2.2.

We note that $M_L = M_S = 0$ occurs three times. One of these corresponds to $L = S = 0$, a singlet S state ($2S + 1 = 1$). We also notice that along the column where $M_S = 0$, M_L takes on the values 0, ±1 and ±2. This corresponds to $L = 2$ and $S = 0$ or a singlet D state. Finally, we are left with a single occurrence of M_L and M_S running from -1 to 1. This corresponds to $L = S = 1$ and a triplet P state. Thus, we have ^1S, ^3P and ^1D states. Now we need to assign J values to these states. For the ^1S and ^1D states, the only choices are $J = 0$ and 2, respectively. For the ^3P state, we have $J = 0, 1$ and 2. To summarize, the allowed terms are $^1S_0, ^3P_{0,1,2}$ and $^1D_2$. It is straightforward to show that our selections account for all the M_J values in Table 2.1 and that there are indeed 15 states. Finally, we state a general property of LS coupling that allows one

Tab. 2.1 The 15 unique selections of quantum numbers for equivalent p^2 electrons in *LS* coupling.

$m_{l_1}^{m_{s_1}}$	$m_{l_2}^{m_{s_2}}$	M_L	M_S	M_J
$-1^{-1/2}$	$-1^{+1/2}$	-2	0	-2
	$0^{-1/2}$	-1	-1	-2
	$0^{+1/2}$	-1	0	-1
	$1^{-1/2}$	0	-1	-1
	$1^{+1/2}$	0	0	0
$-1^{+1/2}$	$0^{-1/2}$	-1	0	-1
	$0^{+1/2}$	-1	1	0
	$1^{-1/2}$	0	0	0
	$1^{+1/2}$	0	1	1
$0^{-1/2}$	$0^{+1/2}$	0	0	0
	$1^{-1/2}$	1	-1	0
	$1^{+1/2}$	1	0	1
$0^{+1/2}$	$1^{-1/2}$	1	0	1
	$1^{+1/2}$	1	1	2
$1^{-1/2}$	$1^{+1/2}$	2	0	2

Tab. 2.2 Occurrences of M_L and M_S values.

M_L \ M_S	–1	0	1
–2		1	
–1	1	2	1
0	1	3	1
1	1	2	1
2		1	

to quickly verify that the selections are correct when coupling two equivalent electrons,

$$L + S \textit{ must be an even number.}$$

It is important to note that had we chosen oxygen to study instead of carbon, we would have had to consider a $2p^4$ configuration. However, this would correspond to two vacancies or holes in the $2p$ configuration and would lead to exactly the same *LS* terms. All the *LS* terms for coupling all combinations of same-shell p, d and f electrons are given in Ref. [2].

2.2.2
jj Coupling

When the second sum in Eq. (2.1) is larger than the last, that is the spin–orbit interaction of both electrons is stronger than the electrostatic interaction between the electrons, *jj* coupling is most appropriate. In other words, the two electrons in question are sufficiently far from each other that the electrostatic interaction between them is weak and thus only their total angular momenta, j, are constants of the motion. This is most often the case when the excited electron is in a high-n state and thus spends most of its time far from the others confined to the core near the nucleus. Consequently, for each electron (or in general for the excited electron and the core electrons) we have $\vec{l}_i + \vec{s}_i = \vec{j}_i$ and $\vec{j}_1 + \vec{j}_2 = \vec{J}$. The possible values for $\vec{J}$ lie between $\left|\vec{j}_1\right| + \left|\vec{j}_2\right|$ and $\left|\vec{j}_1 - \vec{j}_2\right|$. The usual designation for a *jj*-coupled term is

$$(j_1, j_2)_J . \tag{2.11}$$

Again, if the term has odd parity, a superscript $^\circ$ is placed to the right of the right parenthesis. As in *LS* coupling, we can write a state vector,

$$\left|\gamma; (l_1 m_1, s_1 m_1)\, j_1 m_1\, (l_2 m_2, s_2 m_2)\, j_2 m_2; J M_J\right\rangle . \tag{2.12}$$

The application of Pauli's exclusion principle in *jj* coupling leads to no two electrons having the same value for n, l, j, and m_j simultaneously.

As an example, we will look at two equivalent p electrons again. We will keep track of the quantum numbers for each electron in Table 2.3. The following *jj*–coupled terms can be extracted from Table 2.3: $(1/2, 1/2)_0$, $(1/2, 3/2)_{1,2}$ and $(3/2, 3/2)_{0,2}$. One should notice that there are again five states with two having $J = 0$ and 2, and one with $J = 1$. The reader is encouraged to create a table similar to Table 2.2 to count the occurrences. The fact that we obtain the same J values as we did with *LS* coupling means that regardless of how we couple the angular momenta, J is a *good* quantum number, i.e., a constant of the motion.[4]

2.2.3
Intermediate or Pair Coupling

The intermediate coupling scheme, which also involves excited configurations, occurs when neither of the last two terms in Eq. (2.1) clearly dominates. There are two limiting cases of this coupling scheme, *jl* coupling (also known as *jK* coupling) and *LK* coupling (sometimes called *Ls* coupling, not to be confused with *LS* coupling discussed in Section 2.2.1). In either case, the spin of

4) The reader is referred to a standard quantum mechanics text for a proper discussion of quantities that are "constants of the motion."

Tab. 2.3 The 15 unique selections of quantum numbers for equivalent p^2 electrons in *jj* coupling.

j_1	m_1	j_2	m_2	M_J	J
$+\frac{1}{2}$	$-\frac{1}{2}$	$+\frac{1}{2}$	$+\frac{1}{2}$	0	0
$+\frac{1}{2}$	$-\frac{1}{2}$	$+\frac{3}{2}$	$-\frac{3}{2}$	-2	2
			$-\frac{1}{2}$	-1	2, 1
			$+\frac{1}{2}$	0	2, 1
			$+\frac{3}{2}$	$+1$	2, 1
$+\frac{1}{2}$	$+\frac{1}{2}$	$+\frac{3}{2}$	$-\frac{3}{2}$	-1	2, 1
			$-\frac{1}{2}$	0	2, 1
			$+\frac{1}{2}$	$+1$	2, 1
			$+\frac{3}{2}$	$+2$	2
$+\frac{3}{2}$	$-\frac{3}{2}$	$+\frac{3}{2}$	$-\frac{1}{2}$	-2	2
			$+\frac{1}{2}$	-1	2
			$+\frac{3}{2}$	0	2, 0
$+\frac{3}{2}$	$-\frac{1}{2}$	$+\frac{3}{2}$	$+\frac{1}{2}$	0	0
			$+\frac{3}{2}$	$+1$	2
$+\frac{3}{2}$	$+\frac{1}{2}$	$+\frac{3}{2}$	$+\frac{3}{2}$	2	2

the excited electron is coupled last after all the other angular momenta are coupled to give an intermediate angular momentum called K. The name pair coupling comes from the fact that coupling s (= 1/2) to K always produces a pair of levels with total J differing by 1; see Fig. 2.3. In the jl-coupling limit, the order of strength of the interactions is (1) the spin–orbit interaction of the inner or more tightly bound electron, (2) the electrostatic interaction between the two electrons and (3) the spin–orbit interaction of the outer or less tightly bound electron. In this case, j of the inner electron (which we will call j_1) couples to the orbital angular momentum of the outer (excited) electron to give K. Finally, s of the outer electron is coupled to give the total J. The state vector for this limit is given by

$$\left|\gamma; \left[(l_1, s_1)\, j_1, l_2\right] K s_2; J M_J\right\rangle . \tag{2.13}$$

In the other limit, the relative strengths of (1) and (2) are reversed. In this case, the two orbital angular momenta couple to give L, which couples to s of the inner electron to give K. The state vector for this limit is given by

$$\left|\gamma; \left[(l_1, l_2)\, L, s_1\right] K s_2; J M_J\right\rangle . \tag{2.14}$$

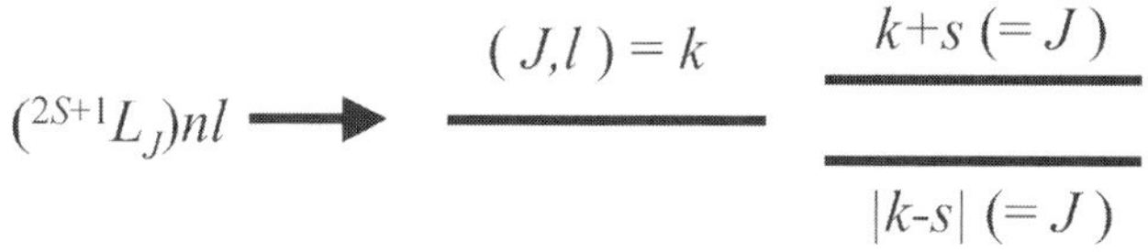

Fig. 2.3 This is an example of KL coupling involving a $^{2S+1}L_J$ core and an nl electron. Coupling the spin to k, splitting the k degeneracy to produce two different values of J, gives this scheme its alternate name of *spin doubling*.

As an example, we will look at Xe, which has 54 electrons in closed shells and whose ground-state configuration is $5p^6\,{}^1S_0$. The low-lying excited levels of Xe obey jl coupling. If a $5p$ electron is promoted to the $6s$ subshell (see below), four levels will arise, one each with $J = 2, 0$ and two with $J = 1$. In the last three columns of the table below, the values for j, jl and J are underlined.

Configuration	$\underline{j}$: Couple l & s for inner electrons	$\{\underline{jl}\}$: Couple j to outer l	$\underline{J}$: Couple $\{jl\}$ to outer spin
$5p^5\,6s$	$5p^5\,(^2P_{\underline{3/2}})$	$[3/2,0]\ \underline{3/2}$	$[3/2,1/2]\ \underline{2}$
$5p^5\,6s$	$5p^5\,(^2P_{\underline{3/2}})$	$[3/2,0]\ \underline{3/2}$	$[3/2,1/2]\ \underline{1}$
$5p^5\,6s$	$5p^5\,(^2P_{\underline{1/2}})$	$[1/2,0]\ \underline{1/2}$	$[1/2,1/2]\ \underline{1}$
$5p^5\,6s$	$5p^5\,(^2P_{\underline{1/2}})$	$[1/2,0]\ \underline{1/2}$	$[1/2,1/2]\ \underline{0}$

Figure 2.4 shows the transition from LS to jj coupling.

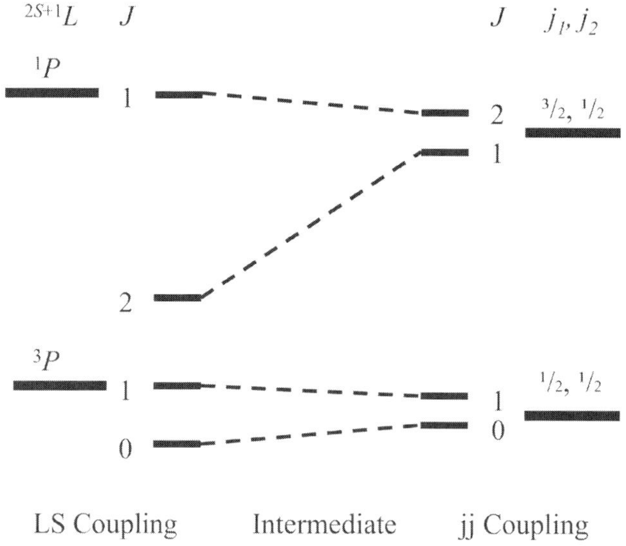

Fig. 2.4 An example of the transition from LS coupling to jj coupling involving two electrons with orbital angular momenta of $l_1 = 1$ and $l_2 = 0$.

2.2.4
Recoupling Between Coupling Schemes

One finds in practice that there is a need to express the wavefunction in one coupling scheme in terms of another coupling scheme. This would be the case, for example, for an electron in a Rydberg state where the electron interacts strongly with the other electrons in the system when it is near the nucleus where *LS* coupling might be appropriate, but only weakly when it is far from the nucleus where *jj* coupling is appropriate.

Consider optical transitions in a two-electron system. If we prepare the system in a well-defined state of J and M far from the nucleus, the state vector takes the form

$$|[(l_1 s_1) j_1 (l_2 s_2) j_2] JM\rangle .$$

The transitions occur near the nucleus so that we need state vectors of the form

$$|[(l_1 l_2) L (s_1 s_2) S] JM\rangle .$$

Since both coupling schemes form a complete set, we can express the former in terms of a series of the latter. The numerical coefficients, the Clebsch–Gordon coefficients (see Appendix D), can be determined from

$$\langle [(l_1 l_2) L (s_1 s_2) S] JM | [(l_1 s_1) j_1 (l_2 s_2) j_2] JM\rangle .$$

As a specific example, consider two nonequivalent p electrons (pp') with total angular momentum $J = 1$; *LS* coupling leads to the following possible states:

$${}^3S_1,\ {}^1P_1,\ {}^3P_1 \text{ and } {}^3D_1$$

while *jj* coupling leads to

$$\left(\frac{1}{2}, \frac{1}{2}\right)_1, \left(\frac{1}{2}, \frac{3}{2}\right)_1, \left(\frac{3}{2}, \frac{1}{2}\right)_1, \left(\frac{3}{2}, \frac{3}{2}\right)_1 .$$

We can build a transformation matrix by calculating the Clebsch–Gordon coefficients for each case:

	$\left(\frac{3}{2},\frac{3}{2}\right)_1$	$\left(\frac{3}{2},\frac{1}{2}\right)_1$	$\left(\frac{1}{2},\frac{3}{2}\right)_1$	$\left(\frac{1}{2},\frac{1}{2}\right)_1$
$^3D_1$	$-\frac{1}{3}\sqrt{\frac{2}{3}}$	$-\frac{1}{3}\sqrt{\frac{5}{6}}$	$\frac{1}{3}\sqrt{\frac{5}{6}}$	$\frac{2}{3}\sqrt{\frac{5}{3}}$
$^3P_1$	0	$\frac{1}{\sqrt{2}}$	$\frac{1}{\sqrt{2}}$	0
$^1P_1$	$\frac{1}{3}\sqrt{5}$	$\frac{1}{3}$	$-\frac{1}{3}$	$\frac{1}{3}\sqrt{2}$
$^3S_1$	$\frac{1}{3}\sqrt{\frac{10}{3}}$	$-\frac{2}{3}\sqrt{\frac{2}{3}}$	$\frac{2}{3}\sqrt{\frac{2}{3}}$	$-\frac{1}{3\sqrt{3}}$

From the table it is clear that if the system were initially prepared in the *jj*-coupled state $(1/2, 1/2)_1$ then the *LS*-coupled wavefunction would have the character

$$|\psi_{LS}\rangle = \frac{1}{3}\left(2\sqrt{\frac{5}{3}}|^3D_1\rangle + \sqrt{2}|^1P_1\rangle - \frac{|^3S_1\rangle}{\sqrt{3}}\right).$$

In this case, although the dominant *LS*-coupled term is the $^3D_1$ term, the $^1P_1$ and $^3S_1$ terms are non negligible.

2.3 Fine Structure

The energy shifts associated with the electrostatic interaction and the spin–orbit interaction have opposite signs. The electrostatic interaction leads to a positive shift in the energy levels, i.e., the levels are less bound. The magnitude of the shift depends on both the total spin and the total orbital angular momentum. In particular, it has been found empirically that for the ground configuration and for configurations with equivalent electrons that *the lowest-energy state will be associated with the term of highest multiplicity (greatest S). The term of lowest energy of a given multiplicity will be that with the greatest L.* This behavior is often referred to as Hund's rule. For both our carbon, $2p^2$, and oxygen, $2p^4$, examples, Hund's rule implies that the lowest state will be the ^3P state and that the highest state will be the ^1S state. This is exactly what one finds.[5]

We saw in Section 1.4.2 that the spin–orbit interaction led to a splitting within a shell; not all levels with the same n quantum number have the same

5) See *Atomic Energy Levels: Vol. I*, NSRDS-NBS 35, which can be found on the NIST web site [9].

energy. It is generally true that levels within nonsinglet LS terms, known as multiplets, are split by the spin–orbit interaction. This *fine-structure splitting* is proportional to J, the *Landé interval rule*. This can be shown by calculating $\Delta E_{\gamma LS;J}$, which is given by the matrix element of the various multiplets over the spin–orbit interaction, $\zeta(r)\overrightarrow{L}\cdot\overrightarrow{S}$, from Eq. (1.90):

$$\Delta E_{\gamma LS;J} = \left\langle \gamma; LM_L, SM_S; JM_J \right| \zeta(r)\overrightarrow{L}\cdot\vec{S}\left| \gamma; LM_L, SM_S; JM_J \right\rangle . \tag{2.15}$$

As was done in the fine-structure calculation for hydrogen, we express $\vec{L}\cdot\vec{S}$ as $J^2 - L^2 - S^2$ and write

$$\Delta E_{\gamma LS;J} = \xi(\gamma LS)\left[J(J+1) - L(L+1) - S(S+1)\right]. \tag{2.16}$$

The spacing between levels J and $J-1$ will be

$$\Delta E_{\gamma LS;J} - \Delta E_{\gamma LS;J-1} = \xi(\gamma LS)\,J. \tag{2.17}$$

The quantity $\xi(\gamma LS)$ is given by

$$\xi(\gamma LS) = \frac{\hbar^2}{2m_e^2c^2}\left\langle \frac{1}{r}\frac{dU(r)}{dr} \right\rangle, \tag{2.18}$$

where $U(r)$ is the effective potential in which the electrons move (see Eq. (2.2)). When $\xi(\gamma LS) > 0$ the splitting of the multiplet is referred to as normal, where the level with lowest J will have the lowest energy. When $\xi(\gamma LS) < 0$ the splitting of the multiplet is referred to as inverted, where the level with highest J will have the lowest energy. A configuration that has a subshell that is less than 50% full will have a normal multiplet while one that is more than 50% full will have an inverted multiplet. Configurations that are 50% full follow no general tend. Carbon is an example of a normal multiplet while oxygen is an example of an inverted multiplet.

There are some important properties of $\Delta E_{\gamma LS;J}$ that should be realized.

1. $\Delta E_{\gamma LS;J}$ has no M dependence, which implies that there is no preferred direction in space – the fine structure is spherically symmetric.

2. Sum rule:

$$\sum_{J=|L-S|}^{L+S} (2J+1)\,\Delta E_{\gamma LS;J} = 0. \tag{2.19}$$

   This implies that the center of gravity of the multiplet,

$$\overline{E}_{\gamma LS;J} = \frac{\sum_J (2J+1)\,\Delta E_{\gamma LS;J}}{\sum_J (2J+1)} = 0, \tag{2.20}$$

   coincides with the unsplit term.

Example. Find the multiplet splitting for $L \neq 0$, $S = 1/2$. Since $J = L \pm 1/2$, the energy for the $L + 1/2$ level will be given by

$$\Delta E_{\gamma LS;L+1/2} = \xi\left(\gamma LS\right) \cdot L, \tag{2.21}$$

while that for the $L - 1/2$ level will be given by

$$\Delta E_{\gamma LS;L-1/2} = -\xi\left(\gamma LS\right) \cdot \left(L+1\right). \tag{2.22}$$

For a normal splitting, $\xi\left(\gamma LS\right) > 0$.

Further Reading

1 W. C. Martin and W. L. Wiese, *Atomic Spectroscopy: A Compendium of Basic Ideas, Notation, Data and Formulas*, Physical Reference Data, National Institute of Standards and Technology [http://physics.nist.gov/Pubs/AtSpec/].

2 R. D. Cowan, *The Theory of Atomic Structure and Spectra*, University of California Press, Los Angeles, CA, 1981.

3 I. I. Sobelman, *Atomic Spectra and Radiative Transitions*, Springer Series in Chemical Physics, Springer, New York, NY, 1979.

4 F. Yang and J. H. Hamilton, *Modern Atomic and Nuclear Physics*, McGraw-Hill, San Francisco, CA, 1996.

5 For additional problems with worked solutions see D. Budker, D. F. Kimball and D. P. DeMille, *Atomic Physics: An Exploration Through Problems and Solutions*, Oxford University Press, New York, NY, 2004.

Problems

2.1 From the energy levels compiled by NIST,[6] estimate the quantum defect, μ, for the s, p and d levels for Li through Ne. How does the trend change for the first ion?

2.2 Using the same data as in the previous problem, estimate $\xi\left(\gamma LS\right)$ for Li through Ne. How does the trend change for the first ion?

6) At the time of writing this book, the link was http://physics.nist.gov/PhysRefData/Handbook/periodictable.htm. If it is no longer active, try to search for "Physical Reference Data" at http://www.nist.gov.

3 Atoms in Static Fields

3.1 External Electric and Magnetic Fields

Thus far, we have looked at the atom in the absence of external fields. External fields will perturb the structure outlined in the previous sections. Weak fields will remove some of the degeneracies while strong fields can cause the atom to fall apart. In this chapter we will study the atom's response to static electric (Stark effect) and magnetic (Zeeman effect) fields. We will treat the response to AC fields, which are responsible for transitions between different states of a system, in Chapter 4.

3.1.1 Stark Effect

When an electric field, F, is applied to an atom in the z direction, a new term, eFz, must be added to the effective potential of Eq. (2.4). Before we look at how the energy levels are perturbed, it is instructive to look at the gross classical behavior associated with this term. This is possible by letting $U(r) \rightarrow V(z)$, where

$$V(z) = -\frac{1}{4\pi\varepsilon_0}\frac{Ze^2}{|z|}. \tag{3.1a}$$

To ease numerical calculations, it is often convenient to use a smoothed version of this one-dimensional potential,

$$V(z) = -\frac{1}{4\pi\varepsilon_0}\frac{Ze^2}{\sqrt{z^2+a^2}}, \tag{3.1b}$$

Light-Matter Interaction: Atoms and Molecules in External Fields and Nonlinear Optics.
W. T. Hill and C. H. Lee

ISBN: 978-3-527-40661-6

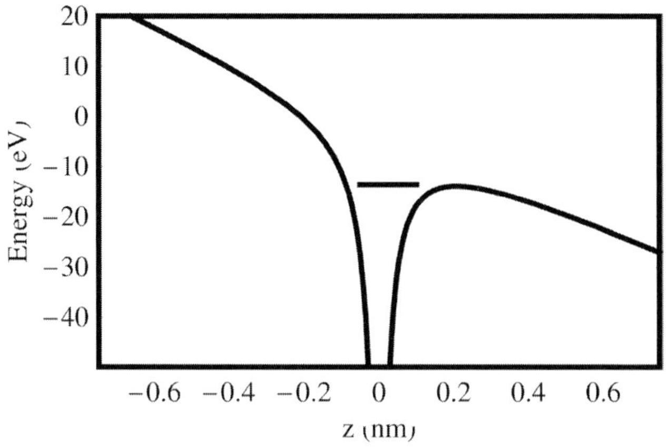

Fig. 3.1 Composite potential, smoothed Coulomb potential of Eq. (3.1b) with $a = 0.01$ plus the potential due to a static electric field, $-eFz$, with $F = 3.2 \times 10^8$ V/cm. The horizontal line inside the potential indicates the level of the ground state of H, which is absolutely bound for all negative z but is unbound, as are all higher levels, for positive z.

first introduced by Eberly's group [10].[1] The parameter a ensures that the potential is finite everywhere and can be used to scale the system's response to that of the actual three-dimensional atom. An examination of the properties of the atom associated with this potential can be found in Ref. [11]. Figure 3.1 shows the composite potential,

$$V'(z) = V(z) - eFz, \tag{3.2}$$

for positive F and small a. Electrons are bound by a potential that extends to ∞ for negative z, but those in states that lie above the barrier at $z_c = \sqrt{Ze/4\pi\varepsilon_0 F}$, determined by setting $\partial V'(z)/\partial z|a \to 0 = 0$, can escape to the right, a process that goes by the name *field ionization*.[2] The barrier height, as $a \to 0$, is given by

$$V_c = V(z_c) - eFz_c = -2e\sqrt{\frac{eFZ}{4\pi\varepsilon_0}}. \tag{3.3}$$

The field, F_c, required to lower this barrier to the level of the ground state can be determined by setting $V'(z_c)$ equal to the ionization potential. For one-electron, hydrogen-like ions the energy for a particular level n is given by

1) It is interesting to note that in honor of Professor Joe Eberly, the Division of Atomic, Molecular and Optical Physics of the American Physical Society jokingly proclaimed on July 28, 2003 that a virtual atom (element) described by this potential would bear the name *Eberlonium*.

2) For the three-dimensional case, z_c locates a saddle point in the Coulomb potential over which the electron escapes.

Eq. (1.15). Thus, the field required for such an *over-the-barrier ionization* of an atom is the simple expression

$$F_c = \frac{e}{4\pi\varepsilon_0 a_0^2} \frac{Z^3}{16n^4} \text{ volts} \tag{3.4a}$$

$$= \frac{Z^3}{16n^4} \text{ a.u.} \tag{3.4b}$$

An electron in $n = 50$ would only require a field of about 130 kV/cm, i.e., 13 kV across a 1-mm gap. The ground state of hydrogen requires an electric field (1/16)th the strength of the atomic field strength, $\sim 3.2 \times 10^8$ V/cm, to ionize via field ionization. Average field strengths of this magnitude can easily be created in the focus of a pulsed laser where the intensity can exceed 1.4×10^{14} W/cm^2. For a general atom, with ionization potential, I_p, the threshold intensity for over-the-barrier field ionization, often called the appearance intensity, is given by [12]

$$I_{app} = \frac{c}{128\pi} \left(\frac{4\pi\varepsilon_0}{e^2} \right)^3 \frac{I_p^4}{Z^2}, \tag{3.5}$$

where c is the speed of light. With I_p in eV, I_{app} will be in W/m^2. Figure 3.2 compares the threshold for field ionization determined by measurement to that predicted by Eq. (3.5). The simple classical theory shows excellent agreement over a wide range of intensities and systems.

Quantum mechanically, when the energy level in Fig. 3.1 is below the barrier, the electron can tunnel through if the laser frequency is not too high. The condition for tunnel ionization is set by the so-called *Keldysh parameter* [13],

$$\gamma = \frac{\omega_L}{\omega_t} = \frac{\omega_L \sqrt{2m_e I_p}}{eF}, \tag{3.6}$$

where ω_L is the laser frequency (typically of the order of 10^{14} Hz), $1/\omega_t$ is the tunneling time and I_p is the ionization potential of the atom. Tunneling occurs when the tunneling time is less than the laser period. That is, the electron must be able to make it through (over) the barrier within one optical cycle. The tunneling time is determined by requiring an electron to acquire enough kinetic energy from the field in a time τ to pass over (through) an energy barrier of height I_o. Thus, $mv^2/2 = I_o$, where $v = a\tau$, with the acceleration given by $a = eF$. This leads to the Keldysh expression $\tau = 1/\omega_t \simeq \sqrt{2mI_o}/eF$. Clearly, tunneling occurs when $\gamma \ll 1$. Consequently, field ionization induced with laser fields is often called *AC tunneling ionization*. In the other extreme, $\gamma \gg 1$, multiphoton ionization dominates. Multiphoton ionization will be discussed in Chapters 4 and 9.

Perelomov, Popov and Terent'ev [14] developed an expression for the tunnel ionization rate for hydrogen. Ammosov, Delone and Krainov [15] extended

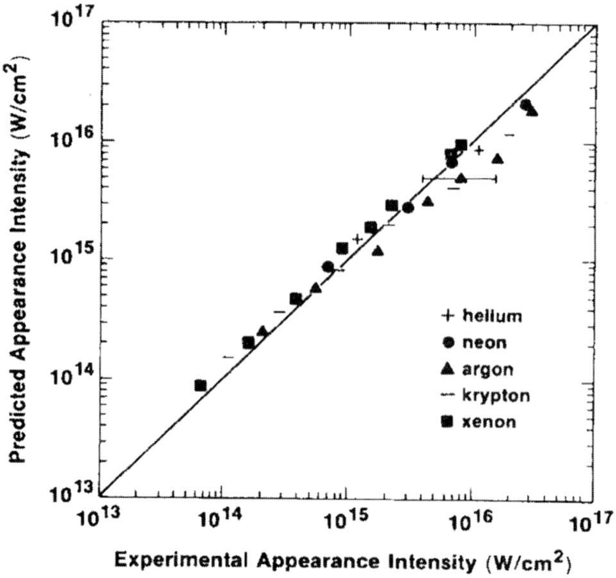

Fig. 3.2 A comparison between experimentally determined and theoretically predicted appearance intensities associated with (Eq. (3.5)) for increasing stages of ionization from the lower left to the upper right. (Reprinted figure with permission from S. Augst, D. Strickland, D. D. Meyerhofer, S. L. Chin and J. H. Eberly, Phys. Rev. Lett. **63**, 2212 (1989). Copyright (1989) by the American Physical Society.)

this to complex atoms. Commonly referred to as the ADK theory, for $n^* \gg l$ this rate in Hz can be written as [3]

$$w = \frac{1}{\tau_{a_0}} \left(\frac{3e}{\pi}\right)^{3/2} \frac{Z^2}{3n^{*3}} \frac{2l+1}{2n^*-1} \left[\frac{4eZ^3}{(2n^*-1)n^{*3}F}\right]^{2n^*-3/2} \times \exp\left[-\frac{2Z^3}{3n^{*3}F}\right], \tag{3.7}$$

where τ_{a_0} is the atomic time (see Appendix A), n^* is the effective quantum number (Eq. (2.6), which in atomic units can be expressed as $Z/\sqrt{2I_p}$ where I_p of the system is also in atomic units), $e = 2.71828\ldots$, l is the orbital angular momentum of the state and F is the field strength in atomic units. The ADK expression fits experimental data better at higher intensities where γ is smaller [19]. Modifications to Eq. (3.7) for molecular systems are discussed in Ref. [20].

For sufficiently weak fields, the low-lying states in an atom will not be ionized but perturbed, where eFz is the perturbation Hamiltonian. Since $z \to -z$

3) This form of the expression is the same as that given in Eq. (2) of Ref. [16] but multiplied by a factor of the atomic frequency, $1/\tau_{a_0}$, to give the expression dimensions of Hz. Note that there is an error in Eq. (3) of Ref. [16]; $C_{n,l}$ should be squared (see Refs. [17,18]).

under inversion, eFz must couple states of opposite parity. Since the l states are degenerate in the nonrelativistic treatment of hydrogen and one-electron ions for the same principal quantum number (i.e., $n > 1$), they will exhibit a linear Stark shift. By linear, we mean that the shifts are calculable in first-order perturbation theory,

$$\Delta E^{(1)} = \langle \psi | eFz | \psi \rangle. \tag{3.8}$$

Since the l degeneracy is lifted by electrostatic interaction in multielectron systems, second-order perturbation theory must be applied,

$$\Delta E^{(2)} = \sum_{\psi_i'} \frac{\langle \psi | eFz | \psi' \rangle \langle \psi' | eFz | \psi \rangle}{E_\psi - E_{\psi'}}. \tag{3.9}$$

The second-order process is usually referred to as the quadratic Stark effect. We will look at both in this section.

3.1.1.1 Linear Stark Effect

Schrödinger [21], Epstein [22] and Walker [23] showed that the linear Stark problem is nicely solved in parabolic coordinates,[4] where the Hamiltonian is written as

$$H' = \frac{e}{2}(\xi - \eta)F. \tag{3.10}$$

In parabolic coordinates the perturbation energy is given by

$$\begin{aligned} \Delta E^{(1)} &= \frac{e}{2} F \langle \psi_{n_1', n_2', m'} | (\xi - \eta) | \psi_{n_1, n_2, m} \rangle \\ &= \frac{3}{2} n (n_1 - n_2) \frac{eFa_o}{Z}, \end{aligned} \tag{3.11a}$$

where

$$n = n_1 + n_2 + |m| + 1. \tag{3.11b}$$

Figure 3.3 shows an example for $n = 2$. We notice that when $n_1 = n_2 = 0$ and $m = \pm(n-1) = \pm 1$, $\Delta E^{(1)} = 0$, reflecting the fact that Eq. (3.11a) is independent of m. The two extreme components corresponding to $m = 0$ ($n_1 = n - 1$, $n_2 = 0$ and $n_1 = 0$, $n_2 = n - 1$) shift in opposite directions, with a separation of these two extremes $3n(n-1)ea_oF/Z$. For $n_1 \neq n_2$, the degeneracy between n_1 and n_2 is removed. Table 3.1 summarizes the allowed values for $n_1 - n_2$ for various values of m.

4) See Appendix B.9 for a discussion of parabolic coordinates and the Schrödinger equation.

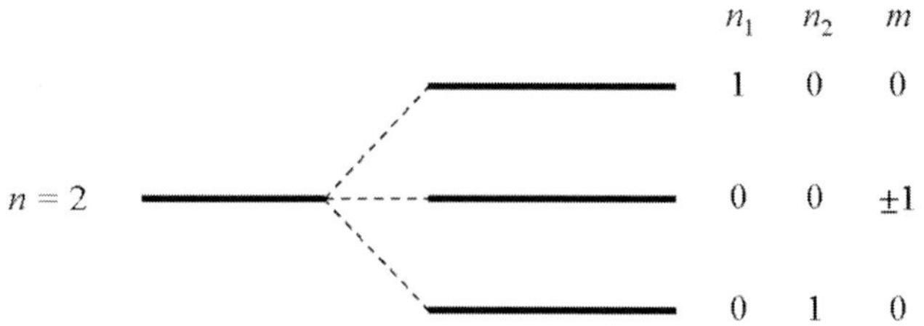

Fig. 3.3 Linear Stark splitting for $n = 2$. The shift for a field of magnitude 5.0×10^5 V/cm is ~ 0.016 eV.

It is possible to express parabolic wavefunctions, $u_{n_1 n_2 m}$, in terms of a linear combination of spherical wavefunctions, ψ_{nlm}, and vice versa. As an example, consider $n = 1$. This value of the parabolic n corresponds to the spherical principal quantum number 2, leading to

$$\begin{aligned} u_{100} &= \mathrm{C}\mathrm{e}^{-ar}\left[1 + ar(1 + \cos\theta)\right], \\ u_{010} &= \mathrm{C}\mathrm{e}^{-ar}\left[1 + ar(1 - \cos\theta)\right], \\ u_{00\pm1} &= \mathrm{C}\mathrm{e}^{-ar} r \sin\theta \mathrm{e}^{\pm im\phi}. \end{aligned}$$

After some manipulation, this leads to

$$\begin{aligned} \psi_{2p0} &= C(u_{010} - u_{100}), \\ \psi_{2s0} &= C(u_{010} + u_{100}). \end{aligned}$$

To see what this means, let us prepare the system in the ψ_{2s0} state at $t = 0$,

$$\Psi(t) = C\left[u_{010}\mathrm{e}^{-i(E_{n=2}-\Delta)t/\hbar} + u_{100}\mathrm{e}^{-i(E_{n=2}+\Delta)t/\hbar}\right]. \tag{3.12}$$

This can be written as

$$\begin{aligned} \Psi(t) &= \left[C(u_{010} + u_{100})\cos(\Delta t/\hbar) + iC(u_{010} - u_{100})\sin(\Delta t/\hbar)\right]\mathrm{e}^{-iE_{n=2}t/\hbar} \\ &= \left[\psi_{2s0}\cos(\Delta t/\hbar) + i\psi_{2p0}\sin(\Delta t/\hbar)\right]\mathrm{e}^{-iE_{n=2}t/\hbar}. \end{aligned}$$

The probability density for the system is

$$|\Psi(t)|^2 = |\psi_{2s0}|^2\cos^2(\Delta t/\hbar) + |\psi_{2p0}|^2\sin^2(\Delta t/\hbar), \tag{3.13}$$

Tab. 3.1 Allowed values for $n_1 - n_2$ for selected values of m.

m	$n_1 - n_2$
0	$n-1, n-3, \ldots, -n+1$
1	$n-2, n-4, \ldots, -n+2$
3	$n-3, n-5, \ldots, -n+3$
$\vdots$	$\vdots$
$n-1$	0

and the system oscillates between the 2*s* and 2*p* states with a period $\tau \sim 2\pi\hbar/\Delta$. For a field of $\sim$ 300 V/cm, $\tau \sim 4.3 \times 10^{-10}$ s. Note that this value is of the order of the radiative lifetime of atomic states, which means that the character of a state will change during a transition. This oscillation causes the 2*s* state to pick up some 2*p* character, allowing forbidden electric dipole transitions (see Section 4.4) to occur.

For Rydberg states, the energy levels are closely spaced. For sufficiently large fields, the manifolds begin to overlap. This so-called *crossing field* (F_c), where the upper extreme from the lower level meets the lower extreme from the upper level, is given by the Inglis–Teller limit $F_c = (3n^5)^{-1}$ [24].

The orbital angular momentum, L, in the presence of an external electric field is not a constant of the motion. However, m is still a good quantum number. This gives some insight into why L is not a good quantum number in a diatomic molecule. The presence of the electric field from the other nucleus distorts the spherical symmetry. While L is not conserved, Λ ($\equiv |m|$) is a good quantum number. The linear case, however, only applies to hydrogen, which has energy-degenerate states of different parity. For other systems, including molecules, we must go to the next order to determine the Stark shifts. As we discuss in the next section, the degeneracy between states of different m is partially lifted, so states of different $|m|$ (i.e., Λ in molecules) will have different energies in general. We will discuss diatomic molecules in Chapter 5.

3.1.1.2 **Quadratic Stark Effect**

The linear Stark shift is absent in the low-lying states of atoms other than hydrogen where the l degeneracy is broken (by the electrostatic interaction, for example). In this case, the second-order term is the source of the lowest-order correction. The quadratic Stark shift was first calculated in parabolic coordinates by Wentzel [25], Walker [23] and Epstein [22], and is given by

$$\Delta E^{(2)} = -\frac{n^4}{16}\left[17n^2 - 3(n_1 - n_2)^2 - 9m^2 + 19\right]\frac{F^2 a_0^3}{Z^4}. \tag{3.14}$$

We note that the degeneracy is now broken between states with different $|m|$ values but a $\pm m$ degeneracy still persists.

In spherical coordinates, Khadjavi et al. [26] showed that the shift takes the form

$$\Delta E^{(2)} = -\frac{1}{2}\left[\alpha_0 + \alpha_2 \frac{3M_J^2 - J(J+1)}{J(2J-1)}\right] F^2, \tag{3.15}$$

where α_0 and α_2 are the so-called scalar and tensor polarizabilities, respectively, which take the form

$$\alpha_0 = -\frac{2}{3}\sum_{J'} \frac{e^2|\langle J||r||J'\rangle|^2}{(2J+1)[E(J) - E(J')]} \tag{3.16a}$$

and

$$\alpha_2 = 2\left[\frac{10J(2J-1)}{3(2J+3)(J+1)(2J+1)}\right]^{1/2} \sum_{J'} \frac{e^2|\langle J||r||J'\rangle|^2}{[E(J)-E(J')]} \times (-1)^{J+J'+1} \left\{ \begin{array}{ccc} J & J' & 1 \\ 1 & 2 & J \end{array} \right\}. \quad (3.16b)$$

These polarizabilities can also be measured experimentally.

3.1.2 Zeeman Effect

When an atom is placed in a magnetic field, B, similar to those used in neutral atom magnetooptical traps or magnetic traps, the degeneracy associated with each J level, i.e., the magnetic sublevels labeled by the magnetic quantum number, m_J, is removed. In this so-called Zeeman effect, each J levels splits into $2J+1$ components, which follows directly from the expectation values of the Hamiltonian for the magnetic interaction,

$$H_{mag} = -\vec{\mu} \cdot \vec{\mathbf{B}}, \quad (3.17)$$

where $\vec{\mu}$ $(= \vec{\mu}_j)$ is the effective magnetic moment of the electron in the $\vec{\mathbf{j}}$ direction as discussed in Appendix A.13.1.

The energy associated with each level is found by taking matrix elements of the Hamiltonian with states of the form $|\gamma LSJM_J\rangle$, which leads to

$$\Delta E = \langle -\vec{\mu} \cdot \vec{\mathbf{B}} \rangle = g_J \mu_B B \langle \vec{\mathbf{J}}_z \rangle = g_J \mu_B B M_J, \quad (3.18)$$

where g_J, the *Landé g* factor, [5] is given by[6]

$$g_J = 1 + \frac{J(J+1) - L(L+1) + S(S+1)}{2J(J+1)}, \quad (3.19)$$

in the LS coupling scheme. Figure 3.4 shows an example of the Zeeman effect for an alkali atom. Since each fine-structure level splits into $2j+1$ levels with a shift from the zero magnetic field level proportional to $g_j m_j$, we label each level by m_j and $g_j m_j$. In a field of 10 G, for example, the splitting between the magnetic sublevels is of the order of 14 MHz $\times g_J$. The energy spacing

5) This quantity also goes by the name g factor. It is the ratio between the quantum and classical gyromagnetic ratios, the ratio between the magnetic moment and the angular momentum. There is also confusion in the literature as some authors call the g factor the gyromagnetic ratio.

6) It is helpful to have an understanding of magnetic moments and the origin of Eq. (3.19). In Appendix A.13 we review the quantum mechanical treatment of the magnetic moments for the electron and nucleons, and derive Eq. (3.19).

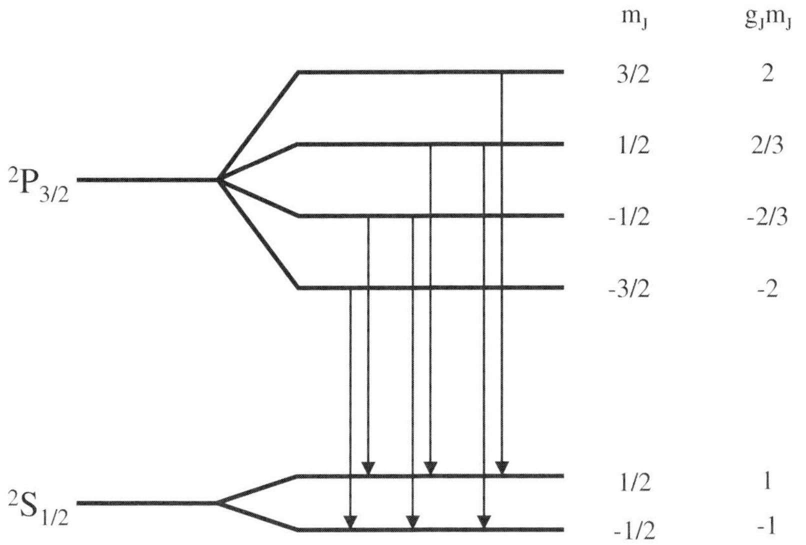

Fig. 3.4 Zeeman effect for the fine structure of alkali atoms.

determined in this section only applies for weak fields. When the applied field is strong, that is when the shifts become comparable to or larger than the spin–orbit splitting, L and S decouple and B will act on the orbital and spin contributions to the magnetic moment separately. The splittings for this case have been worked out in several places. See Sobelman [27], for example.

3.2 Hyperfine Structure

The Zeeman effect discussed in the previous section is appropriate when the spin of the nucleus, I, is zero. For nonzero I, we find substructure within the fine structure known as hyperfine structure. This splitting of the J levels even in the absence of an applied external field can be traced to the dipole moment and quadrupole tensor of the nucleus and their interaction with the magnetic and inhomogeneous electric fields produced by the electrons and the nucleus. In this section we will look at both.

3.2.1 Magnetic Interaction

The Hamiltonian for the hyperfine interaction has a structure similar to that used to calculate the splittings for the Zeeman Hamiltonian,

$$H_{hf} = -\vec{\mu}_I \cdot \vec{\mathbf{B}}_{el}, \tag{3.20}$$

where $\vec{\mu}_I = g_I \mu_N \vec{\mathbf{I}}$ is the magnetic moment of the nucleus. There are several directions we need to consider. The nuclear magnetic moment points in the direction of the nuclear spin, $\vec{\mathbf{I}}$. Since the electrons are in a state of total angular momentum $\vec{\mathbf{J}}$, the magnetic field they produce will be directed along $\vec{\mathbf{J}}$. Consequently, Eq. (3.20) can be represented by

$$H_{hf} = A\mathbf{I} \cdot \mathbf{J}, \tag{3.21}$$

where A depends on the details of the structure and interaction as we will see below. The position of a level one would observe in the laboratory is related to the average value of this Hamiltonian because neither $\vec{\mathbf{I}}$ nor $\vec{\mathbf{J}}$ are constants of the motion. These two vectors couple to give a new vector $\vec{\mathbf{F}} = \vec{\mathbf{I}} + \vec{\mathbf{J}}$ that is a constant of the motion. The appropriate state in which the Hamiltonian is to be averaged will be $|\gamma(LS)JIFM_F\rangle$, leading to an expression similar to what we had for the fine structure,

$$\Delta E_F^{\mu} = \langle H_{hf} \rangle = \frac{A}{2}[F(F+1) - J(J+1) - I(I+1)]. \tag{3.22}$$

The expression for A depends on the value of L. When $L \neq 0$, for example, $A \propto \langle 1/r^3 \rangle$. To indicate this l dependence, we will add the subscript l to A, A_l. When $L = 0$, and the electrons are in s states that can penetrate the nucleus, $A_s \propto |\psi_s(0)|^2$, where $\psi_s(0)$ is the amplitude of the electronic wavefunction at the nucleus (proportional to the Fermi contact potential). When the hyperfine interaction is taken into account, the fine structure J levels split into $2J+1$ sublevels when $I \geq J$ or $2I+1$ sublevels when $J > I$. As an example, we show the hyperfine structure of two fine-structure levels for a one-electron alkali system in Fig. 3.5. In particular, we show the splitting of the ground $^2S_{1/2}$ and the excited $^2P_{3/2}$ states in ^{85}Rb, which has a nuclear spin $I = 5/2$. Each hyperfine level is shifted away from the J level by the right-hand side of Eq. (3.22) and the spacing between adjacent levels is

$$\Delta E_{F,F-1} = \Delta E_F - \Delta E_{F-1} = AF. \tag{3.23}$$

3.2.2 Explicit Expression for A_l

To write an explicit expression for A_l, we must look more closely at the magnetic field generated by the electrons. This field has two contributions, one from the orbital motion and the other from the magnetic moment (spin) of the electron. As was done to generate an expression for the electron in the field of the nucleus, we start by writing classical expressions and then converting the appropriate quantities to operators. From the Biot–Savart law, we know that

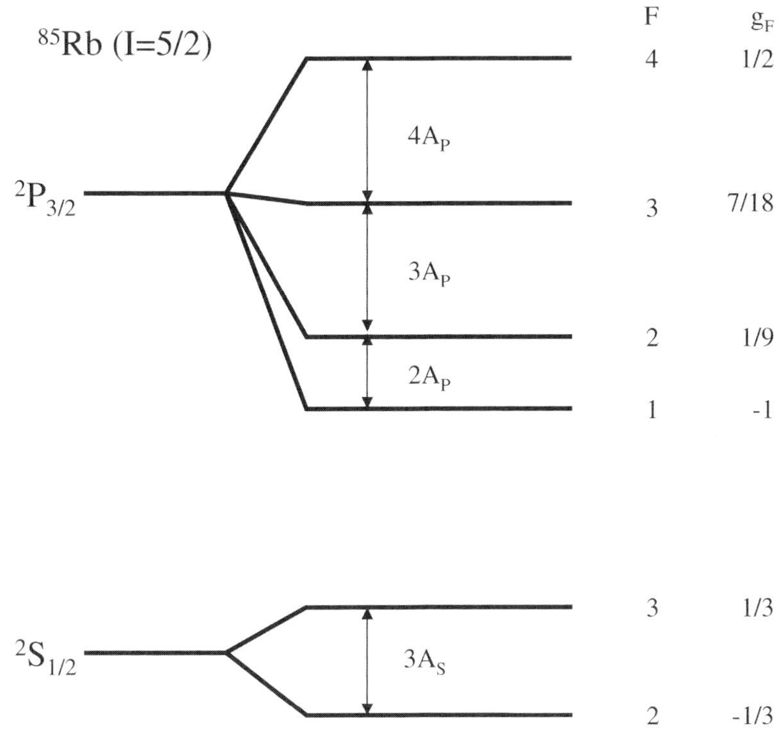

Fig. 3.5 Magnetic hyperfine structure of the cooling states of ^{85}Rb. Each level is labeled by its F quantum number as well as its nuclear g factor, g_F.

the magnetic field of a charge in motion is given by

$$\vec{B}_{el,L} = \frac{\mu_o}{4\pi}\frac{e\vec{v}\times\vec{r}}{r^3}, \tag{3.24}$$

where μ_o is the permeability of the vacuum and $\vec{r}$ is the position of the nucleus relative to the electron. In terms of the angular momentum, which of course is an operator in quantum mechanics, we have

$$\vec{B}_{el,L} = -\frac{\mu_o}{2\pi}\mu_B\frac{\vec{L}/\hbar}{r^3}, \tag{3.25}$$

where μ_B is the Bohr magneton. The field from a classical magnetic dipole is

$$\vec{B}_{el,S} = \frac{\mu_o}{4\pi}\mu_B\frac{1}{r^3}\left[-\vec{\mu} + \frac{3(\vec{\mu}\cdot\vec{r})\vec{r}}{r^2}\right]. \tag{3.26}$$

Substituting $-2\mu_B\vec{S}/\hbar$ for the magnetic moment of the electron (see Appendix A.13.1), we can rewrite this quantum mechanically as

$$\vec{B}_{el,S} = \frac{\mu_0}{2\pi}\mu_B\frac{1}{r^3}\left[\vec{S} - \frac{3(\vec{S}\cdot\vec{r})\vec{r}}{r^2}\right]\frac{1}{\hbar}. \tag{3.27}$$

Combining the two fields gives

$$\vec{B}_{el} = -\frac{\mu_0}{2\pi}\mu_B\frac{1}{r^3}\left[\vec{L} - \vec{S} + \frac{3(\vec{S}\cdot\vec{r})\vec{r}}{r^2}\right]\frac{1}{\hbar}. \tag{3.28}$$

The Hamiltonian in Eq. (3.22) can be written as

$$H_{hf} = -\frac{\mu_0}{2\pi}g_I\mu_B^2\frac{m_e}{m_p}\frac{1}{r^3}\left[I\cdot\left(\vec{L}-\vec{S}\right) + I\cdot\vec{r}\frac{3\,(I\cdot\vec{r})\left(\vec{S}\cdot\vec{r}\right)}{r^2}\right]\frac{1}{\hbar}, \tag{3.29}$$

where m_e/m_p is the electron to proton mass ratio. If we let

$$a_l = \frac{\mu_0}{2\pi}g_I\mu_B^2\frac{m_e}{m_p}\frac{1}{r^3}, \tag{3.30}$$

the energy shifts for fine-structure levels where $L \neq 0$ will be given by

$$\Delta E_F = \langle H_{hf}\rangle = \langle a_l\rangle(-1)^{J+I+F}\langle I\|\|I\rangle\begin{Bmatrix} J & I & F \\ I & J & 1 \end{Bmatrix} \times\left[\langle SLJ\|L\|SLJ\rangle - \sqrt{8\pi}\langle SLJ\|\left(X_K\right)\|SLJ\rangle\right], \tag{3.31}$$

where $\langle SLJ\|\left(X_K\right)\|SLJ\rangle = \langle SLJ\|\left(Y_2\times S_1\right)\|SLJ\rangle$, the reduced matrix element of one of the spherical harmonics ($Y_{l=2,m}$) and the electronic spin, respectively. Equation (E.36) can be used to obtain an expression for this reduced matrix element. It can be shown that

$$\begin{aligned} A_l &= \langle a_l\rangle\frac{l\,(l+1)}{j\,(j+1)} \\ &= \frac{\mu_0}{2\pi}g_I\mu_B^2\frac{m_e}{m_p}\left\langle\frac{1}{r^3}\right\rangle\frac{l\,(l+1)}{j\,(j+1)}. \end{aligned} \tag{3.32}$$

For s states, the major contribution to A_s comes from the electrons penetrating the nucleus, giving

$$A_s = \frac{\mu_0}{2\pi}g_I\mu_B^2\frac{m_e}{m_p}|\psi_s(0)|^2. \tag{3.33}$$

For one-electron systems, one can write an explicit expression for A_l and A_s using the results in Chapter 1. For many-electron systems, only an approximate expression can be written so that numerical calculations are required. Alternatively, A_l can be extracted from experiments.

3.2.3 Hyperfine Zeeman Effect

In the presence of a weak external magnetic field, each F level splits into $2F+1$ sublevels. The shift from the zero-field level is given by $g_F M_F \mu_B B$. Using arguments similar to those employed in deriving g_j in Appendix A.13.1, it is straightforward to show that g_F is approximately given by

$$g_F \simeq g_J \frac{F(F+1) + J(J+1) - I(I+1)}{2F(F+1)}. \tag{3.34}$$

Figure 3.6 is an example in the case of ^{85}Rb.

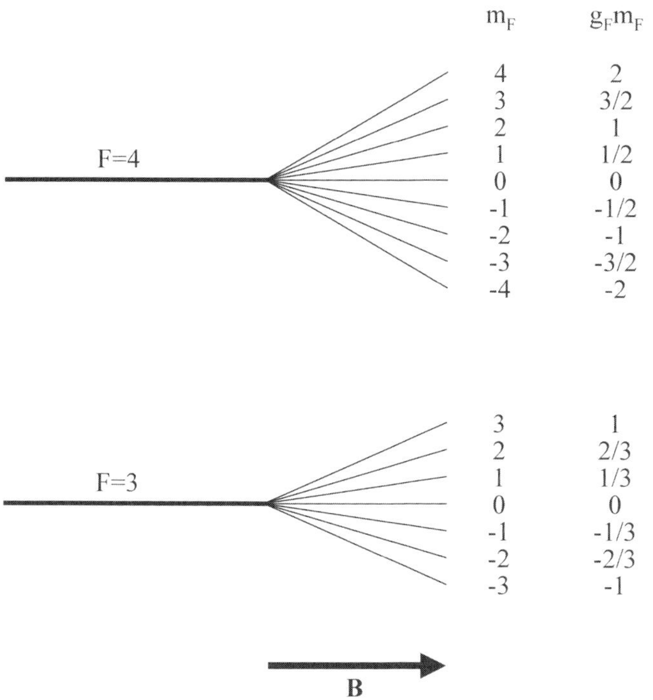

Fig. 3.6 Zeeman effect for the hyperfine structure of ^{85}Rb with nuclear spin 5/2.

3.2.4 Electric Quadrupole Correction

The nucleus, not being a perfect sphere, will also have an electric quadrupole moment, Q. In general, Q will be a tensor (see Appendix A.14). As we mentioned above, the interaction between Q and the inhomogeneous electric field produced by the electrons leads to hyperfine splitting of the fine-structure components as well, albeit smaller than that due to the magnetic dipole. An expression for the energy shift due to the electric quadrupole interaction can be found if we express ΔE_F^{μ} as $AC/2$ in Eq. (3.22). This shift can be shown to be linear in $C(C+1)$ with the total shift given by [27–29]

$$\Delta E^{\mu} + \Delta E^{Q} = \frac{A}{2}C + B[C(C+1) - \frac{3}{4}J(J+1)I(I+1)]. \tag{3.35}$$

The coefficients, A and B, can be determined from experimental measurement of the hyperfine splitting. While outdated, tabulated values can be found in Ref. [30].

Further Reading

1 H. A. Bethe and E. E. Salpeter, *Quantum Mechanics of One- and Two-Electron Atoms*, Plenum, New York, NY, 1977.

2 R. D. Cowan, *The Theory of Atomic Structure and Spectra*, University of California Press, Los Angeles, CA, 1981.

3 I. I. Sobelman, *Atomic Spectra and Radiative Transitions*, Springer Series in Chemical Physics, Springer, New York, NY, 1979.

4 T. F. Gallagher, *Rydberg Atoms*, Cambridge Monographs on Atomic, Molecular and Chemical Physics 3, Cambridge University Press, New York, NY, 1994.

5 R. F. Stebbings and F. B. Dunning, *Rydberg States of Atoms and Molecules*, Cambridge University Press, New York, NY, 1993.

6 W. C. Martin and W. L. Wiese, *Atomic Spectroscopy: A Compendium of Basic Ideas, Notation, Data and Formulas*, Physical Reference Data, National Institute of Standards and Technology [http://physics.nist.gov/Pubs/AtSpec/].

7 For additional problems with worked solutions see D. Budker, D. F. Kimball and D. P. DeMille, *Atomic Physics: An Exploration Through Problems and Solutions*, Oxford University Press, New York, NY, 2004.

8 L. Armstrong, Jr, *Theory of the Hyperfine Structure of Free Atoms*, Wiley-Interscience, New York, NY, 1971.

Problems

3.1 The fine-structure splitting can also be viewed as a magnetic interaction since in the electron's rest frame, the nucleus is seen to be a circulating current. Estimate the order of magnitude of the magnetic field produced by this circulating current for the $3p$ and $4p\ ^2P_{1/2,3/2}$ terms in Na.

3.2 Determine the possible LS and jj terms for a d^2 configuration.

3.3 Follow the steps outlined in Appendix A.13.1 to combine μ_J and μ_I, due to the nuclear spin, to show that

$$\langle \mu_F \rangle = \mu_B g_F \sqrt{F(F+1)} \tag{3.36}$$

and that g_F is given by Eq. (3.34). Describe why Eq. (3.34) is a good approximation.

3.4 Use the relationships in Appendix E to show explicitly that

$$\langle \gamma LSJM_J | \mathbf{J}^2 | \gamma LSJM \rangle = J(J+1). \tag{3.37}$$

3.5 Carry out the steps to show Eq. (3.32) and show that it agrees with Eq. (3.22).

3.6 Calculate the energy-level diagram for the strong magnetic field (so-called Paschen–Back) effect. Draw a diagram of the energy levels for ^{85}Rb as a function of magnetic field.

4
Atoms in AC Fields

4.1
Applied EM Fields

In the last chapter we saw that when an atom is placed in a static electric or magnetic field, its structure changes. Under severe conditions, the distortions can be so great that electrons are no longer bound. In this chapter, we will examine the details of atoms in electromagnetic fields – light. Here, we will focus primarily on light-induced transitions where electrons jump from one state to another. While we will write down equations that govern multiphoton transitions in the perturbative regime, we will not discuss nonlinear phenomena, such as harmonic generation, four-wave mixing, etc., in this chapter. The general features of the nonlinear response of media will be the subjects of Part II. Much of what we will discuss in this chapter will also apply to molecules as well. Thus, the first part of this presentation will be generic formalism for transitions. We will take a semiclassical approach. That is, we will treat the radiation field classically and the atom (and molecule) quantum mechanically. At the end of the chapter, we will look at a few specific atomic examples. Molecular examples will be presented in Chapter 6.

4.1.1
Radiation Hamiltonian

To begin, we write out the radiation Hamiltonian that alters the motion of the electrons in electromagnetic fields,[1]

$$\widehat{H} = \frac{1}{2\mu}\left(\vec{\mathbf{p}} - e\vec{\mathbf{A}}\right)^2 - e\phi + V', \tag{4.1}$$

1) The quantity inside the parentheses is what we added to the Dirac equation in Section 1.4.1.1 that led to spin–orbit interaction.

Light-Matter Interaction: Atoms and Molecules in External Fields and Nonlinear Optics.
W. T. Hill and C. H. Lee

ISBN: 978-3-527-40661-6

in which

$$\begin{aligned}
\mu &= \text{the reduced mass,}\\
e &= \text{electric charge (taken to be a positive quantity),}\\
\vec{\mathbf{p}} &= \frac{\hbar}{i}\overrightarrow{\nabla} = \text{the electron's momentum,}\\
\vec{\mathbf{A}} &= \text{the vector potential,}\\
\phi &= \text{the scalar potential,}\\
V' &= \text{the static electric potential.}
\end{aligned}$$

The $\vec{\mathbf{E}}$ and $\vec{\mathbf{B}}$ fields can be derived from $\vec{\mathbf{A}}$ using Eq. (1.77). To make the problem more tractable, we let $\widehat{H} = \widehat{H}_o + \widehat{H}_{rad}$, where $\widehat{H}_o$ describes the motion of the electron in the absence of the radiation field and $\widehat{H}_{rad}$, the radiation Hamiltonian, describes the interaction between the field and an isolated electron in the absence of any other interaction (i.e., the Coulomb interaction of the atomic potential). Expanding the right-hand side of Eq. (4.1) with the aid of the definition of $\vec{\mathbf{p}}$ gives

$$\widehat{H}_o = -\frac{\hbar^2}{2\mu}\nabla^2 + V, \tag{4.2a}$$

where we have let $V = -e\phi + V'$ and

$$\widehat{H}_{rad} = \frac{e}{2\mu}\left[2i\hbar\vec{\mathbf{A}}\cdot\overrightarrow{\nabla} + i\hbar\left(\overrightarrow{\nabla}\cdot\vec{\mathbf{A}}\right) + e\vec{\mathbf{A}}\cdot\vec{\mathbf{A}}\right]. \tag{4.2b}$$

The $\overrightarrow{\nabla}$ inside the parentheses means that it operates only on $\vec{\mathbf{A}}$. We will primarily be interested in electronic transitions that electrons make in the presence of this field. There are two instructive limits to consider first – the weak- and strong-field limits. In the weak-field limit, $|\vec{\mathbf{p}}| \gg \left|e\vec{\mathbf{A}}\right|$, in which case only the first term need be considered in Eq. (4.2b) and we will seek solutions to the Schrödinger equation that are perturbations to wavefunctions that are eigenstates of $\widehat{H}_o$ that we discussed in Chapter 1. At higher intensities both the first and third terms are important and must be kept. As we will show in the next section, the second term in Eq. (4.2b) can be set to zero through a gauge transformation. This is because $\overrightarrow{\nabla}$ acts only on $\vec{\mathbf{A}}$, which is just a complex number. As the strength of $\widehat{H}_{rad}$ increases, perturbation theory breaks down and new nonperturbative approaches must be employed. At extreme intensities, the Schrödinger equation will no longer be valid and one must use a relativistic equation, such as the Dirac equation (Eq. (1.66)), to describe the electron's motion. We point out that when $|\vec{\mathbf{p}}| \ll \left|e\vec{\mathbf{A}}\right|$, the tables are turned and $\widehat{H}_o$ becomes a perturbation of $\widehat{H}_{rad}$. In this chapter we will limit our discussion of transitions and electron dynamics to the nonrelativistic regime.

4.1.2 Coulomb or Radiation Gauge

Since the second term in Eq. (4.2b) is just a complex number, we can simplify our calculations by choosing a gauge where this number is zero. This is possible with a gauge transformation of the first kind:

$$\vec{\mathbf{A}} \rightarrow \vec{\mathbf{A}}' = \vec{\mathbf{A}} + \vec{\nabla}\chi, \tag{4.3a}$$

$$\phi \rightarrow \phi' = \phi - \partial\chi/\partial t. \tag{4.3b}$$

We are free to choose $\chi(\vec{\mathbf{r}}, t)$ to be any value, so we will choose it such that $\vec{\nabla} \cdot \vec{\mathbf{A}} = 0$. It is straightforward to show by direct substitution into Eq. (1.77) that the Coulomb gauge leaves the electric and magnetic fields unchanged. It is also straightforward to show that the Schrödinger equation is invariant under this gauge transformation as well, since

$$\psi \rightarrow \mathrm{e}^{ie\chi/\hbar}\psi.$$

Applying the gauge transformation to Eq. (4.2b) transforms the radiation Hamiltonian into

$$\widehat{H}_{rad} = \frac{i\hbar e}{\mu}\vec{\mathbf{A}} \cdot \vec{\nabla} + \frac{e^2}{2\mu}\vec{\mathbf{A}} \cdot \vec{\mathbf{A}}. \tag{4.4}$$

In the presence of the atomic nucleus and the radiation field, the time-dependent Schrödinger equation for the electron can be written as

$$i\hbar\frac{\partial}{\partial t}\Psi\left(\vec{r}, t\right) = \left[\widehat{H}_o + \widehat{H}_{rad}\right]\Psi\left(\vec{r}, t\right). \tag{4.5}$$

The remainder of this chapter is concerned with solutions to and consequences of this equation.

4.2 Free-Electron Wavefunction

As we have seen in Chapter 1, exact solutions can be found to Eq. (4.5) when $\widehat{H}_{rad} = 0$. Exact solutions are also possible when $\widehat{H}_o \rightarrow -\hbar^2\nabla^2/2\mu$. These solutions take the form

$$\psi(\vec{r}, t) = \mathrm{e}^{i\vec{\mathbf{k}}\cdot\vec{\mathbf{r}}} \exp\left\{-\frac{i}{2\mu\hbar}\int_t \left[\hbar\vec{\mathbf{k}} - e\vec{\mathbf{A}}(t')\right]^2 dt'\right\}. \tag{4.6}$$

We can evaluate the integral for various forms of light. For linearly polarized light, $\vec{\mathbf{A}}$ can be written as

$$\vec{\mathbf{A}}(t) = \vec{\varepsilon}_z A_0(t)\cos(\omega t + \varphi), \tag{4.7a}$$

while for circularly polarized light it takes the form

$$\vec{\mathbf{A}}(t) = A_0(t)\left[\vec{\varepsilon}_x \cos(\omega t + \varphi) - \vec{\varepsilon}_y \sin(\omega t + \varphi)\right], \tag{4.7b}$$

where $A_0(t)$ is a time-dependent amplitude. With Eq. (4.7a) the solution takes the form

$$\begin{aligned}\psi(\vec{\mathbf{r}}, t) = \exp\Bigg\{ &-\frac{i}{\hbar}\left[\frac{\hbar^2 k^2}{2\mu} + \frac{e^2 E_0^2}{4\mu\omega^2} + \frac{e^2 E_0^2}{8\hbar\omega^3\mu}\sin 2(\omega t + \varphi)\right.\\ &\left. - \left(\frac{eE}{\mu\omega^2}\right)\vec{\varepsilon}_z \cdot \vec{\mathbf{k}}\sin(\omega t + \varphi)\right] + i\vec{\mathbf{k}}\cdot\vec{\mathbf{r}}\Bigg\}.\end{aligned} \tag{4.8}$$

In this expression, $e^2E_0^2/4\mu\omega^2$ is the so-called ponderomotive energy, U_p, and $eE/\mu\omega^2$ is the average vibration radius, α. In terms of the intensity of the light in W/cm^2 and the frequency in Hz, the approximate value for U_p in eV is

$$U_p = 5.42 \times 10^{16}\frac{I}{\omega^2}. \tag{4.9}$$

The approximate value for α in nm is

$$\alpha = 4.88 \times 10^{23}\frac{\sqrt{I}}{\omega^2}. \tag{4.10}$$

4.3 Radiative Transitions

To solve Eq. (4.5) when the atomic potential does not vanish, one must resort to a variety of approximation techniques appropriate to the relative sizes of $\widehat{H}_o$ and $\widehat{H}_{rad}$. In this section we are interested in the nature of single and multiphoton transitions in atomic and molecular systems. Our goal is to generate expressions for transition rates and cross sections. We find that for many applications making a Dyson expansion of the transition operator is convenient at intensities where the lowest-order term describes the physics well. Given the transition amplitude, the rates and cross sections we seek are just the amplitude squared as we average over a transition cycle. This is perhaps most straightforward when the wavefunction is expressed in the interaction picture (see, for example, Eq. (3.17) of Ref. [6] or a standard quantum mechanics text),

$$\Psi(t) = e^{i\widehat{H}_o t/\hbar}\psi(t), \tag{4.11}$$

where $\psi(t)$ is an eigenfunction of the unperturbed states, $|i\rangle$, with eigenvalue E_i. Note that, since we are interested primarily in the time dependence of the process, we dropped the explicit reference to space in the wavefunction.

When we substitute this wavefunction into Schrödinger's equation, Eq. (4.5), we find that $\psi(t)$ must satisfy

$$i\hbar\frac{\partial}{\partial t}\psi(t) = e^{i\widehat{H}_o t/\hbar}\widehat{H}_{rad}e^{-i\widehat{H}_o t/\hbar}\psi(t) = \widehat{V}(t)\psi(t). \tag{4.12}$$

The matrix elements of this interaction will be

$$\langle j|\widehat{V}(t)|i\rangle = e^{i(\omega_j-\omega_i)t}\langle j|\widehat{H}_{rad}(t)|i\rangle, \tag{4.13}$$

where we have made use of the fact that $\widehat{H}_o|i\rangle = \hbar\omega_i|i\rangle$. We note that the matrix element of our new interaction differs from the original interaction by a phase. To solve Eq. (4.12) we will seek integral solutions of the form

$$|\psi_j(t)\rangle = |j\rangle + \left(-\frac{i}{\hbar}\right)\int_{-\infty}^{t} dt_1\widehat{V}(t_1)|\psi_j(t_1)\rangle. \tag{4.14}$$

One should view $|\psi_j(t)\rangle$ as a state of the system that evolves in time. In general, we will be interested in transitions from an initial state $|i\rangle$ to a final state $|f\rangle$. At this point, we have not specified if the states are bound or free.

The transition amplitude will be a projection of $\langle f|$ out of $|\psi_j(t)\rangle$,

$$\langle f|\psi_j(t)\rangle \equiv \langle f|\widehat{T}(t,t')|i\rangle, \tag{4.15}$$

where $\widehat{T}(t,t_1)$ is a unitary operator that takes the system from state $|i\rangle$ at time t_1 to state $|\psi_i(t)\rangle$ at time t,

$$|\psi_i(t)\rangle = \widehat{T}(t,t_1)|i\rangle. \tag{4.16}$$

Now substituting this into Eq. (4.14) leads to

$$\widehat{T}(t,t_1) = 1 - \frac{i}{\hbar}\int_{t_1}^{t} dt_1'\widehat{V}(t_1')\widehat{T}(t_1',t_1). \tag{4.17}$$

In general, we can make the Dyson expansion

$$\widehat{T}(t,t_1) = 1 + \sum_{n=1}^{\infty}\left(-\frac{i}{\hbar}\right)^n\int_{t_1}^{t} dt_1'\int_{t_1}^{t_1'} dt_2'\cdots\int_{t_1}^{t_{n-1}'} dt_n'\widehat{V}(t_1')\cdots\widehat{V}(t_n'). \tag{4.18}$$

We are interested in n-photon processes. Thus, we define $\widehat{T}^{(n)}(t,t_1)$ such that

$$\widehat{T}^{(n)}(t,t_1) = \left(-\frac{i}{\hbar}\right)^n\int_{t_1}^{t} dt_1'\int_{t_1}^{t_1'} dt_2'\cdots\int_{t_1}^{t_{n-1}'} dt_n'\widehat{V}(t_1')\cdots\widehat{V}(t_n'). \tag{4.19}$$

These are particularly useful for cases where the field is weak enough that the lowest-order term describes the physics well. When higher-order terms are required, this approach quickly becomes inconvenient to use. The lowest-order n-photon transition amplitude, however, would be given by

$$T_{if}^{(n)}(t,t_1) = \langle f|\, \widehat{T}^{(n)}(t,t_1)\, |i\rangle \,. \tag{4.20}$$

Typically, only certain terms within $T_{if}^{(n)}(t,t_1)$ exist to describe the transition from $|i\rangle$ to $|f\rangle$. For example, if the two levels have opposite parity, only the odd terms will exist; the rest vanish. In order to perform a calculation to estimate $T_{if}^{(n)}(t,t_1)$, we must write an explicit expression for $\widehat{V}(t)$. The choices include the length, velocity and acceleration forms. We will work with the length form, which usually gives the best results with approximate wavefunctions. In the length form we have

$$\widehat{V}(t) = \mathrm{e}^{i\widehat{H}_0 t/\hbar}\left[\vec{\mathbf{d}}\cdot\vec{\mathbf{F}}(t)\right]\mathrm{e}^{-i\widehat{H}_0 t/\hbar}, \tag{4.21}$$

where $\vec{\mathbf{F}}(t)$ is the electric field and is expressed as

$$\vec{\mathbf{F}}(t) = \frac{1}{2}\left[\vec{\mathbf{F}}^{*}\mathrm{e}^{i\omega t} + \vec{\mathbf{F}}\mathrm{e}^{-i\omega t}\right], \tag{4.22}$$

while $\vec{\mathbf{d}}$ is the dipole moment and takes the form

$$\vec{\mathbf{d}} = -e\vec{\mathbf{r}}. \tag{4.23}$$

The dot product between $\vec{\mathbf{d}}$ and $\vec{\mathbf{F}}$ will select the component of $\vec{\mathbf{d}}$ along the polarization axis, $\vec{\epsilon}$, of the laser field. Thus, we will denote $\vec{\mathbf{d}}\cdot\vec{\mathbf{F}}$ as $d^{\epsilon}F$ and write $\widehat{V}(t)$ as

$$\widehat{V}(t) = \mathrm{e}^{i\widehat{H}_0 t/\hbar}\left[\frac{d_{\epsilon}F^{*}}{2}\mathrm{e}^{i\omega t} + \frac{d_{\epsilon}F}{2}\mathrm{e}^{-i\omega t}\right]\mathrm{e}^{-i\widehat{H}_0 t/\hbar}. \tag{4.24}$$

We are now in a position to look at n-photon transitions.

4.3.1
One-Photon Transitions

The transition matrix element for a one-photon transition takes the form

$$T_{if}^{(1)}(t,t_1) = -\frac{i}{\hbar}\int_{t_1}^{t} dt_1' \left[\mathrm{e}^{i(\omega_0+\omega)t_1'}\frac{F^{*}d_{if}^{\epsilon}}{2} + \mathrm{e}^{i(\omega_0-\omega)t_1'}\frac{F d_{if}^{\epsilon}}{2}\right], \tag{4.25}$$

where $\omega_0 = \left(E_f - E_i\right)/\hbar$ and

$$d_{if}^{\epsilon} = \langle f|\, d^{\epsilon}\, |i\rangle \,. \tag{4.26}$$

To perform the integration, we recognize that the interaction will take place over the length of the pulse, which is typically very long compared with the atomic time scale ($\sim 10^{-17}$ s, Eq. (A.22)). This allows us to let $t \to \infty$ and $t_1 \to -\infty$. Then, using the definition of the Dirac delta function, Eq. (B.7), we write

$$T_{if}^{(1)}(\infty, -\infty) = -\frac{2\pi i}{\hbar} d_{if}^{\epsilon} \left[\frac{F^*}{2} \delta(\omega_0 + \omega) + \frac{F}{2} \delta(\omega_0 - \omega) \right]. \tag{4.27}$$

The first term on the right contributes when $\omega = -\left(E_f - E_i\right)/\hbar$, which corresponds to the transition $f \to i$, stimulated emission, while the second term contributes when $\omega = \left(E_f - E_i\right)/\hbar$, which corresponds to the transition $i \to f$, stimulated absorption. We note that there is no spontaneous emission. In this semiclassical treatment, spontaneous emission must be added artificially and treated phenomenologically with the Einstein A coefficient (see Chapter 2 of Volume 1). Spontaneous emission emerges naturally when the radiation field is quantized as done in Chapter 5 of Ref. [6]. The photons that permeate the vacuum are said to stimulate spontaneous emission.

4.3.2 Two-Photon Transitions

For $n = 2$, two-photon transitions, $T_{if}^{(n)}(t, t_1)$ takes the form

$$\begin{aligned} T_{if}^{(2)}(t, t_1) = \left(-\frac{i}{\hbar}\right)^2 \sum_j \int_{t_1}^{t} dt_1' e^{i(\omega_{fj} - \omega)t_1'} \left(\frac{d_{jf}^{\epsilon} F^*}{2} + \frac{d_{jf}^{\epsilon} F}{2} \right) \\ \times \int_{t_1}^{t_1'} dt_2' e^{i(\omega_{ji} - \omega)t_2'} \left(\frac{d_{ij}^{\epsilon} F^*}{2} + \frac{d_{ij}^{\epsilon} F}{2} \right), \end{aligned} \tag{4.28}$$

where $\omega_{fj} = \omega_f - \omega_j$ and $\omega_{ji} = \omega_j - \omega_i$. The summation is a result of inserting a complete set between $\widehat{V}(t_1')$ and $\widehat{V}(t_2')$. We solve the integrals from right to left. We first let $t_1 \to -\infty$, assume everything vanishes at $-\infty$ and solve the t_2' integral. We then let $t \to \infty$ and solve the t_1' integral. This gives

$$\begin{aligned} T_{if}^{(2)}(\infty, -\infty) = \frac{2\pi i}{\hbar^2} \sum_j \Bigg[\delta(\omega_o + 2\omega) \left(\frac{F^*}{2}\right)^2 \frac{d_{jf}^{\epsilon} d_{ij}^{\epsilon}}{\omega_{ji} + \omega} \\ + \delta(\omega_o - 2\omega) \left(\frac{F}{2}\right)^2 \frac{d_{jf}^{\epsilon} d_{ij}^{\epsilon}}{\omega_{ji} - \omega} \\ + \delta(\omega_o) \left|\frac{F}{2}\right|^2 \left(\frac{d_{jf}^{\epsilon} d_{ij}^{\epsilon}}{\omega_{ji} + \omega} + \frac{d_{jf}^{\epsilon} d_{ij}^{\epsilon}}{\omega_{ji} - \omega} \right) \Bigg], \end{aligned} \tag{4.29}$$

where $\omega_0 = \omega_f - \omega_i$. The first two terms correspond to two-photon emission and absorption, respectively, while no real photons are absorbed nor emitted in the last two terms. These terms describe dynamic Stark shifts, if the initial and final states are the same or a mixing between degenerate states.

Following this same prescription, it is possible to write down an expression for an arbitrary number of photons, at least to lowest order. The reader is encouraged to compare the one- and two-photon expressions with those of Chapter 9.

4.3.3
Transition Rate: Fermi's Golden Rule

What one observes in the laboratory is not the transition amplitude but quantities related to the transition rate, which is proportional to the modulus squared of the amplitude. In lowest order, we define the transition rate for an n-photon absorption as the probability, $P_{if}^{(n)}$,[2] per unit time as we let time approach infinity,

$$R_{if}^{(n)} \equiv \lim_{t\to\infty} \frac{1}{t} \left| T_{if,a}^{(n)} (t/2, -t/2) \right|^2 , \tag{4.30}$$

where the subscript a stands for absorption and is associated with the second term in Eq. (4.27) or (4.29). An equivalent expression can be written for stimulated emission.

At first glance, the structure of the rate appears to be a bit odd, mathematically, because it contains a delta function squared. We see this explicitly from the one-photon rate ($n = 1$),

$$R_{if}^{(1)} = \lim_{t\to\infty} \frac{1}{t} \left| \left(-\frac{2\pi i}{\hbar} \right) \left(\frac{d_{if}^{\epsilon} F}{2} \right) \delta (\omega_0 - \omega) \right|^2 . \tag{4.31}$$

One of the delta functions, however, is associated with the line shape, as we discuss below, while the other is proportional to the period and can be eliminated. To see this, we write Eq. (4.31) as

$$R_{if} = \frac{2\pi}{\hbar^2} \left| \frac{d_{if}^{\epsilon} F}{2} \right|^2 \delta (\omega_0 - \omega) \left(\lim_{t\to\infty} \frac{1}{t} \int_{-t/2}^{t/2} dt' e^{i(\omega_0 - \omega)t'} \right) ,$$

2) We point out that one must be careful with probabilities,

$$P_{if}^{(n)} = \int R_{if}^{(n)} (t') dt',$$

because they grow with time and eventually tend towards 1, even for very weak transitions if one waits long enough. Thus, if one is not careful, one will find oneself in a situation where the probability is greater than 1.

where we used Eq. (B.8) to express $2\pi\delta(\omega_o - \omega)$ as the integral and combined the two limits. Since the entire expression for R_{if} vanishes when $\omega \neq \omega_o$, the integral reduces to t, which leads to the one-photon transition rate

$$R_{if} = \frac{2\pi}{\hbar^2}\left|\frac{d^{\epsilon}_{if}F}{2}\right|^2 \delta\left(\omega_0 - \omega\right). \tag{4.32}$$

Equation (4.32) is often referred to as Fermi's Golden Rule. The steps for the more general n-photon rate can be obtained in a similar way.

The one-photon rate can be written in several useful forms. In terms of the resonant Rabi frequency, Ω_0, for example, it takes the form

$$R_{if} = \frac{\pi}{2}\Omega_0^2\delta\left(\omega_0 - \omega\right), \tag{4.33}$$

where

$$\Omega_0 = \frac{d^{\epsilon}_{if}F}{\hbar}, \tag{4.34}$$

which is the same as Eq. (2.9) of Ref. [6]. It is often convenient to express the rate in terms of the intensity of the incident field, $I_\omega = \hbar\omega\eta_\omega$, where η_ω is the photon flux at frequency ω. Typically, the intensity is expressed with dimensions of $\mathrm{W/cm^2}$ and the flux with dimensions of $\mathrm{cm^{-2}\ s^{-1}}$. The rate then takes the form

$$R_{if} = \frac{4\pi^2\alpha}{\hbar}I_\omega\left|r^{\epsilon}_{if}\right|^2\delta\left(\omega_0 - \omega\right) \tag{4.35a}$$

$$= 4\pi^2\alpha\eta_\omega\omega\left|r^{\epsilon}_{if}\right|^2\delta\left(\omega_0 - \omega\right), \tag{4.35b}$$

where $|r^{\epsilon}_{if}| = |d^{\epsilon}_{if}/e|$ and α is the fine-structure constant. Finally, if we let the matrix element have units of Bohr radii, a_o, we obtain

$$R_{if} = 4\pi^2 a_0^2\alpha\eta_\omega\omega\left|\widetilde{r}^{\epsilon}_{if}\right|^2\delta\left(\omega_0 - \omega\right). \tag{4.35c}$$

We distinguish the matrix element in units of Bohr radii by the symbol $\left|\widetilde{r}^{\epsilon}_{if}\right|$. Since $|\widetilde{r}^{\epsilon}_{if}|$ is of order 1, we estimate the order of magnitude of the transition rate in the case where neither the initial nor the final state is degenerate to be

$$\sim 8 \times 10^{-18}\eta\omega\delta(\omega - \omega_o). \tag{4.35d}$$

The rate given in Eq. (4.35c) is not well defined because of the presence of the delta function. Literally, it says that the transition rate is zero for all ω except for $\omega = \omega_o$, where it is infinite! Dirac pointed out that since the state represented by the delta function is infinitely narrow, the probability for detecting such a line would be of the order zero [31,32]. Thus, to measure the probability for a transition, one must integrate this rate over an energy or frequency range in the vicinity of the line. In addition, at least one of the states involved in a real transition is usually not infinitely narrow, but has a definite shape and width. Consequently, as mentioned above, the delta function should be replaced by the line shape, $L(\omega - \omega_o)$.

The line shape can be due to one or more of a variety of effects such as the natural lifetime of the state, collisions during the transition or perhaps the translational energy of the system. We will consider, as an example, the natural lifetime, which leads to a Lorentzian line profile that was defined in Eq. (1.50) of Ref. [6] as

$$L(\omega - \omega_o) = \frac{2}{\pi} \frac{\gamma_{\frac{1}{2}}}{(\omega - \omega_o)^2 + \gamma_{\frac{1}{2}}^2}, \tag{4.36}$$

where in this case $\gamma_{\frac{1}{2}}$ is the half width at half maximum and is equal to half the radiative lifetime between the two states.[3] The line shape obeys

$$\int_0^\infty L(\omega - \omega_o) d\omega = 1.$$

Replacing $2\pi\delta(\omega_0 - \omega)$ in Eq. (4.35c) with $L(\omega - \omega_o)$, the rate is written as

$$R_{if} = 2\pi a_0^2 \alpha \eta_\omega \omega \left| \widetilde{r}_{if}^{\epsilon} \right|^2 L(\omega - \omega_o), \tag{4.37}$$

which corresponds to the transition rate for a monochromatic source at ω.

While we have derived Eq. (4.37) assuming a Lorentzian profile, a similar expression holds for other profiles such as a Gaussian shape (Doppler-broadened line) or even for Voigt profiles (a convolution of a Gaussian with a Lorentzian as discussed in Appendix B.10).

4.3.3.1 **Degeneracy**

The transition rate given in Eq. (4.37) corresponds to the case where there is no degeneracy. We know from our study of the structure of atoms that a state of angular momentum J will be $(2J + 1)$-fold degenerate. The appropriate rate in the degenerate case for polarized light is obtained by averaging over the initial states and summing over the final states to give

$$R_{if} = \frac{2\pi}{g_i} \alpha \eta \omega \left| \widetilde{r}_{if}^{\epsilon} \right|^2 L(\omega_0 - \omega), \tag{4.38}$$

3) The radiative lifetime is given by the Einstein A coefficient coupling the two states. Compare Eq. (4.36) with Eq. (2.33) of Ref. [6].

where $g_i = 2J_i + 1$, the number of degenerate initial states of different M_i, and $|\widetilde{r}_{if}^{\epsilon}|^2$ now takes the form

$$\left|\widetilde{r}_{if}^{\epsilon}\right|^2 = \sum_{M_i,M_f} \left|\left\langle \gamma_f J_f M_f \left| r^{\epsilon} \right| \gamma_i J_i M_i \right\rangle\right|^2. \tag{4.39}$$

An additional sum over the polarization components is required when the light is unpolarized.

4.3.3.2 Narrow and Broad Sources

When the transition is induced with a laser that has a line width small compared with $\gamma_{\frac{1}{2}}$,

$$L(\omega - \omega_0) \simeq \frac{2}{\pi\gamma_{\frac{1}{2}}},$$

on resonance, and the rate takes the form

$$R_{if}(\omega_0) = 4a_0^2\alpha\eta\frac{\omega_0}{g_i\gamma_{\frac{1}{2}}}\left|\widetilde{r}_{if}^{\epsilon}\right|^2. \tag{4.40}$$

For a source broad compared with the width of the line, one needs to integrate over the line. Assuming that the intensity is flat over the line, one can write

$$R_{if}^{broad} = \frac{2\pi}{g_i}a_0^2\alpha\eta_{\omega_0}\left|\widetilde{r}_{if}^{\epsilon}\right|^2\int_0^{\infty} d\omega\omega L(\omega - \omega_0). \tag{4.41}$$

4.3.4 Transition Strength: Absorption

Although the rate is often what is needed when deciding the feasibility of an experiment, it depends on the strength of the light source. It is often useful to have an expression for the strength of the transition that depends only on the atomic system. The common ways to express the strength of a transition are the *line strength*, the *cross section* and the *oscillator strength*. We will look at each in this section.

4.3.4.1 Line Strength

Exploiting the Wigner–Eckart theorem (Appendix E.3) we see that $|\widetilde{r}_{if}^{\epsilon}|^2$, from Eqs. (4.38) and (4.39), is related to the reduced matrix element for the transition. Specifically, as shown in Chapter 9 of [27], a single polarization component, ϵ, can be written as

$$\left|\widetilde{r}_{if}^{\epsilon}\right|^2 = \frac{\left|\widetilde{r}_{if}\right|^2}{3} = \frac{\left|\left\langle \gamma_f J_f \left|\left|\widetilde{r}\right|\right| \gamma_i J_i \right\rangle\right|^2}{3}. \tag{4.42}$$

The modulus squared of the reduced matrix element, $\left\langle \gamma_f J_f \left|\left|\vec{r}\right|\right| \gamma_i J_i \right\rangle$, defines the line strength,

$$S \equiv e^2 a_0^2 \left| \left\langle \gamma_f J_f \left|\left|\vec{r}\right|\right| \gamma_i J_i \right\rangle \right|^2 , \tag{4.43}$$

which is a sum over all transitions and all three directions. The line strength gives the probability for a transition from a specific initial level to all $2J_f + 1$ levels of the final state. The line strength has units of C^2 cm^2.

4.3.4.2 **Cross Section**

The cross section is defined by dividing the transition rate of Eq. (4.38) by the photon flux,[4]

$$\sigma(\omega) = \frac{R_{if}}{\eta_\omega} = \frac{2\pi}{g_i} a_0^2 \alpha \omega \left| \vec{r}_{if}^{\,\epsilon} \right|^2 L(\omega - \omega_o). \tag{4.44}$$

The cross section as defined has units of area, typically cm^2.

4.3.4.3 **Oscillator Strength**

The oscillator strength, introduced in Section 2.2.5 of Volume 1 [6], can be thought of as a fudge factor to bring the classical cross section into agreement with the quantum mechanical cross section for absorption. Specifically, it is the dimensionless number by which the classical cross section must be multiplied to give the quantum cross section for a particular transition. Classically, the absorption cross section for a damped oscillating electron can be expressed as

$$\sigma^{cl}(\omega) = \frac{1}{4\pi\varepsilon_0} \frac{\pi e^2}{m_e c} L(\omega - \omega_o), \tag{4.45}$$

where e, m_e, c and ε_0 have their usual meanings. The absorption strength is given by the total or integrated cross section (sometimes called the *spectral* cross section),

$$\begin{aligned} \sigma_T^{cl} &= \int_0^\infty d\omega \sigma(\omega) \\ &= \frac{1}{4\pi\varepsilon_0} \frac{\pi e^2}{m_e c}. \end{aligned} \tag{4.46}$$

The oscillator strength is then defined as

$$\sigma_T = \sigma_T^{cl} f = \frac{1}{4\pi\varepsilon_0} \frac{\pi e^2}{m_e c} f = 2.65 \times 10^{-2} f \ \text{cm}^2/\text{s}. \tag{4.47}$$

4) One should understand that while we have been using $L(\omega - \omega_o)$ to indicate a Lorentzian, in Eq. (4.44) it is meant to convey the appropriate line shape (Lorentzian, Gaussian, etc.).

Comparing Eqs. (4.43)–(4.45) and (4.47) along with the definition of the fine-structure constant (Appendix A.3), an explicit expression for the absorption oscillator strength, f_{if}, takes the form

$$f_{if} = \frac{2m_e a_0^2 \omega}{g_i \hbar} \left| \tilde{r}_{if}^{\epsilon} \right|^2 = \frac{2m_e \omega}{3\hbar e^2} \frac{S}{g_i}. \tag{4.48}$$

Sometimes the strength of the transition is given in terms of gf, where for absorption and emission we have respectively

$$gf = (2J_f + 1) f_{if}, \tag{4.49a}$$

$$= -(2J_i + 1) f_{fi}. \tag{4.49b}$$

The oscillator strength defined by Eq. (4.48) is related to the integrated cross section (see Eq. (4.47)) and assumes that the transition strength is confined to a small region of frequency space. In essence, it is only appropriate for transitions between bound states. For transitions to continuum states, it is more appropriate to talk about the oscillator strength density, df/dE. Expressions for df/dE are discussed in Ref. [33].

4.3.5 Transition Strength: Emission

There are corresponding expressions for the downward transition – emission – to quantities presented in the preceding section. We give the oscillator strength here, but the rest can be determined from Table 2.1 of Volume 1 [6]. The oscillator strength for stimulated emission takes the form

$$f_{fi} = -\frac{2m_e \omega}{3\hbar e^2} \frac{S}{g_f}, \tag{4.50}$$

where $g_f = 2J_f + 1$. Note that f_{fi} is usually expressed as a negative value. The meaning of Eq. (4.50) is the probability for emission from a specific upper level to all the lower levels. The oscillator strength is also related to the lifetime[5] of the state through

$$A_{fi} = -\frac{e^2 \omega^2}{2\pi \varepsilon_0 m c^3} f_{fi} \tag{4.51a}$$

$$= \frac{g_i}{g_f} \frac{e^2 \omega^2}{2\pi \varepsilon_0 m c^3} f_{if}. \tag{4.51b}$$

5) The lifetime of the state refers to how long the electron will stay in an excited state before it spontaneously emits a photon and makes a transition to a lower state. Spontaneous emission occurs in the absence of an applied external field and is due to photons in the vacuum stimulating a transition. This interpretation follows naturally when the radiation field is quantized.

Combining Eqs. (4.50) and (4.51a) (or Eqs. (4.48) and (4.51b)) allows us relate the lifetime to the line strength as well,

$$A_{fi} = \frac{1}{3\pi\varepsilon_0}\frac{\omega^3}{\hbar c^3}\frac{S}{g_f}. \tag{4.51c}$$

4.4
Selection Rules for Atomic Transitions

There are three types of transitions commonly observed in atomic systems: the electric dipole (so-called E1), the magnetic dipole (M1) and the electric quadrupole (E2) transitions. Of these, the electric dipole is by far the most common and strongest. The other two are often referred to as electric-dipole forbidden. Since the E1 transitions are the most common, we will work out the selection rules for them in some detail and only give the final results for the other two.

4.4.1
Electric Dipole (E1) Transitions

From Eq. (4.39) we know that the transition rate for a dipole transition is proportional to a matrix element of the form $< f|r|i >$. It is not necessary to calculate this matrix element for all possible states because most of them will be zero. What we will do in this section is to determine which are nonzero. There are a few ways we can approach this problem. We will limit ourselves to two. First, we will use our knowledge of the symmetry of the wavefunction to work out much of the detail. We will then calculate the nonzero matrix element more formally with the aid of the Wigner–Eckart theorem. (See Appendix E.3 for a discussion of the Wigner–Eckart theorem.)

We know that in coordinate space, the integral represented by the matrix element represents a volume integration over 4π. Thus, we know that if the integrand is odd with respect to inversion, the matrix element will be identically zero. We have three parts of the matrix element to consider – the wavefunctions of the initial and final states and r. Looking first at r, we know that under inversion $r \to -r$. Such a function has odd parity. The first requirement for a nonzero matrix element is that $< f|$ and $|i >$ have opposite parity. We learned in Section 1.3.1 that the parity of the state is given by l. Consequently, for a nonzero matrix element we must have

$$\Delta l = \pm 1. \tag{4.52}$$

We can say a few more things about the selection rules from what is missing from Eq. (4.39). Specifically, the matrix element does not involve the principal quantum number, n, or the spin. Thus, there is no restriction on the principal quantum number nor on the spin quantum number.

The fact that there is a restriction on the change in value of l would imply that there should be a restriction on L and perhaps J as well. To explore this we need to turn to the Wigner–Eckart theorem and the properties of the $n - j$ symbols discussed in Appendix D. To that end, we express $\vec{r}$ as a spherical vector. In Cartesian coordinates the components of $\vec{r}$ are

$$r_1 = x,$$
$$r_2 = y,$$
$$r_3 = z,$$

while in spherical coordinates they are

$$r_0 = z,$$
$$r_{+1} = -\frac{1}{\sqrt{2}}(x + iy),$$
$$r_{-1} = +\frac{1}{\sqrt{2}}(x - iy).$$

Expressing r in spherical coordinates and using the Wigner–Eckart theorem, Eq. (E.14), the transition matrix is seen to be related to the line strength, a 3-j symbol and a phase,[6]

$$\langle \gamma jm | r_q | \gamma' j' m' \rangle = (-1)^{j-m} \begin{pmatrix} j & 1 & j' \\ -m & q & m' \end{pmatrix} \langle \gamma j \| r \| \gamma' j' \rangle , \tag{4.53}$$

where $q = 0, \pm 1$ and the last factor is the reduced matrix element, which is independent of the magnetic components (see Appendix E.3) and gives the probability for a transition from a specific initial level to all $2j + 1$ levels of the final state.[7] From the properties of the 3-j symbol (see Appendix D.2), we know that the matrix element will be zero unless j, 1 and j' add as vectors and $-m + q + m' = 0$. In other words, we have the two additional selection rules,

$$\Delta j - 0, \pm 1, \tag{4.54a}$$
$$\Delta m = 0, \pm 1, \tag{4.54b}$$

with two exceptions,

$$j' = 0 \nrightarrow j = 0, \tag{4.54c}$$
$$m' = 0 \nrightarrow m = 0 \text{ when } \Delta j = 0. \tag{4.54d}$$

6) The reader should note that in Appendix E.3, the rank of the tensor operator is also included in the subscript in the general expression of the matrix element. We drop the rank on r to avoid confusion with the $+1$ component.

7) We note that the reduced matrix element in the Wigner–Eckart theorem is not in atomic units, so $S \equiv |e \langle \gamma j \| r \| \gamma' j' \rangle|^2$.

In LS coupling, one can show that

$$\Delta L = 0, \pm 1, \tag{4.54e}$$

$$\Delta S = 0, \tag{4.54f}$$

$$L' = 0 \nrightarrow L = 0. \tag{4.54g}$$

4.4.2 Magnetic Dipole (M1) Transitions

The magnetic dipole operator, which has a form similar to Eq. (3.17), couples states of the same parity. The Wigner–Eckart theorem (Eq. (E.14)) can be used to determine the selection rules as we did for E1 transitions. The resulting selection rules are

$$\Delta l = 0, \tag{4.55a}$$

$$\Delta j = 0, \pm 1, \tag{4.55b}$$

$$\Delta m = 0, \pm 1, \tag{4.55c}$$

again with two exceptions,

$$j' = 0 \nrightarrow j = 0, \tag{4.55d}$$

$$m' = 0 \nrightarrow m = 0 \text{ when } \Delta j = 0. \tag{4.55e}$$

In LS coupling,

$$\Delta L = 0, \tag{4.55f}$$

$$\Delta S = 0, \tag{4.55g}$$

$$\Delta J = \pm 1. \tag{4.55h}$$

4.4.3 Electric Quadrupole (E2) Transitions

The electric quadrupole operator is a second-rank tensor $\widehat{\mathbf{Q}}_{2,q}$, which is commonly written in Cartesian coordinates as $Q_{ij} = er_i r_j$ and couples states of the same parity. Using the Wigner–Eckart theorem (Eq. (E.14)), the selection rules for transitions are determined to be

$$\Delta l = 0, \pm 2, \tag{4.56a}$$

$$\Delta j = 0, \pm 1, \pm 2, \tag{4.56b}$$

$$\Delta m = 0, \pm 1, \pm 2, \tag{4.56c}$$

again with the exceptions,

$$j' = 0 \nrightarrow j = 0, \quad (4.56d)$$

$$j' = 1/2 \nrightarrow j = 1/2, \quad (4.56e)$$

$$j' = 0 \nrightarrow j = 1. \quad (4.56f)$$

In *LS* coupling,

$$\Delta L = 0, \pm 1, \pm 2, \quad (4.56g)$$

$$\Delta S = 0. \quad (4.56h)$$

4.5 Atomic Spectra

Before we conclude our discussion of atoms, it is instructive to look at the simplest, unique features in the spectra of one- and two-electron atoms. A more comprehensive review is beyond the scope of this introductory text. There are a variety of sources to which the interested reader can turn for additional examples. For a few classic examples, see *Photoabsorption, Photoionization, and Photoelectron Spectroscopy* by Berkowitz and *Atomic and Molecular Spectroscopy: Basic Aspects and Practical Applications* by Svanberg listed under further reading at the end of this chapter and references therein. Here, we will look at *Rydberg series*, *autoionization* and *above-threshold ionization*.

4.5.1 Rydberg Series

From the selection rules of the previous section, it is clear that from the ground state, transitions to any state n are possible. However, the strength of the transition will vary with n. In the absence of strong interactions between different series, the strength will decrease as $1/n^{*3}$ for a so-called hydrogenic system. Figure 4.1 shows this reduction theoretically for H. This trend, however, is violated in general due to a variety of interactions between series. We will look at a classic example later in this chapter.

The oscillator strength defined in Section 4.3.4.3 strictly refers to bound–bound transitions, where f_{if} is linked to the integrated (over the absorption line) cross section. For bound–continuum transitions, photoionization, the spectrum is no longer composed of discrete lines but, rather, a continuum. To discuss the strength, it is more appropriate to define a spectral density of the oscillator strength, df/dE. Thus, in Fig. 4.1, df/dE for bound–bound transitions is plotted as histograms, where the area corresponds to f. Such a figure allows us to compare the strengths of the discrete and continuum portions of

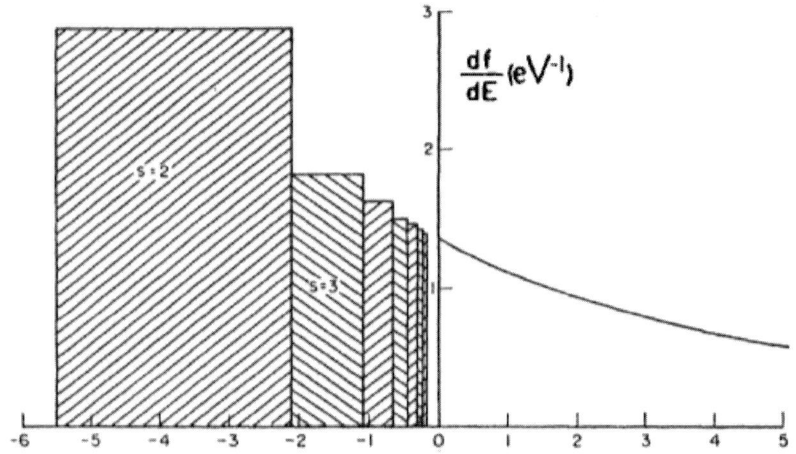

Fig. 4.1 The theoretical oscillator strength distribution, df/dE, to within a proportionality constant for bound-bound (histogram) and bound-continuum transitions from the ground state in H taken from Ref. [33]. The horizontal axis is in eV. The oscillator strengths for the bound states correspond to the area of the histogram. It is clear that the distribution in the discrete region continues smoothly into the continuum. (Reprinted figure with permission from U. Fano and J. W. Cooper, Rev. Mod. Phys. **40**, 441 (1968). Copyright (1968) by the American Physical Society.)

the spectrum on an equal footing. The key observations one should make from this figure are that f varies with n and df/dE continues smoothly from the discrete into the continuum. The latter is true in general even when the intensity does not fall as $1/n^{*3}$. The reason for this behavior is due the fact that the Rydberg series and its adjoining continuum (a channel) are intimately related, as we pointed out in Section 1.5. Typically, there is more than one channel that can be accessed from a particular level. When the Rydberg series converge to the same ionization limit, the members march in lock-step, without crossing each other. In the absence of any other perturbation, their intensities follow a simple, smooth progression similar to that displayed in Fig. 4.1.[8] This simple structure is common in one-electron (e.g. alkali) systems.

The spectrum is richer in two-electron systems. Specifically, it is common for two or more Rydberg series to occur in the same energy region but converge to different ionization thresholds. In this case, the spacing between levels is not the same and members of the two series periodically cross. This intermingling of the levels leads to an interaction between the series, particularly

8) While the interaction between Rydberg series may be small, more exotic interactions can change the simple behavior observed in Fig. 4.1 dramatically even for one-electron systems. See Ref. [33] for examples.

if the energies of two levels are nearly degenerate, resulting in energy shifts of the bound states away from their *proper* locations as dictated by Eq. (2.7). At the same time, the strength of the lines can vary rather dramatically near the crossing points.

The classic example is the Xe isoelectronic sequence (same electronic configuration). Part of the Ba^{2+} spectrum near the lowest ionization limit is shown in Fig. 4.2. From the ground state, a $5p^6\ {}^1S_0$ configuration, transitions to configurations of the form $5p^5nl$ consist of five distinct channels,[9] three of which converge to the first limit producing Ba^{3+} in its ${}^2P_{3/2}$ state with the other two converging to the second limit producing Ba^{3+} in its ${}^2P_{1/2}$ state. The spectrum in Fig. 4.2a is plotted as a function of the effective quantum (ν_2, see Eq. (2.6)) associated with the upper limit. On this scale, levels that are part of the Rydberg series converging to the upper limit are equally spaced. Consider the levels marked ns' and nd', which are equally spaced and do not cross each other. On the other hand, the spacing between levels marked ns and nd decreases as the effective quantum number increases causing these unprimed levels to cross the primed levels periodically; crossing occurs near $\nu_2 = 4.8$, 5.8 and 6.8. One notices that the intensity of the lines increases significantly near these crossing points. We see this again in the bottom graph of $\log gf$ (see Eq. (4.49) and Fig. 4.2c), which compares the calculated values of the oscillator strength for this Ba^{2+} case (the lines) against the ideal values for the hydrogenic case ($\propto 1/n^{*3}$, indicated by the dots). The connection between the Rydberg series and its continuum is clearly seen as well, in that the same intensity anomaly in the discrete region is repeated in the continuum – the physics associated with the perturbation in the two cases is essentially the same.

Figure 4.2b contains a plot of the positions of the bound resonances relative to the upper limit (ν_2) vs their positions relative to the lower limit (ν_1) mod 1, a so-called Lu–Fano plot [36–38]. In general, the unprimed levels lie on curves that are more or less horizontal while the primed levels lie on vertical curves. The interaction between the channels is strongest where the curves cross. The stronger the interaction, the larger the avoided crossings between the nominally vertical and horizontal curves. Resonances that fall close to where the curves would cross in the absence of any interaction experience significant energy shifts. ν_1 for the $19d$ level, for example, is shifted to lower values relative to other members of the sequence. At the same time, we notice that $20s$ is also shifted, but by much less than $19d$, indicating that the interaction between the nd and nd' channels is much stronger than the interaction between the ns and ns' channels. The differences in the interaction are more acute above the threshold (near $\nu_2 = 7.8$) in the autoionization (see next section) region.

9) Only four of the Rydberg series are observed in Fig. 4.2 because two of those converging to the lower limit are nearly degenerate and unresolved.

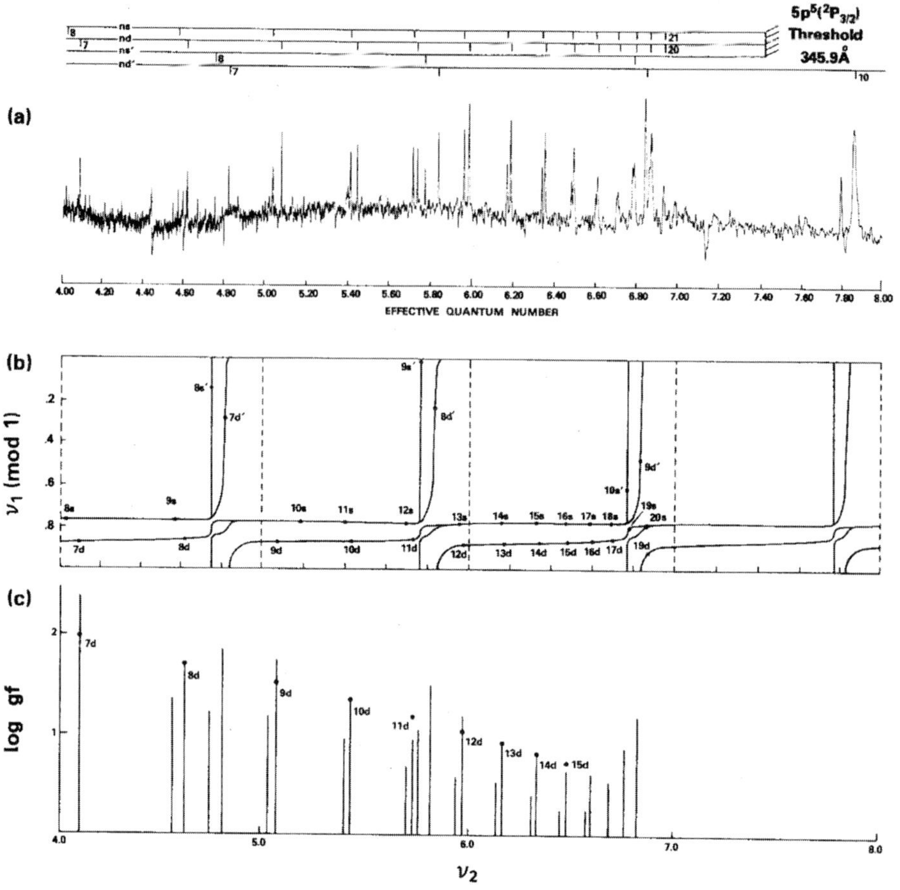

Fig. 4.2 An example of interacting Rydberg series in a 2-electron system, Ba^{2+}, taken from Ref. [34]: (a) photoabsorption spectrum plotted as a function of ν_2 (Eq. (2.6)), (b) Lu-Fano plot of position of the resonances relative to the first and second limits (i.e., $\nu_1(\mod 1)$ vs. ν_2) and (c) bar graph of calculated oscillator strengths (gf) for the bound nd series with the length of the lines representing the theoretical value, including full interactions, and the dots representing the hydrogenic values falling as $1/\nu_1^3$. (Reprinted figure with permission from W. T. Hill, III, J. Sugar and T. B. Lucatorto and K. T. Cheng, Phys. Rev. A **36**, 1200 (1987). Copyright (1987) by the American Physical Society.)

The stronger interaction between the nd and nd' channels compared with the interaction between the ns and ns' channels makes the nd' resonance wider than the ns' resonance. Figures 4.3 and 4.4 show that as electrons are removed along the isoelectronic sequence, the interaction is reduced. The intimate connection between the interaction in the bound and continuum regions is again seen. The wider autoionization resonances in Fig. 4.3 lead to larger avoided crossings about the bound levels in the Lu–Fano plots in Fig. 4.4.

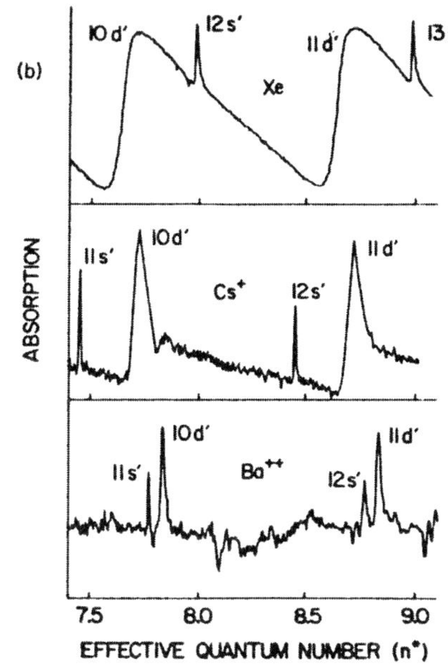

Fig. 4.3 Autoionization widths along the Xe-isoelectronic sequence taken from [35]. The spectrum of Xe, Cs^+ and Ba^{2+} are shown respectively from top to bottom between $\nu_2 = 7.5$ and 9.0. The change in the interactions is seen in three ways: a change in the widths and shapes of the nd' resonances and a shift in the position of the ns' resonances. (Reprinted figure, Fig. 3b) on page 188, from W. T. Hill, III and C. L. Cromer, "Laser-Driven Ionization and Photoabsorption Spectroscopy of Atomic Ions," in *Lasers, Spectroscopy and New Ideas: A Tribute to Arthur L. Schawlow*, Opt. Sci. Series Vol **54**, 183 (1987), with kind permission from Springer Science and Business Media. Copyright (1987) by Springer-Verlag.)

4.5.2
Autoionization

Autoionization is an interaction between a bound state that belongs to a closed channel (energy below the ionization threshold) and a continuum state (belonging to an open channel) that is energy degenerate. Thus, the bound state is only *quasibound* because the interaction leads to its decay into the continuum state and a free electron. The stronger the interaction the faster the decay and the wider, in energy space ($\Delta E \Delta t \sim \hbar$), the resonance. The nd' resonances are wider than the ns' resonances in Fig. 4.2, indicating that they autoionize faster.

The formal treatment of autoionization was done first by Fano [39], whose steps we follow here. The analysis begins by assuming eigenstates of our Hamiltonian describing the state of the system in the absence of excitation. The eigenstates will include bound ($|n\rangle$), quasibound ($|\phi\rangle$) and continuum ($|E\rangle$) states. In the autoionization region (above the first ionization threshold)

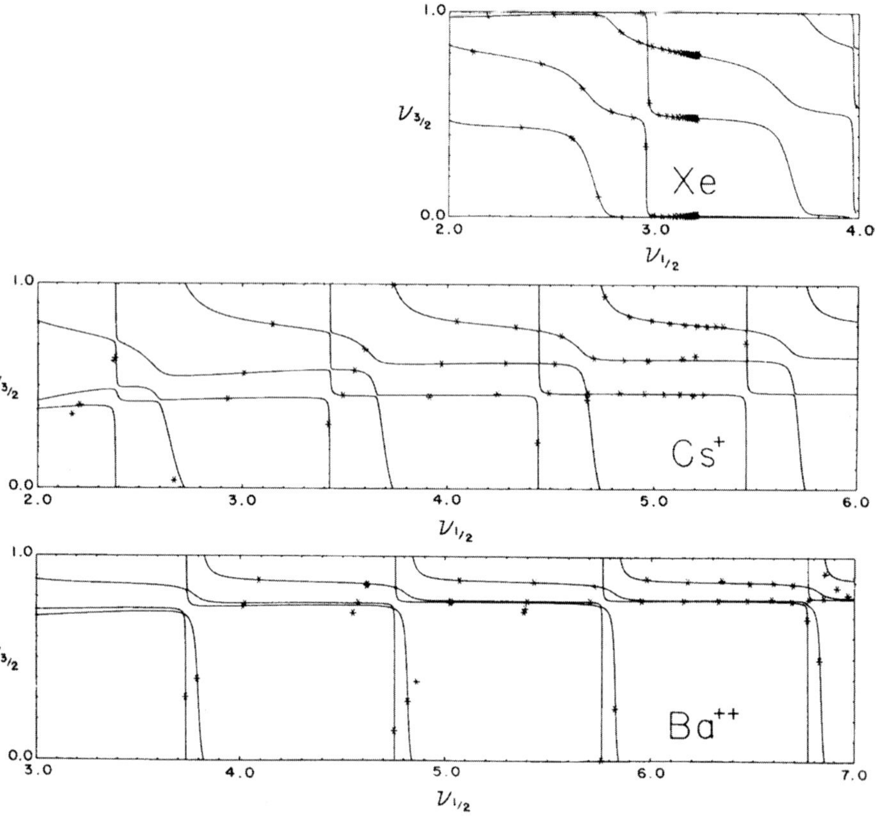

Fig. 4.4 Lu-Fano plots of the bound Rydberg resonances associated with the autoionization resonances in Fig. 4.3 taken from [35]. As electrons are removed from the system, the interactions decrease (smaller avoided crossings) and the resonances shift. (Reprinted figure, Fig. 4 on page 190, from W. T. Hill, III and C. L. Cromer, "Laser-Driven Ionization and Photoabsorption Spectroscopy of Atomic Ions," in *Lasers, Spectroscopy and New Ideas: A Tribute to Arthur L. Schawlow*, Opt. Sci. Series Vol **54**, 183 (1987), with kind permission from Springer Science and Business Media. Copyright (1987) by Springer-Verlag.)

we have

$$\langle \phi | \widehat{H}_o | \phi' \rangle = E_n \delta_{\phi\phi'}, \tag{4.57a}$$

for the quasibound state and

$$\langle E | \widehat{H}_o | E' \rangle = E\delta(E - E'), \tag{4.57b}$$

for the continuum state. We define the matrix element of $\widehat{H}_o$ between the quasibound and continuum states,

$$V = \langle E | \widehat{H}_o | \phi \rangle, \tag{4.57c}$$

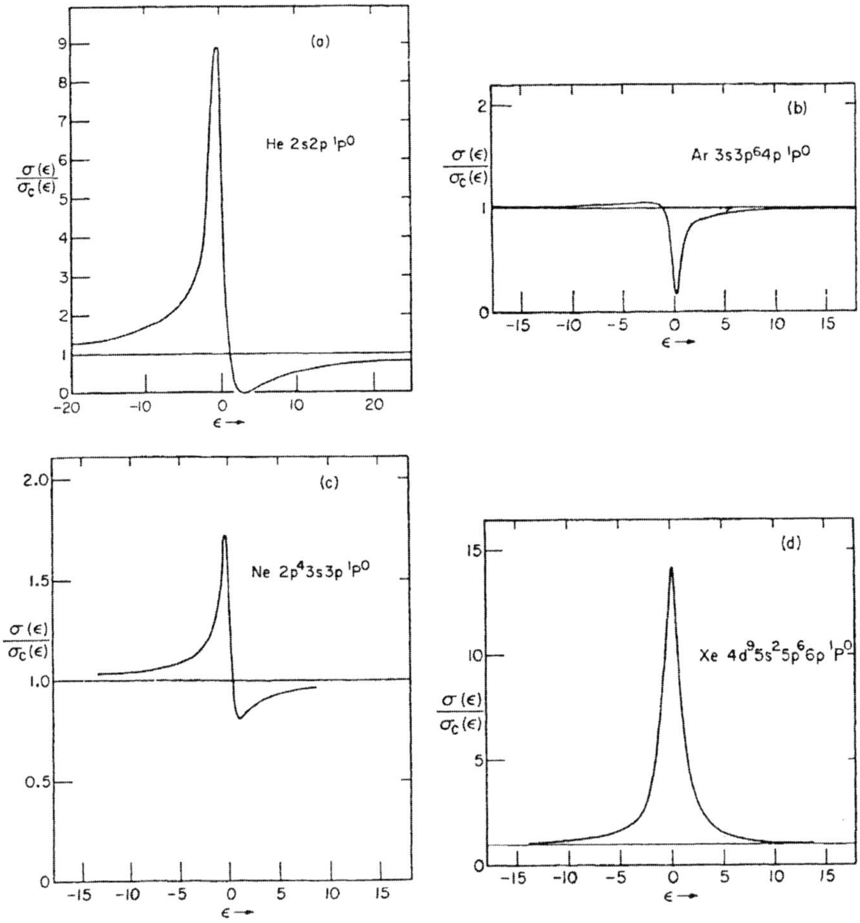

Fig. 4.5 Various observed autoionization line profiles from [33]: (a) He ($q = -2.8, \rho^2 = 1$) $2s2p\,^1$P^o, (b) Ar ($q = -0.22, \rho^2 = 0.86$) $3s3p^6$ $4p\ ^1$P^o, (c) Ne ($q = -2.0, \rho^2 = 0.17$) $2p^4\ 3s3p\ ^1P^o$ and (d) Xe ($q \sim 200, \rho^2 \sim 0.0003$) $4d^9 5s^2 5p^6 6p\ ^1$P^o. Here, $\rho = \langle E|\phi\rangle$ and what is plotted is $\rho^2[\sigma(\epsilon) - 1] + 1$. (Reprinted figure with permission from U. Fano and J. W. Cooper, Rev. Mod. Phys. **40**, 441 (1968). Copyright (1968) by the American Physical Society.)

which indicates the strength of the interaction between the two. We next define a transition matrix element from the initial (bound) state to the quasi-bound and to the continuum states,

$$t^b = \langle \phi|\widehat{T}|i\rangle \tag{4.58a}$$

and

$$t^c = \langle E|\hat{T}|i\rangle. \tag{4.58b}$$

The eigenstate of the system above threshold with eigenvalue E will be a superposition of the continuum and quasibound states,

$$|\Psi(E)\rangle = a|\phi(E)\rangle + \int dE' b|E(E')\rangle, \tag{4.59}$$

where a (a function of E) and b (a function of E and E') are coefficients. It can be shown [39] that the transition matrix element and cross section will take the form

$$t(\epsilon) = \frac{q+\epsilon}{\sqrt{1+\epsilon^2}} t^c, \tag{4.60a}$$

$$\sigma(\epsilon) = \frac{(q+\epsilon)^2}{1+\epsilon^2} \sigma^c, \tag{4.60b}$$

where

$$\sigma^c \equiv |t^c|^2, \tag{4.60c}$$

$$q \equiv \frac{t^b}{\pi V^* t^c}, \tag{4.60d}$$

$$\epsilon \equiv \frac{E - E_r}{\pi |V|^2}. \tag{4.60e}$$

In Eqs. (4.60), E_r is the energy where the enhancement in the continuum occurs (the resonance) and q is real and measures the relative strength of transitions to the quasibound and continuum states. The magnitude and sign of q tend to control the shape of the resonance while $|V|^2$ is related to its width. Clearly, these parameters are different for each member of the isoelectronic sequence of Xe shown in Fig. 4.3. It is possible to fit the profiles to determine the parameters in each case. Several additional shapes along with the parameters are displayed in Fig. 4.5.

4.5.3 Photoionization with Intense Lasers

Thus far, we have only looked at weak-field transitions involving one-photon transitions between bound and continuum states and two-photon transitions between bound states. At higher intensities, n-photon transitions, where n can become quite large, are possible between bound states as well as to continuum states. Transitions to continuum states is referred to as multiphoton ionization (MPI). It is straightforward, albeit tedious, to calculate the n-photon transition rate by following the procedure used to determine the one- and two-photon rates in Section 4.3. While the lowest order is usually sufficient, if this perturbative approach is valid it is also possible to calculate higher orders. Diagrammatic approaches have been developed to aid bookkeeping [40]. In lowest order, the n-photon rate can be written as [40,41]

$$R_{if}^{(n)} = \frac{2\pi}{\hbar}\left(\frac{2e^2}{\varepsilon_0 c}\right)^n I^n \times \left| \sum_{i,j,\ldots,k} \frac{\langle E|r|k\rangle\langle k|r|j\rangle\cdots\langle i|r|g\rangle}{(E_k - E_g - (n-1)\hbar\omega)\cdots(E_i - E_g - \hbar\omega)} \right| \times \delta(E - E_g - n\hbar\omega), \quad (4.61)$$

where $|g\rangle$ is the initial state with the associated energy E_g and $|E\rangle$ is the continuum state with the associated energy E. Energy conservation leads to photoelectron energies given by

$$E = n\hbar\omega - E_g\,, \quad (4.62)$$

where n is the minimum number of photon to exceed the ionization threshold.

At sufficiently low intensities, Eqs. (4.61) and (4.62) do a reasonable job describing the photoelectron energy spectrum. At intensities of 10^{12}–10^{13} W/cm^2, Eq. (4.62) begins to fail because a new phenomenon, called

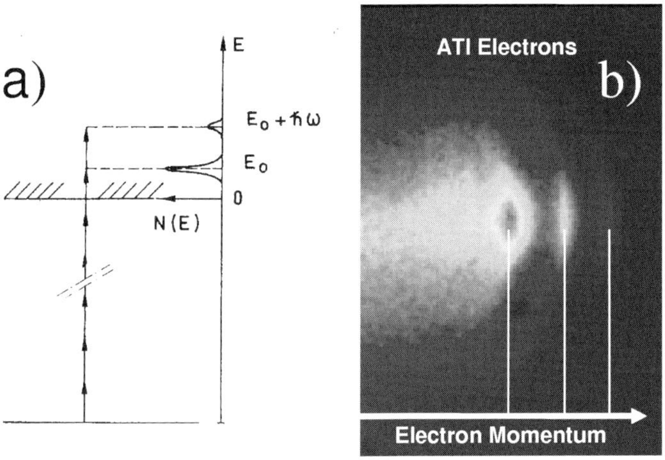

Fig. 4.6 Above-threshold ionization: (a) energy level schematic taken from Ref. [42] and (b) photoelectron momentum spectrum showing the characteristic energy spacing corresponding to $\hbar\omega_l$. Note, on a momentum scale, $p_e = \sqrt{2m_e\hbar\omega_l}$. (Reprinted figure, Fig. 4.6a, with permission from P. Agostini, F. Fabre, G. Mainfray, and G. Petite, Phys. Rev. Lett. **42**, 1127 (1979). Copyright (1979) by the American Physical Society.)

above-threshold ionization (ATI), becomes active. The central feature of ATI is that electrons do not cease to absorb energy from the laser field once $n\hbar\omega$ exceeds the ionization threshold, as they do at lower intensities. Rather, as shown in Fig. 4.6, the electrons can leave the atom having absorbed an integral number of additional or excess photons [42].

In order for the ATI spectrum to conserve energy, one must modify Eq. (4.62) such that

$$E = (n + s)\hbar\omega - E_g, \tag{4.63}$$

where s is an integer. We can understand why the spectrum contains peaks spaced by the photon energy when we recognize that some of the intermediate states in Eq. (4.61) will be continuum states. The summations, which resulted from inserting complete sets, involving the continuum states are really integrals. Thus, the rates peak every $\hbar\omega$ because of the energy denominators involving those continuum states.

Further Reading

1 F. H. Faisal, *Theory of Multiphoton Processes*, Plenum, New York, NY, 1987.

2 I. I. Sobelman, *Atomic Spectra and Radiative Transitions*, Springer Series in Chemical Physics, Springer, New York, NY, 1979.

3 H. A. Bethe and E. E. Salpeter, *Quantum Mechanics of One- and Two-Electron Atoms*, Plenum, New York, NY, 1977.

4 R. D. Cowan, *The Theory of Atomic Structure and Spectra*, University of California Press, Los Angeles, CA, 1981.

5 J. Berkowitz, *Photoabsorption, Photoionization and Photoelectron Spectroscopy*, Academic, New York, NY, 1979.

6 S. Svanberg, *Atomic and Molecular Spectroscopy: Basic Aspects and Practical Applications*, Springer, Berlin, 1991.

7 R. F. Stebbings and F. B. Dunning, *Rydberg States of Atoms and Molecules*, Cambridge University Press, New York, NY, 1993.

8 T. F. Gallagher, *Rydberg Atoms*, Cambridge Monographs on Atomic, Molecular and Chemical Physics 3, Cambridge University Press, New York, NY, 1994.

9 J.-P. Connerade, *Highly Excited Atoms*, Cambridge Monographs on Atomic, Molecular and Chemical Physics 9, Cambridge University Press, New York, NY, 1998.

10 G.W.F. Drake (ed.), *Atomic, Molecular, & Optical Physics Handbook*, American Institute of Physics, Woodbury, NY, 1996.

11 http://physics.nist.gov/PhysRefData/contents.html. This web site, maintained by the US National Institute of Standards and Technology (NIST), contains an assortment of tables and links to the latest atomic spectroscopic information and fundamental constants. The link was valid at the time of publication.

12 For additional problems with worked solutions see D. Budker, D. F. Kimball and D. P. DeMille, *Atomic Physics: An Exploration Through Problems and Solutions*, Oxford University Press, New York, NY, 2004.

Problems

4.1 Show by direct substitution that $E' = E$ and $B' = B$ when the Coulomb gauge is made.

4.2 Show by direct substitution that the Schrödinger equation is invariant under the Coulomb gauge transformation.

4.3 Show that the Lorentzian function (Eq. (4.36)) is properly normalized for optical frequencies.

4.4 Estimate the relative magnitudes of Γ_o and Γ_D (see Appendix B.10) that determine when the Lorentzian and Doppler profiles dominate the line shape. Determine the peak of the absorption when $\Gamma_o = \Gamma_D$.

4.5 Using the equations given in Appendix B.10, show that when $\omega = \omega_o$ the line shape takes the form

$$S(\omega_o) = \left(\frac{4\ln 2}{\pi}\right)^{1/2} \frac{e^{b^2}}{\Gamma_D}\ \texttt{erfc}(b), \tag{4.64}$$

where the *complementary error function* is given by

$$\texttt{erfc}(x) = \frac{2}{\sqrt{\pi}} \int_x^{\infty} \mathrm{e}^{-u^2} du. \tag{4.65}$$

4.6 Using the result of Problem 4.5, estimate the peak absorption magnitudes in the limit where the Lorentzian and Doppler widths dominate.

4.7 Carry out the steps to determine the transition amplitude for two-photon absorption, $\left\langle f \left| \widehat{T}^{(2)}(\infty, \infty) \right| \right\rangle$, and then write an expression for the transition rate. Estimate the rate for two-photon absorption between the $2s\,^2S$ and $3d\,^2D$ states in Li. Assume that the dominant intermediate state is the $2p\,^2P^{\circ}$ state. You can use the values in *Atomic Transition Probabilities: Hydrogen Through Neon*, NSRDS-NBS 4, Vol. 1 (which can be found online at Ref. [9]) to do this problem.

4.8 Starting from the Hamiltonian

$$\widehat{H} = \frac{1}{2\mu}\left[\vec{p} - e\vec{A}(t)\right]^2 - e\phi + V', \tag{4.66}$$

where $\vec{p}$ is the electron momentum, μ the reduced mass, e the elementary charge, $\vec{A}$ the vector potential, ϕ the scalar potential and V $(= -e\phi + V')$ the static potential in which the electron moves, complete the steps to show that the radiation Hamiltonian takes the form

$$\widehat{H}_{rad} = \frac{i\hbar e}{\mu}\vec{A}(t)^2 \cdot \vec{\nabla} + \frac{e}{2\mu}\vec{A}(t) \cdot \vec{A}(t).$$

Make a gauge transformation of the second kind,

$$\psi(\vec{r},t) = e^{i\frac{e}{\hbar}\vec{r}\cdot\vec{A}(t)}\Psi(\vec{r},t),$$

and show that the radiation Hamiltonian can also take the form

$$\widehat{H}_{rad} = -e\vec{r}\cdot\vec{E}(t).$$

4.9 Carry out the steps to verify the M1 and E2 selection rules of Eqs. (4.55) and (4.56).

4.10 Use the Wigner–Eckart theorem to determine the selection rules for a two-photon transition in one-electron atoms.

4.11 Derive the restrictions on ΔL, ΔJ and ΔS in LS coupling for one-photon and two-photon transitions in the dipole approximation.

4.12 Show that the following pairs of equations lead to the same physics:

a) Eqs. (4.27) and (9.37) and

b) Eqs. (4.29) and (9.49).

5
Diatomic Molecules

A diatomic molecule is an atom with
one atom too many.
– Arthur L. Schawlow

In this section we will look at the simplest case of two atoms working in concert – the diatomic molecule. The collective motion of the two atoms will lead to symmetries and electronic states related to but distinct from atomic states. Just as in atoms, we will be able to define electronic configurations for the electrons from which molecular states and potential curves can be derived. One of the major differences between an atom and a diatomic molecule is that the states in which the electrons move will depend upon an additional parameter, the internuclear separation, R, between the two atomic nuclei. This parameter R will, hence, couple the electronic motion to the nuclear motion. The nuclei can vibrate and rotate, introducing new structure. Electronic states will not be the result of a single configuration, in general, which will again produce configuration interaction. While we will not explore them in this book, this interaction leads to interesting radiationless decay channels – *autoionization* and *predissociation*.[1]

We will employ a simple model to classify and analyze these systems. This model has proven itself over the years to be able to capture most of the essence of the physics of the diatomic system. The cartoon in Fig. 5.1 helps us to picture the treatment. Specifically, the atoms are treated as heavy masses in close proximity ($\sim$ 0.1 nm) attached to each other via massless springs. The energy of the system will have two parts, an electronic part and a nuclear part. The nuclear part will be determined by a combination of vibrating and rotating dumbbells. The fact that the atoms can rotate and vibrate at the same time means that the vibrations will not be exactly harmonic nor will the moment of inertia remain fixed during rotation. The electrons, which are shared between the nuclei, will move in a distorted, cylindrically symmetric potential established by the Coulomb fields of the two atomic nuclei and the other electrons.

1) Readers interested in these phenomena are encouraged to read Ref. [43].

Light-Matter Interaction: Atoms and Molecules in External Fields and Nonlinear Optics.
W. T. Hill and C. H. Lee

ISBN: 978-3-527-40661-6

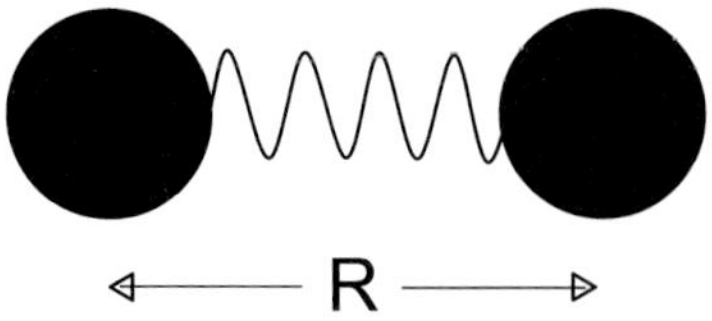

Fig. 5.1 The classical model of a diatomic molecule, two atoms separated by a distance R, that we will employ in our discussion in this section. The two atoms are assumed to be attached as if by a massless spring allowing them to vibrate. In addition, the system is free to rotate about an axis perpendicular to $\vec{\mathbf{R}}$ that contains the center of mass.

5.1 The Hamiltonian

We begin by looking at the Hamiltonian for the general case, which takes the form

$$\widehat{H} = \underbrace{-\frac{\hbar^2}{2}\left(\frac{\nabla_1^2}{M_1}+\frac{\nabla_2^2}{M_2}\right)}_{A} \underbrace{-\frac{\hbar^2}{2m_e}\sum_{i=1}^{N}\nabla_i^2}_{B} + \underbrace{\frac{Z_1Z_2e^2}{4\pi\varepsilon_0 R}+\frac{e^2}{4\pi\varepsilon_0}\sum_{i<j}^{N}\frac{1}{r_{ij}}}_{C}$$
$$\underbrace{-\frac{e^2}{4\pi\varepsilon_0}\sum_{i=1}^{N}\left(\frac{Z_1}{\left|\vec{r}_i-\vec{R}_1\right|}+\frac{Z_2}{\left|\vec{r}_i-\vec{R}_2\right|}\right)}_{D} + \zeta, \quad (5.1)$$

where A is the nuclear kinetic energy, B the electronic kinetic energy, C the electrostatic interaction, D the Coulomb potential and ζ is composed of smaller relativistic interaction terms, such as spin–orbit, spin–spin, etc., that we will not consider explicitly. For a neutral diatomic molecule, $N = Z_1 + Z_2$. In A, $\nabla_{1,2}^2$ acts only on the nuclear coordinates, R_1 and R_2, while in B, ∇_i^2 acts on the electronic coordinates, r_i; $R = |R_1 - R_2|$ is the internuclear separation and r_{ij} is the distance between electrons i and j. In the center-of-mass coordinates, expression A can be written as

$$\frac{\hbar^2}{2\left(M_1+M_2\right)}\nabla_{cm}^2 + \frac{\hbar^2}{2\mu}\nabla^2, \quad (5.2)$$

where the first term corresponds to the translation of the entire molecule, which does not affect the internal energy, and the second describes the relative motion of the two nuclei; ∇ only operates on R and the nuclear reduced mass is represented by μ $(= M_1M_2/\left(M_1+M_2\right))$. In center-of-mass coordinates the

Hamiltonian takes the form

$$\widehat{H} = -\frac{\hbar^2}{2\mu}\nabla^2 - \frac{\hbar^2}{2m_e}\sum_{i=1}^{N}\nabla_i^2 + V(r_i; R_1, R_2), \tag{5.3}$$

where to first order

$$V(r_i; R_1, R_2) = \frac{Z_1 Z_2 e^2}{4\pi\varepsilon_0 R} + \frac{e^2}{4\pi\varepsilon_0}\sum_{i<j}^{N}\frac{1}{r_{ij}}$$
$$-\frac{e^2}{4\pi\varepsilon_0}\sum_{i=1}^{N}\left(\frac{Z_1}{\left|\vec{r}_i - \frac{M_2}{M_1+M_2}\vec{R}\right|} + \frac{Z_2}{\left|\vec{r}_i + \frac{M_1}{M_1+M_2}\vec{R}\right|}\right). \tag{5.4}$$

In Eq. (5.4) we have assumed that M_1 is to the left of the center of mass and M_2 is to the right. Applying this Hamiltonian to Eq. (1.2) yields

$$\frac{\hbar^2}{2\mu}\nabla^2|\psi\rangle + \frac{\hbar^2}{2m_e}\sum_{i=1}^{N}\nabla_i^2|\psi\rangle + (E - V)|\psi\rangle = 0. \tag{5.5}$$

5.2 Born–Oppenheimer Approximation

We will seek a separable solution (known as the Born–Oppenheimer separation) of the form

$$|\psi\rangle = |\psi_N\rangle\,|\psi_{el}\rangle, \tag{5.6}$$

where ψ_N and ψ_{el} are the nuclear and electronic wavefunctions, respectively. These wavefunctions are solutions to

$$\frac{\hbar^2}{2m_e}\sum_{i=1}^{N}\nabla_i^2|\psi_{el}\rangle + (E_{el} - V_{el})|\psi_{el}\rangle = 0 \tag{5.7a}$$

and

$$\frac{\hbar^2}{2\mu}\nabla^2|\psi_N\rangle + (E - E_{el} - V_N)|\psi_N\rangle = 0. \tag{5.7b}$$

Equation (5.7a) describes the motion of the electrons when the nuclei are fixed in space. In the center-of-mass coordinates,

$$V_{el}(r_i; R_1, R_2) = \frac{e^2}{4\pi\varepsilon_0}\sum_{i<j}^{N}\frac{1}{r_{ij}}$$
$$-\frac{e^2}{4\pi\varepsilon_0}\sum_{i=1}^{N}\left(\frac{Z_1}{\left|\vec{r}_i - \frac{M_2}{M_1+M_2}\vec{R}\right|} + \frac{Z_2}{\left|\vec{r}_i + \frac{M_1}{M_1+M_2}\vec{R}\right|}\right), \tag{5.8}$$

which is a function of the positions of all the electrons, r_i, as well as the internuclear separation, R. This implies that $|\psi_{el}\rangle$ and E_{el} both depend on R. Equation (5.7b) describes the motion of the nuclei in a field produced by the electrons, E_{el}, and the electrostatic interaction with each other, $V_N = Z_1 Z_2 e^2/4\pi\varepsilon_0 R$. Thus, E is the total energy of the molecule because it depends on both nuclear and electronic motion. We define $U(R) = E_{el} + V_N$, the average electronic potential in which the nuclei move, as

$$U(R) = \frac{Z_1 Z_2 e^2}{4\pi\varepsilon_0 R} + \langle\psi_{el}| \left\{ \frac{e^2}{4\pi\varepsilon_0} \sum_{i<j}^{N} \frac{1}{r_{ij}} - \frac{e^2}{4\pi\varepsilon_0} \sum_{i=1}^{N} \left(\frac{Z_1}{\left|\vec{r}_i - \frac{M_2}{M_1+M_2}\vec{R}\right|} + \frac{Z_2}{\left|\vec{r}_i + \frac{M_1}{M_1+M_2}\vec{R}\right|} \right) \right\} |\psi_{el}\rangle \,. \quad (5.9)$$

Putting all this together, Eq. (5.5) becomes

$$\underbrace{\frac{\hbar^2}{2\mu}\left(\nabla^2|\psi_N\rangle\right)|\psi_{el}\rangle + (E - E_{el} - V_N)|\psi_N\rangle|\psi_{el}\rangle}_{\text{A}}$$
$$\underbrace{+\frac{\hbar^2}{2\mu}\left[2\left(\vec{\nabla}|\psi_N\rangle\right)\cdot\left(\vec{\nabla}|\psi_{el}\rangle\right) + |\psi_N\rangle\nabla^2|\psi_{el}\rangle\right]}_{\text{B}}$$
$$\underbrace{+\frac{\hbar^2}{2m_e}\sum_{i=1}^{N}|\psi_N\rangle\nabla_i^2|\psi_{el}\rangle + (E_{el} - V_{el})|\psi_N\rangle|\psi_{el}\rangle}_{\text{C}} = 0, \quad (5.10)$$

where, again, ∇ only operates on quantities inside the parentheses.

Equation (5.10) is subject to the constraints of Eqs. (5.7a) and (5.7b), so A and C are identically zero. It is straightforward to see that B is nearly zero as well. Consider $\hbar/2\mu\nabla|\psi_{el}\rangle$, which acts on the nuclear coordinate, R. Specifically, it measures the variation of the electronic wavefunction as R changes. Since $-i\hbar\nabla|\psi_{el}\rangle$ acts over nearly the same dimensions as $-i\hbar\nabla_i|\psi_{el}\rangle$, it is $\sim p_{el}$. Thus, B is of order $m_e/\mu E_{el}$, which is very small. To a good approximation we can ignore B. Setting B to zero is called the Born–Oppenheimer approximation. Essentially, the approximation relies on the fact that in most cases, the electrons move much faster than the nuclei and thus adjust very rapidly to any small changes in R – the nuclei are assumed to be fixed in space when estimating the electron dynamics. This is generally true for low-lying bound states.

5.3 Nuclear Equation

We will now look at the nuclear equation (Eq. (5.7b)) in some detail. As mentioned at the beginning, the nuclear dynamics consists of vibration and rotation. In general,

$$|\psi_N\rangle = |\psi_{vib}\rangle \, |\psi_{rot}\rangle \,. \tag{5.11}$$

Writing ∇^2 in spherical coordinates and substituting Eq. (5.11), the nuclear equation becomes

$$-\frac{\hbar^2}{2\mu}\left[\frac{1}{R}\frac{\partial}{\partial r}\left(R^2\frac{\partial}{\partial r}\right) - \frac{\widehat{\kappa}^2}{R^2}\right]|\psi_{vib}\rangle\,|\psi_{rot}\rangle = (E - U(R))\,|\psi_{vib}\rangle\,|\psi_{rot}\rangle\,, \tag{5.12}$$

where

$$\widehat{\kappa}^2 = \frac{1}{\sin\vartheta}\frac{\partial}{\partial\vartheta}\left(\sin\vartheta\frac{\partial}{\partial\vartheta}\right) + \frac{1}{\sin^2\vartheta}\frac{\partial^2}{\partial\varphi^2} \tag{5.13}$$

is the angular momentum operator. When R is constant (say at $R = R_e$, the equilibrium separation) we have the so-called *rigid rotator* and the equation reduces to

$$\frac{\hbar^2}{2\mu R_0^2}\widehat{\kappa}^2\,|\psi_{rot}\rangle = (E - U(R_e))\,|\psi_{rot}\rangle\,. \tag{5.14}$$

The solutions are spherical harmonics, $|\psi_{rot}\rangle = Y_{Km}(\vartheta,\varphi)$ (see Appendix B.7), and the rotational energy is[2]

$$E_{rot} = (E - U(R_e)) = \frac{K(K+1)\hbar^2}{2\mu R_e^2}. \tag{5.15}$$

Usually, the rotational energy will be written in wavenumbers (see Appendix A.2), E_{rot}/hc. Thus, we define the rotational term energy, $F(K)$, which is generally taken to be positive, as

$$F(K) = \frac{E_{rot}}{hc} = B_e K(K+1)\,. \tag{5.16}$$

Here, B_e is the rotational constant,

$$B_e = \frac{h}{8\pi^2\mu c R_e^2} = \frac{h}{8\pi^2 c I_e}, \tag{5.17a}$$

2) We point out that in this section, we use K to denote the quantum number for rotation, as was done in Ref. [44]. This is not the agreed upon convention used by contemporary books and papers, however. The international agreed upon symbol for the rotational quantum number is R, as we show in Table 5.2. We use K in this tutorial environment to avoid possible confusion with the internuclear separation, which is also denoted by R.

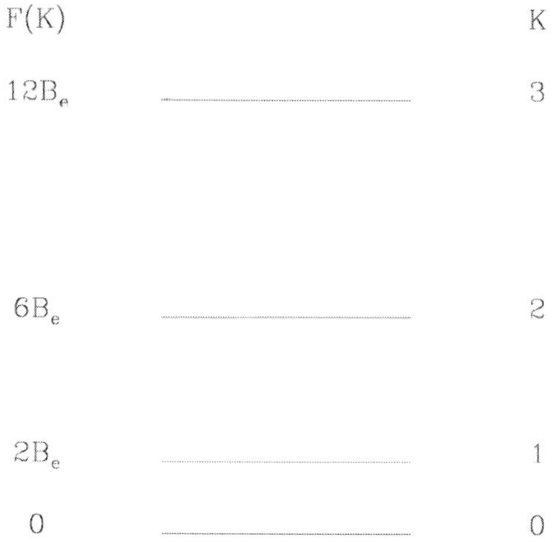

Fig. 5.2 Rotational energy level structure.

where

$$I_e = \mu R_e^2 = \frac{M_1 M_2}{M_1 + M_2} R_e^2 \tag{5.17b}$$

is the moment of inertia. The spacing between adjacent levels,

$$\Delta F(K) = F(K+1) - F(K) \tag{5.18a}$$

$$= B_e \left[(K+1)(K+2) - K(K+1) \right] \tag{5.18b}$$

$$= 2B_e (K+1), \tag{5.18c}$$

is not constant but increases with increasing K as shown in Fig. 5.2.

5.3.1 Harmonic Approximation of $U(R)$

If we relax the restriction that R is fixed at R_e then we must solve Eq. (5.12) to determine how R deviates from R_e. To that end, we make the substitution

$$|\psi_{vib}\rangle = \frac{\xi(R)}{R}, \tag{5.19a}$$

which leads to

$$|\psi_N\rangle = \frac{\xi(R)}{R} Y_{Km}(\vartheta, \varphi). \tag{5.19b}$$

Equation (5.12) then becomes

$$\frac{d^2}{dr^2}\xi(R) + \frac{2\mu}{\hbar^2}\left[E - U(R) - \frac{K(K+1)\hbar^2}{2\mu R^2}\right]\xi(R) = 0. \tag{5.20}$$

To solve this equation we must know something about $U(R)$, the average potential in which the nuclei move. In general there are no analytic solutions for $U(R)$, so approximations must be made. If we assume that R does not deviate too much from its equilibrium value, then we can expand $U(R)$ into a Taylor series about R_e,

$$U(R) = U(R_e) + (R - R_e) \left.\frac{dU(R)}{dR}\right|_{R=R_e} + \frac{1}{2}(R - R_e)^2 \left.\frac{d^2U(R)}{dR^2}\right|_{R=R_e} + \cdots . \tag{5.21}$$

For bound states, $U(R)$ represents a potential well with its minimum located at $U(R_e)$, so the second term vanishes. Without loss of generality we can choose an energy scale such that $U(R_e) = 0$, so the first term also vanishes. As a result the potential will be harmonic, described by

$$U(R) = \frac{1}{2}(R - R_e)^2 \left.\frac{d^2U(R)}{dR^2}\right|_{R=R_e} = \frac{k_s}{2}(R - R_e)^2 \tag{5.22}$$

to lowest order, near R_e. Our radial equation now becomes

$$\frac{d^2}{dr^2}\xi(R) + \frac{2\mu}{\hbar^2}\left[E - \frac{k_s}{2}(R - R_e)^2 - \frac{K(K+1)\hbar^2}{2\mu R^2}\right]\xi(R) = 0. \tag{5.23}$$

When $K = 0$, the equation reduces to

$$\frac{d^2}{dr^2}\xi(R) + \frac{2\mu}{\hbar^2}\left[E - \frac{k_s}{2}(R - R_e)^2\right]\xi(R) = 0, \tag{5.24}$$

which describes a harmonic oscillator with energy

$$E_{vib} = \hbar\sqrt{\frac{k_s}{\mu}}\left(v + \frac{1}{2}\right). \tag{5.25}$$

The wavefunctions are given by

$$\xi(R) = N_v e^{-\alpha R^2/2} H_v\left(\sqrt{\alpha}R\right), \tag{5.26}$$

where H_v are the Hermite polynomials and $\alpha = \sqrt{k_s\mu}/\hbar$. We identify the quantized vibrational energy as

$$E(v) = E_{vib} = hc\omega_e\left(v + \frac{1}{2}\right) \tag{5.27a}$$

or, in wavenumbers,

$$G(v) = \frac{E(v)}{hc} = \omega_e\left(v + \frac{1}{2}\right), \tag{5.27b}$$

where

$$\omega_e = \frac{1}{2\pi c}\sqrt{\frac{k_s}{\mu}} = \frac{\alpha\hbar}{2\pi c}. \tag{5.27c}$$

We note that the levels are equally spaced. However, the lowest level, $v = 0$, does not sit at the bottom of the well, but has zero-point energy, $\omega_e/2$.

It is straightforward to see that the rotational energy is considerably smaller than the vibrational energy. Consider H_2 where $R_e = 0.074$ nm and $\mu c^2 \approx 469$ MeV. If we express $\hbar c$ as 197 eV nm then the rotational energy is

$$E_{rot} = \frac{(\hbar c)^2}{2\mu c^2 R_e^2} = 7.6 \times 10^{-3}\ eV \tag{5.28}$$

and

$$\frac{E_{rot}}{hc} \to 60.9\ \text{cm}^{-1}. \tag{5.29}$$

Using Eq. (5.57) (see Problem 5.1), we have for the spring constant

$$k_s = 2\left(\frac{e}{4\pi\varepsilon_0 R_e}\right)\frac{e}{R_e^2} = 7.1 \times 10^3\ eV/\text{nm}^2. \tag{5.30}$$

For $v = 0$, the vibrational energy is

$$E_{vib} = \frac{\hbar c}{2}\sqrt{\frac{k_s}{\mu c^2}} = 0.38\text{eV} \tag{5.31}$$

and

$$\frac{E_{vib}}{hc} \to 3100\ \text{cm}^{-1}, \tag{5.32}$$

about 50 times larger.[3]

5.3.2 Beyond the Harmonic Approximation of $U(R)$

Our harmonic oscillator potential, which is parabolic, is unrealistic because it implies that the system remains bound as $R \to \infty$. We know that at some point the system separates into two individual atoms. We can improve our approximation by carrying out our expansion of $U(R)$ to higher orders – cubic, quartic, etc. For the first correction we have

$$E(v) = hc\left[\omega_e\left(v + \frac{1}{2}\right) - \omega_e x_e\left(v + \frac{1}{2}\right)^2\right], \tag{5.33}$$

3) The actual vibrational energy is about 2200 cm^{-1}.

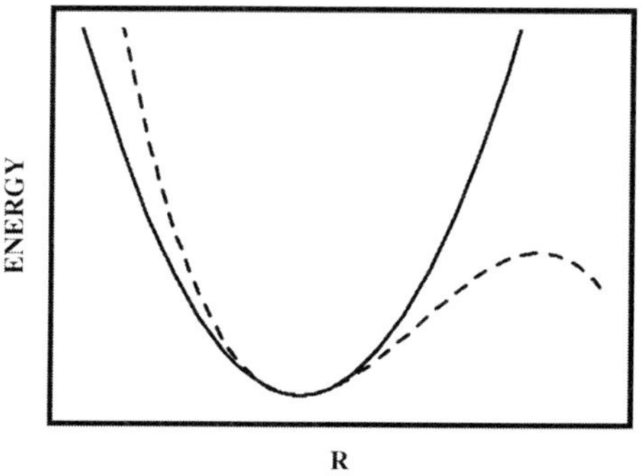

Fig. 5.3 This figure shows the effect of adding a cubic term (dashed curve) to the harmonic oscillator potential.

where typically $\omega_e > \omega_e x_e > 0$.[4] It should be recognized that just adding a cubic term can lead to unphysical behavior as well (see Fig. 5.3). Consequently, depending upon how far out one wants to determine the curve, additional terms must be included. Huber and Herzberg [45] provided comprehensive tables that include ω_e and $\omega_e x_e$ for numerous systems, while Herzberg [44] tabulated values for $\omega_e y_e$ ($\ll \omega_e x_e$) in some cases. Constants can be found online as well [46]. Thus, we redefine $G(v)$ as

$$G(v) = \omega_e \left(v + \frac{1}{2}\right) - \omega_e x_e \left(v + \frac{1}{2}\right)^2 + \omega_e y_e \left(v + \frac{1}{2}\right)^3 + \cdots . \tag{5.34}$$

5.3.3 Vibrating Rotator

Thus far, we have ignored the R dependence for rotation. That is, we have decoupled rotation from vibration. This too is unphysical because, since we are modeling the molecule as two masses attached by a spring, the atoms will experience a centrifugal force as they rotate, which will tend to increase R as the energy increases. This must be the case to conserve angular momentum. This can be taken into account by adding terms that are nonlinear in $K(K+1)$ to Eq. (5.16),

$$F(K) = B_v K(K+1) - D_v K^2 (K+1)^2 + \cdots . \tag{5.35}$$

An approximate expression for B_v can be written in the form of Eq. (5.17a) with $1/R_e^2$ replaced by an average of $1/R_v^2$,

$$B_v = \frac{h}{8\pi^2 c\mu} \left\langle \frac{1}{R_v^2} \right\rangle . \tag{5.36}$$

4) The sign of the $\omega_e x_e$ term is chosen to make $\omega_e x_e$ positive definite. Thus, this anharmonicity causes the potential curve to flatten out as shown in Fig. 5.3 – the potential becomes softer.

To account for its v dependencc, we write

$$B_v = B_e - \alpha_e \left(v + \frac{1}{2} \right) + \gamma_e \left(v + \frac{1}{2} \right)^2 + \cdots, \tag{5.37}$$

where $\alpha_e > 0$ and $\gamma_e > 0$, so that as $v \to \infty$, $R \to \infty$ and the energy of the system decreases.[5] The subscript e refers to the equilibrium internuclear separation, R_e. In a similar way we write

$$D_v = D_e - \beta_e \left(v + \frac{1}{2} \right) + \cdots. \tag{5.38}$$

Values for B_e, D_e, α_e, β_e and γ_e can all be determined spectroscopically.

Putting all this together, the expression for the energy of a rovibrational state is

$$\nu(v, K) = \nu_e + G(v) + F(K), \tag{5.39}$$

where ν_e is the energy of the bottom of the electronic state (which is zero for the ground state).

Before we end this section it is worthwhile pointing out that Dunham [47] (see also Refs. [44, 48]) introduced a different expression for the energy for the vibrating rotator,

$$T = \sum_{ij} Y_{ij} \left(v + \frac{1}{2} \right)^i K^j (K+1)^j, \tag{5.40}$$

which leads to expressions for the vibrational and rotational constants that deviate from our earlier expressions. These deviations are quite small and usually ignorable except for H_2. Nevertheless, the vibrational–rotational level will be given by [48]

$$\nu(v, K) = \nu_e + Y_{00} + G(v) + F_v(K), \tag{5.41}$$

where

$$Y_{00} = \frac{B_e}{4} + \frac{\alpha_e \omega_e}{12 B_e} + \frac{\alpha_e^2 \omega_e^2}{144 B_e^2} - \frac{\omega_e x_e}{4}, \tag{5.42}$$

$$\nu_e = T_e - G_0(X). \tag{5.43}$$

The last expression gives the zero-point-adjusted electronic term energy. For H_2, $G_0(X) = 2197.3\ \text{cm}^{-1}$ [48].[6]

5) Here again, the sign of the α_e term is chosen to make α_e positive definite. In so doing, B_v reduces as the molecule stretches – the potential becomes softer, again.

6) Note that the last term in Eq. (5.42) is a correction to the expression given in Ref. [44]. See Ref. [48].

5.3.4
Analytic Expression for $U(R)$

The first correction to the harmonic potential leads to an exact expression for the potential curve called a Morse potential that can be written as

$$U(R) = U_{el} + D^o[1 - e^{-a(R-R_e)}]^2, \tag{5.44}$$

where U_{el} is a constant that sets the overall energy of the bottom of the potential well relative to the zero of the energy scale. For the example shown in Fig. 5.4, $U_{el} = 0$. For this potential, at R_e we have

$$U(R_e) = U_{el},$$
$$\left.\frac{dU}{dR}\right|_{R=R_e} = 0,$$
$$\left.\frac{d^2U}{dR^2}\right|_{R=R_e} = 2a^2D^o \equiv k_s.$$

Substituting the Morse function (Eq. (5.44)) into Eq. (5.20), the nuclear equation takes the form

$$\frac{d^2}{dR^2}\xi(R) + \frac{2\mu}{\hbar}\left\{E - U_{el} - D^o\left[1 - e^{-a(R-R_e)}\right]^2 - \frac{K(K+1)\hbar^2}{2\mu R^2}\right\}\xi(R) = 0. \tag{5.45}$$

When $K = 0$, the solution takes the form [49]

$$G(v) = \frac{U_{el}}{hc} + \omega_e\left(v + \frac{1}{2}\right) - \omega_e x_e\left(v + \frac{1}{2}\right)$$

without any higher-order terms, where

$$\omega_e = a\left(\frac{\hbar D^o}{\pi c\mu}\right)^{\frac{1}{2}},$$
$$\omega_e x_e = \frac{\omega_e^2}{4D^o} = \frac{\hbar a^2}{4\pi\mu c}.$$

Measured or tabulated values of ω_e and $\omega_e x_e$ can be used to determine D^o, the dissociation energy, or to draw the potential curve. One is cautioned, however, that $a^2\hbar/4\pi c\mu$ is not equal to experimental values of $\omega_e x_e$ in general.

5.3.5
More Accurate Techniques

While the Morse approach is simple and quick to implement, it is not the most accurate means of estimating the potential. The RKR method (named for

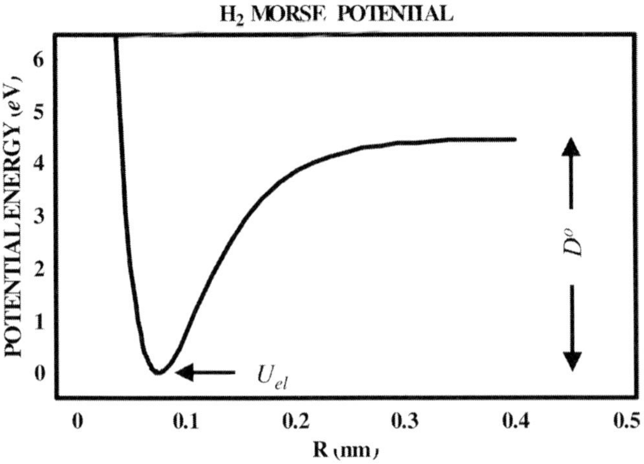

Fig. 5.4 The *Morse potential*, for the ground state of H_2 where U_{el}, the energy of the bottom of the potential well, is 0 and D^o, the dissociation energy – the height of the well, is 4.46 eV in this case.

its inventors Rydberg [50], Klein [51] and Rees [52]) is a powerful technique. This method relies on knowledge of the energy levels and constants obtained from spectroscopy to relate the potential energy curve, U, to the internuclear separation. Specifically, the classical inner and outer turning points for U are given analytically by

$$R_{max} = \left(\frac{f}{g} + f^2\right)^{\frac{1}{2}} + f, \tag{5.46a}$$

$$R_{min} = \left(\frac{f}{g} + f^2\right)^{\frac{1}{2}} - f. \tag{5.46b}$$

The functions f and g, which depend on the energy of the molecule, $E = G_v + B_v J(J+1)$, are defined in terms of partial derivatives of

$$S\,(E,\kappa) = \frac{1}{\pi\sqrt{2\mu}} \int_0^{I'} \sqrt{E - U\,(I,\kappa)}dI, \tag{5.47}$$

where $U(I,\kappa)$ is the sum of the vibrational and rotational energies of the molecule,

$$\frac{\omega_e}{h}I - \frac{\omega_e x_e}{h^2}I^2 + \frac{\omega_e y_e}{h^3}I^3 + \frac{1}{R_e^2}\kappa - \frac{D_e}{B_e^2 R_e^4}\kappa^2 - \frac{\alpha_e}{h\omega_e B_e R_e^2}I\kappa + \cdots,$$

$I = h(v + 1/2)$, $\kappa = h^2 J(J+1)/8\pi^2\mu$ and $I = I'$ when $E = U$; f and g are then

$$f = \frac{\partial S}{\partial E}, \tag{5.48a}$$

and

$$g = -\frac{\partial S}{\partial \kappa}. \tag{5.48b}$$

We identify I above as the action when we recognize that the RKR method is the inverse of the JWKB procedure. In the JWKB procedure $U_{eff}(R)$ is known and E is sought. This involves quantizing the action,

$$\begin{aligned} I &= \oint p dq = h(v + 1/2) \\ &= \oint \sqrt{2\mu \left[E - U_{eff}(R) \right]} dR, \end{aligned} \tag{5.49}$$

where the integration is over a complete cycle – R_{min} to R_{max} and back.

One advantage of the RKR approach is that it is possible to generate potential energy curves without a good value of the dissociation limit. At the same time, one of the drawbacks of this method is that it does not give an estimate of the dissociation limit. Traditionally, a Birge–Sponer extrapolation [53] gives such an estimate. The technique uses the spacing between vibrational levels, which is given by

$$\Delta G_{v+1/2} \equiv G(v + 3/2) - G(v + 1/2) = \omega_e - 2(v + 1)\omega_e x_e, \tag{5.50}$$

up to the first anharmonic term (the cubic term), to determine how many vibrational levels a potential well can hold. Since the spacing decreases linearly with v, it is possible to plot a few values of $\Delta G(v)$ and extrapolate the curve to $\Delta G_{v+1/2} = 0$ to determine v_{max} and hence the onset of dissociation (i.e., D^0). It is well known that near the dissociation limit the plot of the spacing deviates from the extrapolation because of the influence of long-range interatomic potentials. LeRoy and Bernstein [54] and Stwalley [55] have developed methods to correct for these long-range forces, giving more accurate results.

5.4 Electronic States

In this section, we will look at the electronic potentials determined by Eq. (5.34) in more detail. Our primary goals are to determine: (1) the relevant quantum numbers (conserved quantities) for bound electronic motion, (2) the symmetries that distinguish various states and (3) the selection rules for transitions (which we will investigate in the next chapter). The first two will allow electronic configurations similar to what we have for atoms to be constructed while the third will enable molecular spectroscopy.

Before we get to the details, however, we can make a few general comments that will allow us to use what we already know to reduce our workload. First, since the diatomic molecule is composed of two nuclei in close proximity, the spherical symmetry we have for atoms will be broken. As a consequence, it is necessary to revisit angular momentum and angular momenta coupling in cylindrical symmetry. Second, the energies of the electronic states depend on the internuclear separation as shown in Fig. 5.4. We will find that some states support binding, a lowering of the energy as R (the internuclear separation) decreases, while others will not. Equation (5.7a) must be solved to determine the electronic energy, E_{el}. Third, the spacings between the electronic states will be on par with those of the corresponding atomic states (see below) of the constituent atoms and, in general, much larger than the spacing between rotational and vibrational levels. Finally, each electronic state will be associated with a distinct pairing of atomic states. We will exploit this to build molecular configurations (molecular orbitals) from linear combinations of atomic configurations (atomic orbitals), which will provide a rudimentary understanding of molecular structure.

5.4.1
Angular Momenta in Cylindrically Symmetric Fields

Electrons bound to molecules have orbital and spin angular momenta, as they do in atoms, which couple to the molecular rotation and spin to produce a total angular momentum for the system. This coupling can occur in a variety of different ways. Frequently, the molecular analogue to LS coupling is appropriate where all the electronic orbital angular momenta couple to produce $\vec{\mathbf{L}}$ as do the electronic spins to produce $\vec{\mathbf{S}}$. This occurs, as we found with atoms, when the spin–orbit interaction is small compared with the electrostatic interaction. Electronic orbital angular momentum, however, is a good quantity (constant of the motion) in spherically symmetric fields. In our study of the Stark effect (see Sections 3.1.1.1 and 3.1.1.2), we found that $\vec{\mathbf{L}}$ is not constant when an electric field is applied, which breaks the spherical symmetry. Specifically, we found that

- the applied field reduces the spherical symmetry to cylindrical symmetry;
- $\vec{\mathbf{L}}$ is not conserved;
- M_L, the projection of $\vec{\mathbf{L}}$ along the electric field direction, is a good quantum number and
- part of the $2L + 1$ degeneracy for a given n shell is removed, but $\pm M$ states are still degenerate – all states with $L \geq 1$ are doubly degenerate.[7]

7) This remaining degeneracy can be seen by reversing either the sense of the orbital motion of the electrons or the direction of the electric field; $M_L \rightarrow -M_L$, but the energy would be unchanged.

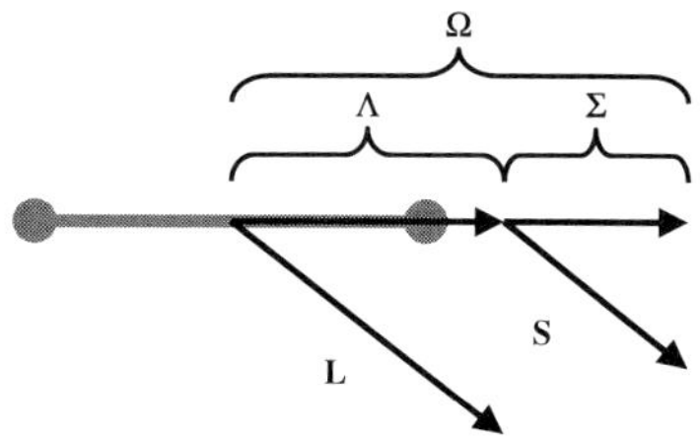

Fig. 5.5 The projections of the orbital and spin angular momenta of the electrons of a diatomic molecule onto the internuclear axis, defining Λ, Σ and Ω.

The electric field produced by the two atoms also has cylindrical symmetry. Thus, the results from the Stark effect provide a framework to begin our analysis.

5.4.1.1 **Orbital Angular Momentum**

Since $\vec{\mathbf{L}}$ is not conserved, it would be better to classify the states according to M_L, which is conserved. However, since there is a degeneracy between the positive and negative values of M_L (see Eqs. (3.14) and (3.15)), it makes more sense to classify the states in terms of $|M_L|$.[8] Thus, we define

$$\Lambda \equiv |M_L|, \tag{5.51}$$

the projection of the orbital angular momentum of the electrons along the internuclear axis as shown in Fig. 5.5. Each Λ corresponds to a distinct electronic state. For each L there will be $L+1$ distinct states (e.g., for $L=2$ the possible projections along the internuclear axis are $\Lambda = 0$, 1 and 2.) Table 5.1 gives the designations of the first few molecular states.

Tab. 5.1 Molecular state designation

	Λ				
	0	1	2	3	...
Molecular designation	Σ	Π	Δ	Φ	...
Atomic analogy	S	P	D	F	...

5.4.1.2 **Spin Angular Momentum**

The aggregate electronic spin, $\vec{\mathbf{S}}$, unlike $\vec{\mathbf{L}}$, remains a good quantum operator in the cylindrical field because the Hamiltonian, due to the fact that the electric field of the nuclei does not involve the electron's spin (see for example Eq. (3.10)). In addition, its projection onto the internuclear axis, $M_S \equiv \Sigma$

8) We point out that building up the molecule at large internuclear separation leads to the same conclusion.

(see Fig. 5.5), is also a good quantum number.[9] Unlike Ω, however, Σ has both positive and negative values. States of different spin are degenerate in Σ states ($\Lambda = 0$ states), as they are for S states in atoms. However, when $\Lambda \neq 0$, the motion of the electrons about the nuclei leads to $2S+1$ values for Σ (LS coupling) since it can be positive or negative (i.e., the projection of S can be parallel or antiparallel to the magnetic field produced by the electrons).

5.4.1.3 **Multiplet Splitting**

The multiple splitting for $\Lambda \neq 0$ is proportional to the interaction of the electronic spin and the magnetic field, which has the form $\vec{\mu} \cdot \vec{\mathbf{H}}_{elec}$ (see, for example, Eqs. (3.17) and (3.20)). This leads to a multiplet splitting that will be similar to the Zeeman effect, which will be proportional to σ. As a first approximation, the term values will obey

$$T_e = T_0 + A\Lambda\Sigma, \tag{5.52}$$

where T_0 is the energy of the term in the absence of spin and A the proportionality constant independent of Λ and Σ. It is clear that unlike the multiplet splitting in atoms, this splitting is equally spaced.

5.4.1.4 **Total Angular Momentum**

Tab. 5.2 Table of standard symbols for angular momentum operators and quantum numbers, total and projections along the internuclear axis, used to describe the structure and dynamics of diatomic molecules. This notation is consistent with that used in Refs. [43, 56]. We point out that $\vec{\mathbf{R}}$ that appears in this table to denote nuclear rotation is identical to $\vec{\mathbf{K}}$ used in Ref. [44] and in earlier sections of this chapter (see footnote on page 101). Entries with "· · ·" indicate no standard notation.

Operator	Description	Total quantum number	Projection quantum number
$\vec{\mathbf{R}}$	Nuclear rotation	R	Zero
$\vec{\mathbf{L}}$	Electronic orbital	L	Λ
$\vec{\mathbf{S}}$	Electronic spin	S	Σ
$\vec{\mathbf{I}}$	Nuclear spin	I	· · ·
$\vec{\mathbf{J}}$	$\vec{\mathbf{R}}+\vec{\mathbf{L}}+\vec{\mathbf{S}}$	J	$\Omega = \|\Lambda+\Sigma\|$
$\vec{\mathbf{F}}$	Total ($\vec{\mathbf{J}}+\vec{\mathbf{I}}$)	F	· · ·
$\vec{\mathbf{J}}_a$	$\vec{\mathbf{L}}+\vec{\mathbf{S}}$	J_a	· · ·
$\vec{\mathbf{N}}$	$\vec{\mathbf{R}}+\vec{\mathbf{L}}=\vec{\mathbf{J}}-\vec{\mathbf{S}}$	N	Λ
$\vec{\mathbf{O}}$	$\vec{\mathbf{J}}-\vec{\mathbf{L}}=\vec{\mathbf{R}}+\vec{\mathbf{S}}$	O	· · ·

9) This choice for the name of the projection is unfortunate and must not be confused with a Σ ($\Lambda = 0$) electronic state.

Tab. 5.3 Nomenclature example for $\Lambda = 1$, Π state.

$S = \frac{3}{2}$	$\rightarrow$	Quartet state
$\Lambda + \Sigma$	$=$	$-\frac{1}{2}, \frac{1}{2}, \frac{3}{2}, \frac{5}{2}$
Four states	$\rightarrow$	${}^4\Pi_{-\frac{1}{2}}, {}^4\Pi_{\frac{1}{2}}, {}^4\Pi_{\frac{3}{2}}, {}^4\Pi_{\frac{5}{2}}$

By definition, the total angular momentum, $\vec{\mathbf{J}}$, in the absence of external fields is always conserved. Regardless of how the electronic angular momenta couple to the internuclear axis, $\vec{\mathbf{J}}$ is a result of coupling of the electronic and the rotational angular momenta. According to Eq. (5.15), rotation is quantized with integer quantum numbers. We used the vector $\vec{\mathbf{K}}$ in Eq. (5.15) to designate pure rotation. The more conventional notation of Table 5.2 for pure rotation is $\vec{\mathbf{R}}$. For the rest of this chapter, we will assume that $\vec{\mathbf{K}} \equiv \vec{\mathbf{R}}$. In Table 5.2, $\vec{\mathbf{J}}$ is due to coupling of $\vec{\mathbf{R}}$, $\vec{\mathbf{L}}$ and $\vec{\mathbf{S}}$, where $\vec{\mathbf{R}}$ is the total angular momentum ignoring spin.

5.4.1.5 **Labeling Nomenclature**

We will define a total electronic angular momentum along the internuclear axis as

$$\Omega = \Lambda + \Sigma, \tag{5.53}$$

which ranges from $\Lambda - |\Sigma|$ to $\Lambda + |\Sigma|$.[10] For light systems $\vec{\mathbf{L}}$ and $\vec{\mathbf{S}}$ couple separately to the internuclear axis as shown in Fig. 5.5. Following a labeling system similar to that used to define *LS*-coupled terms in atoms, the nomenclature for electronic states is

$$^{2S+1}\Lambda_\Omega .$$

Table 5.2 is an example for a $\Lambda = 1$ state. The independent coupling of $\vec{\mathbf{L}}$ and $\vec{\mathbf{S}}$ to the internuclear axis is just one example of how they can couple. In some systems, $\vec{\mathbf{L}}$ couples to $\vec{\mathbf{R}}$, the rotation vector, to form a resultant that then couples to $\vec{\mathbf{S}}$ to form the total angular momentum $\vec{\mathbf{J}}$.

10) This convention agrees with contemporary papers (see Refs. [56] and [43] for example) but differs from that of Ref. [44] where Ω is defined as $|\Lambda + \Sigma|$. The advantage of the latter definition is that Ω is positive definite. The disadvantage is that for some cases, distinct states with the same values for Λ and S can have the same value for Ω. This will not be the case with the definition we will use.

5.4.2 Angular Momenta Coupling: Hund's Cases

There are several possible ways in which the various angular momenta can couple to produce $\vec{\mathbf{J}}$. Typically, these are organized into five major idealized categories called Hund's cases (a)–(e). We point out that while real molecules can fall between cases or change from one case to another as the nuclear rotation increases or the system becomes more excited, Hund's cases provide a good approximation for grouping or a starting point for analysis. We will review four of them: cases (a)–(d). Table 5.4 summarizes the different ways in which the various angular momenta can couple in these four cases. A more thorough discussion of all five cases can be found in Herzberg's book on diatomic molecules [44].

Tab. 5.4 Summary of Hund's cases (a)–(d).

Case	Precoupling to axis	Order of coupling
(a)	$\vec{L} \to \vec{\Lambda}$ & $\vec{S} \to \vec{\Sigma}$	$\vec{\Lambda}, \vec{\Sigma} \to \vec{\Omega}, \vec{R} \to \vec{J}$
(b)	$\vec{L} \to \vec{\Lambda}$	$\vec{\Lambda}, \vec{R} \to \vec{N}, \vec{S} \to \vec{J}$
(c)	$\vec{L}, \vec{S} \to J_A \to \Omega$	$\Omega, \vec{R} \to \vec{J}$
(d)	Nothing	$\vec{L}, \vec{R} \to \vec{N}, \vec{S} \to \vec{J}$

5.4.2.1 Hund's Case (a)

In case (a) the interaction between the nuclear rotation and the electronic orbital and spin angular momenta is weak. Thus, $\vec{\mathbf{L}}$ and $\vec{\mathbf{S}}$ couple to the internuclear axis separately to give $\vec{\Lambda}$ and $\vec{\Sigma}$, respectively, which then combine to

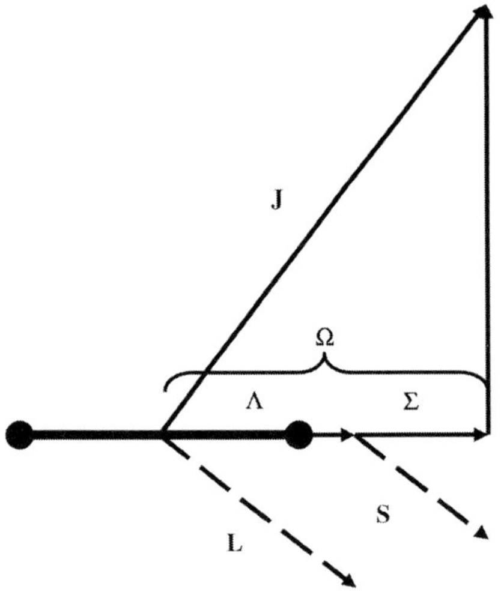

Fig. 5.6 Hund's case (a).

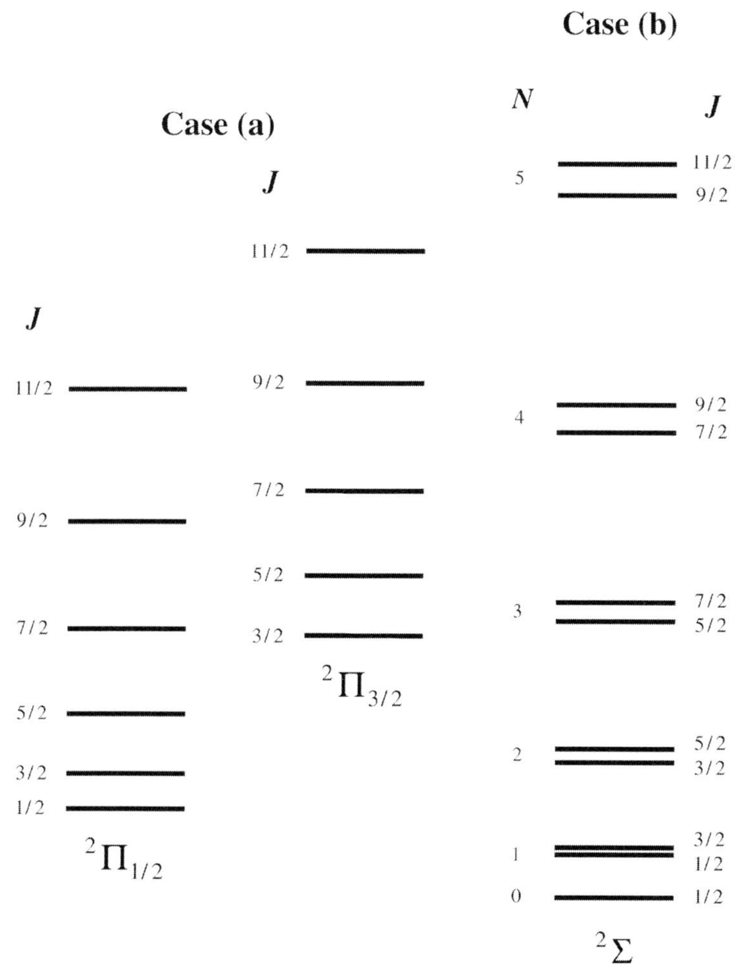

Fig. 5.7 Examples of the rotational structure for $^2\Pi$ (left) and $^2\Sigma$ states in cases (a) and (b), respectively. In case (a), $S = 1/2$, $\Sigma = 1/2$ and $\Lambda = 1$, leading to two states, $^2\Pi_{1/2}$ and $^2\Pi_{3/2}$. Note that the $J = 1/2$ level is missing for the $^2\Pi_{3/2}$ state. In case (b), $S = 1/2$ as well, but there is only one state. However, each R level (rotational level), except the lowest level, splits into two J levels.

give $\vec{\Omega}$. This was the situation in Fig. 5.5 for diatomics composed of light atoms. Figure 5.6 shows that the resulting $\vec{\Omega}$ couples to $\vec{\mathbf{R}}$ to give $\vec{\mathbf{J}}$. For case (a), the rotational energy levels are given by

$$F_v(J) = B_v[J(J+1) - \Omega^2] + \ldots,$$

while J takes on values

$$J = \Omega, \Omega + 1, \Omega + 2, \ldots,$$

and cannot be less than Ω. Since J must be positive, $J \geq \Omega$ demands that some of the values of J may be missing from the spectrum. An example of the rotational structure for a $^2\Pi$ state in this case is shown in Fig. 5.7.

5.4.2.2 **Hund's Case (b)**

When $S \neq 0$ but couples very weakly, or not at all, to the internuclear axis, case (b) applies. Weak coupling is easy to understand when $\Lambda = 0$ (Σ states), especially for the the case of zero nuclear spin. The internal magnetic field with which the electronic spin can interact will be very weak. Consequently, in case (b), $\vec{\mathbf{S}}$ tends to couple to the rotation vector instead of the internuclear axis. Case (b) can also apply in situations where $\Lambda \neq 0$; $\vec{L}$ couples to the internuclear axis but $\vec{S}$ does not. As sketched in Fig. 5.8, $\vec{\Lambda}$ and $\vec{\mathbf{R}}$ couple to give $\vec{\mathbf{N}}$, which takes on integer values $\Lambda + R$,[11]

$$N = \Lambda, \Lambda + 1, \Lambda + 2, \ldots .$$

Finally, $\vec{\mathbf{N}}$ couples to $\vec{\mathbf{S}}$ to produce $\vec{\mathbf{J}}$, which takes on values

$$J = |N - S|, |N - S + 1|, \ldots, (N + S - 1), (N + S).$$

When $N \geq S$, there will be $2S + 1$ J components for each N. The rotational levels for $^2\Sigma$ and $^3\Sigma$ states are given in Table 5.5 and displayed graphically in Fig. 5.7 for a $^2\Sigma$ state.

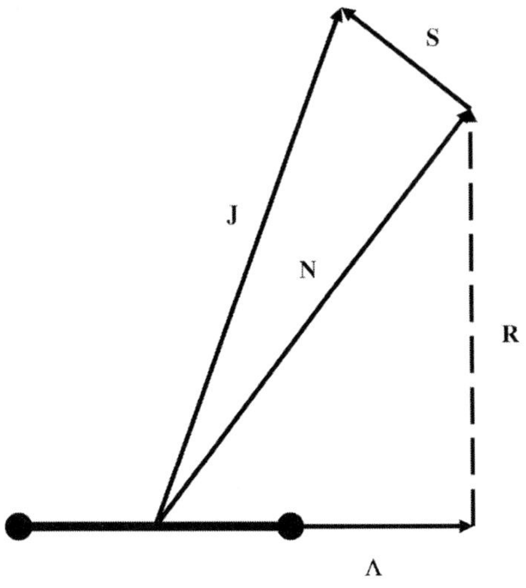

Fig. 5.8 Hund's case (b).

Tab. 5.5 Rotational energy levels in Hund's case (b) for $S = 1/2$ and $S = 1$. The splitting constants γ ($S = 1/2$, $S = 1$) and λ ($S = 1$) are usually very small compared with B_v.

$$S=\tfrac{1}{2}\quad\left\{\begin{array}{lcl} F_{N+\frac{1}{2}}(N) &=& B_vN(N+1)+\frac{1}{2}\gamma N \\ F_{N-\frac{1}{2}}(N) &=& B_vN(N+1)-\frac{1}{2}\gamma(N+1) \end{array}\right.$$

$$S=1\quad\left\{\begin{array}{lcl} F_{N+1}(N) &=& B_v[N(N+1)+(2N+3)]-\lambda+\gamma(N+1) \\ && -\sqrt{(2N+3)^2B_v^2+\lambda^2-2\lambda B_v} \\ F_N(N) &=& B_vN(N+1) \\ F_{N-1}(N) &=& B_v[N(N+1)-(2N-1)]-\lambda-\gamma N \\ && +\sqrt{(2N-1)^2B_v^2+\lambda^2-2\lambda B_v} \end{array}\right.$$

5.4.2.3 **Hund's Case (c)**

For heavy atoms, the spin–orbit interaction becomes larger than the electrostatic interaction and $\vec{\mathbf{L}}$ and $\vec{\mathbf{S}}$ are expected to couple first to give $\vec{\mathbf{J}}_a$, before coupling to the internuclear axis. Its projection now defines $\vec{\boldsymbol{\Omega}}$, which couples to the rotation to give the total angular momentum $\vec{\mathbf{J}}$. This constitutes case (c) and is illustrated in Fig. 5.9. In this case, however, the states are not classified in terms of Λ and Σ any longer because they are no longer conserved. Rather, they are classified by Ω, which is conserved. The energies of the rotational

11) Clearly, $\vec{R} \equiv \vec{N}$ when $\Lambda = 0$.

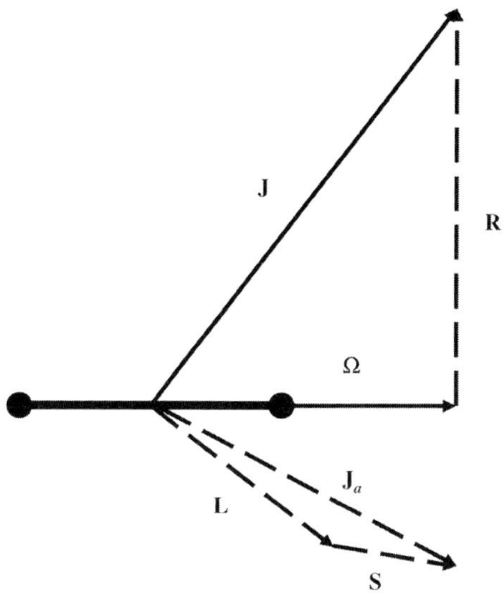

Fig. 5.9 Hund's case (c).

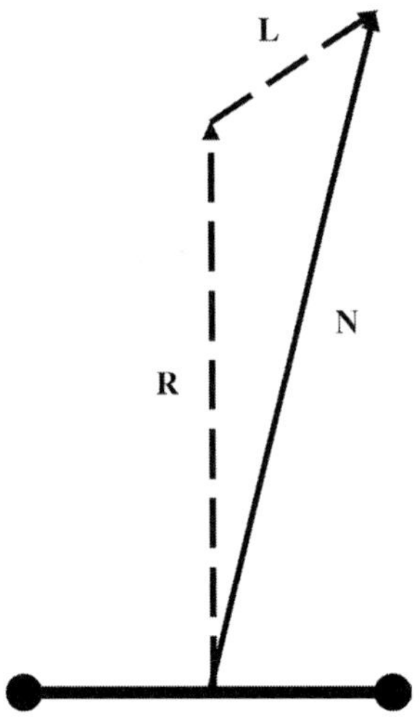

Fig. 5.10 Hund's case (d).

levels and the values over which J ranges are the same as in case (a). We point out that some investigators doubt that this case exists [57].

5.4.2.4 **Hund's Case (d)**

The final case we will mention is that where $\vec{\mathbf{L}}$ couples to the rotation vector, $\vec{\mathbf{R}}$, instead of the internuclear axis, to give $\vec{\mathbf{N}}$. In this case there are $2L + 1$ values for N, which range over

$$N = (R + L), (R + L - 1), \ldots, |R - L|.$$

The rotational energies are given by

$$F_v(R) = B_v R(R + 1),$$

with each R level split into $2L + 1$ components. While the total $\vec{\mathbf{J}}$ results when $\vec{\mathbf{N}}$ is coupled to $\vec{\mathbf{S}}$, this coupling is usually very small and neglected, as we do in Fig. 5.10.

5.4.3 Molecular Symmetries: Electronic Motion

We have already seen that the electrons in diatomic molecules, and all linear molecules for that matter, move in a cylindrically symmetric field. We now want to look at what impact this symmetry has on the electronic states.[12] In

12) We acknowledge that some of the symmetry arguments used in this section were inspired by those presented by G. Herzberg in *Spectra of Diatomic Molecules* [44].

particular, we will determine how the wavefunctions associated with the various states transform under inversion, reflection and exchange of nuclei.

5.4.3.1 **Inversion Symmetry**

In a rectangular system, inversion means that $r \to -r$ or $x \to -x$, $y \to -y$ and $z \to -z$. What does it mean in cylindrical coordinates? We can determine this with the help of Fig. 5.11, where the two positive charges are found along the z-axis, the internuclear axis. If we start at (a), inversions will take us to (c). We can make the trek via two paths. Path 1: rotate the system by π about the y-axis to reach (b) and reflect through the xz-plane. Path 2: reflect through the xz-plane to reach (d) and rotate by π about the y-axis. The rotation takes $z \to -z$ (and $x \to -x$) while the reflection takes $\phi \to -\phi$ (and $y \to -y$). It is clear that applying either sequence of operations a second time brings us back to (a), which means that at best

$$\psi_{el} \Rightarrow \pm\psi_{el}$$

under inversion.[13] The possibility of a phase change to the wavefunction puts restrictions on the values that Λ can assume. To see what these restrictions are, consider two functions in cylindrical coordinates,

$$\chi_+(z_i, \rho_i, \Delta\phi_i) \qquad \text{and} \qquad \chi_-(z_i, \rho_i, \Delta\phi_i),$$

where i labels the electrons and the difference between $\chi_\pm$ is the sign in front of $\Delta\phi_i$. We define

$$\Delta\phi_i = \phi_i - \phi_1,$$

where ϕ_1 is the angle for the first electron and $\Delta\phi_i$ is the relative azimuthal angle between the first electron and all the others.[14] The electronic wavefunction can be defined as

$$\begin{aligned} \psi_{el} &= \chi_+ e^{i\Lambda\phi_1} \\ &= \chi_- e^{-i\Lambda\phi_1}. \end{aligned} \tag{5.54}$$

Both representations of ψ_{el} are eigenfunctions of

$$\hat{L}_z = \frac{\hbar}{i}\frac{\partial}{\partial\phi_1},$$

13) Since the act of rotation cannot change the physics, we could have reflected the system through the yz-plane instead of the xz-plane and then rotated the system about the z-axis by π to get to (b), without introducing any new physics. In fact, we can reflect through any plane containing the z-axis followed by an appropriate rotation about the z-axis and the physics would remain the same.

14) The physics must also not depend on the initial coordinate system so that the value of ϕ_1 is irrelevant.

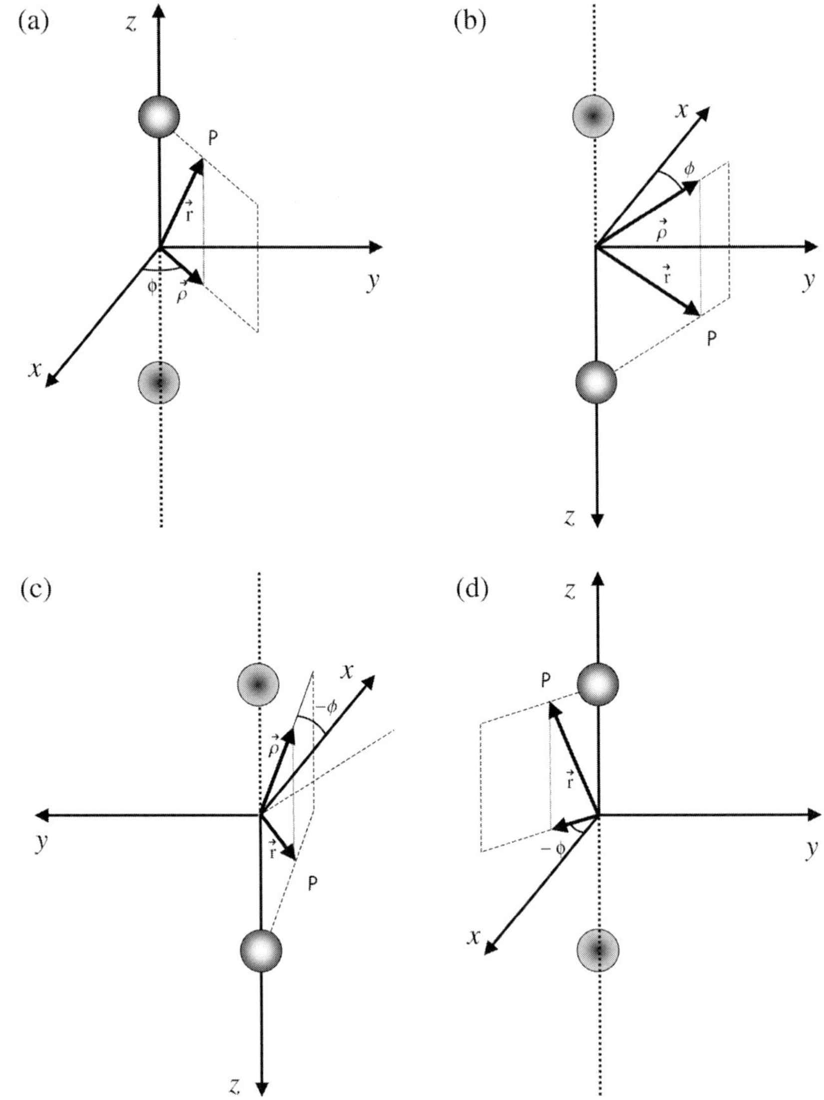

Fig. 5.11 Inversion via rotation and reflection. Path 1: starting at (a), rotate by π about the y-axis to reach (b) followed by reflection through the xz-plane to reach (c). Path 2: starting at (a), reflect through the xz-plane to reach (d) followed by a π rotation about the y-axis to reach (c).

the angular part of Schrödinger's equation. Thus, these functions describe rotations in the opposite sense and Λ represents the projection of the orbital angular momentum onto the Z-axis. To see that Λ must be an integer, assume that it is a half-integer. If we were to rotate the system through 2π, then we should be back to where we started. However, for half-integer Λ,

$$\psi_{el}(\phi_1 = 2\pi) \rightarrow \psi_{el}(\phi_1 = 0)\mathrm{e}^{\pm i\pi} \neq \psi_{el}(\phi_1 = 0).$$

Now consider a reflection through the xz-plane. This takes $\phi \rightarrow -\phi$, which implies that

$$\chi_+\mathrm{e}^{i\Lambda\phi_1} \rightleftarrows \chi_-\mathrm{e}^{-i\Lambda\phi_1}.$$

Since the general solution to Schrödinger's equation takes the form

$$\psi_{el} = a\chi_+e^{i\Lambda\phi_1} + b\chi_-e^{-i\Lambda\phi_1}, \tag{5.55}$$

the electronic wavefunctions fall into two categories – $\Lambda = 0$ and $\Lambda \neq 0$. We will look at the symmetry properties of each category separately.

5.4.3.2 Reflection Symmetry and Σ States ($\Lambda = 0$)

In this case, upon reflection

$$\psi_{el} \rightarrow \pm\psi_{el},$$

which implies that

$$\psi_{el}^{\pm} = \chi_+ \pm \chi_-,$$

to an arbitrary phase. The Σ^+ states correspond to

$$\psi_{el}^{+} = \chi_+ + \chi_-,$$

and do not change sign upon reflection. The Σ^- states correspond to

$$\psi_{el}^{-} = \chi_+ - \chi_-,$$

and do change sign upon reflection. In general, the two states are not energy degenerate and must be treated as independent electronic states. This can be shown as follows. Assume that

$$H|\psi_{el}^{\pm}> = E^{\pm}.$$

Then

$$\begin{aligned}E^+ &= <\psi^+|H|\psi^+> \\ &= \int \chi_+^* H\chi_+ d^3r + \int \chi_+^* H\chi_- d^3r \\ &\quad + \int \chi_-^* H\chi_+ d^3r + \int \chi_-^* H\chi_- d^3r,\end{aligned}$$

while

$$\begin{aligned}E^- &= <\psi^-|H|\psi^-> \\ &= \int \chi_+^* H\chi_+ d^3r + \int \chi_-^* H\chi_- d^3r \\ &\quad - \int \chi_+^* H\chi_- d^3r - \int \chi_-^* H\chi_+ d^3r,\end{aligned}$$

so that $E^+ \neq E^-$.

5.4.3.3 Reflection Symmetry and $\Lambda \neq 0$ States

When $\Lambda \neq 0$, the ensuing positive- and negative-symmetry states (such as $\Pi^\pm$, $\Delta^\pm$, etc.) will be energy degenerate. Following the same approach as above, we have

$$\begin{aligned}E^\pm &= <\psi^\pm|H|\psi^\pm> \\ &= \int \chi_+^* H\chi_+ d^3r + \int \chi_-^* H\chi_- d^3r \\ &\quad \pm \int \chi_+^* H\chi_+ e^{-2i\Lambda\phi_1} d^3r, \\ &\quad \pm \int \chi_-^* H\chi_+ e^{2i\Lambda\phi_1} d^3r.\end{aligned}$$

Since $\chi_\pm$ does not depend on ϕ_1 explicitly, we can break up the integration into cylindrical components, z, ρ and ϕ and carry out the ϕ integral first. The last two terms above become

$$\int_z dz \int_\rho \chi_\pm^* H\chi_\pm \rho d\rho \int_0^{2\pi} e^{\mp 2i\Lambda\phi_1} d\phi_1.$$

Since Λ is an integer,

$$\int_0^{2\pi} e^{\mp 2i\Lambda\phi_1} d\phi_1 = 0,$$

which means that $E^+ = E^-$. As a consequence of this, two-fold degeneracy of the idealized molecule allow us to drop the $\pm$ designation on $\Lambda \neq 0$ electronic states.

5.4.3.4 Exchange of Nuclei

As we mentioned above, when dealing with homonuclear diatomics we need to consider the wavefunction upon the exchange of the identical nuclei. As with reflection, the wavefunction can either remain the same or change sign. When the sign remains the same the electronic state is known as a *gerade state* (g) after the German word for *even*. When the state changes sign with the

Tab. 5.6 Summary of Hund's cases (a)–(d).

Λ	0	1	2	
State	$\Sigma^{\pm}_{g,u}$	$\Pi_{g,u}$	$\Delta_{g,u}$	Homonuclear
	$\Sigma^{\pm}$	Π	Δ	Heteronuclear

exchange of identical nuclei the state is called an *ungerade state* (u, German for *odd*). We summarize the symmetry properties of electronic states in Table 5.6.

5.4.4
Molecular Symmetries: Nuclear Motion

Now that we have a basic understanding of the symmetry of the electronic part of the wavefunction we need to examine the nuclear part. Here, we are concerned with vibration and rotation of the molecule. Recall that the nuclear part of the wavefunction is given by $\psi_N = \psi_{vib} \cdot \psi_{rot}$, where

$$\psi_{vib} = \frac{1}{R_n}\zeta(R_n).$$

But, since ψ_{vib} only depends on $|\vec{R_n}|$ (the internuclear separation)[15] it always has positive symmetry with respect to reflection and inversion. The rotational wavefunction is proportional to the spherical harmonics, $Y_{R,m}(\Omega)$, and thus has the following symmetry under inversion:

$$\begin{array}{ll} \text{even } R & \psi_{rot} \rightarrow +\psi_{rot}, \\ \text{odd } R & \psi_{rot} \rightarrow -\psi_{rot}, \end{array}$$

5.4.5
Molecular Symmetries: Herzberg Bookkeeping Diagram

We are going to want to examine the selection rules for transitions between molecular states. For the most part, we will focus on dipole transitions where the parities of the two states involved must be opposite. For high-resolution studies, we will be interested in transitions between individual rotational levels. Consequently, we need to consider the parity of the rotational levels of electronic states.

We will first consider a heteronuclear system with $\Lambda = 0$. In this case,

$$\begin{array}{lll} \Sigma^{+} & \Rightarrow & \text{even rotational levels have positive symmetry} \\ \Sigma^{-} & \Rightarrow & \text{odd rotational levels have positive symmetry} \end{array}$$

15) We have added the subscript n to help distinguish the internuclear separation from the rotational quantum number.

To specify the symmetry of the rotational levels it is convenient to use a bookkeeping method introduced by Herzberg, which we will refer to as the Herzberg diagram. To demonstrate these diagrams we will consider two Hund's case (b) states where $\vec{L}$ couples to the internuclear axis but $\vec{S}$ does not. In case (b) the good quantum numbers will be

$$\Lambda, R, N \text{ and } J,$$

where

$$N = \Lambda + R,$$
$$\vec{\mathbf{J}} = \vec{\mathbf{N}} + \vec{\mathbf{S}}.$$

The Herzberg diagram for the structure of the levels has three lines and can be written as follows for a singlet ($S = 0$) state:

(top)	$N = R =$	0	1	2	3
(middle)	${}^1\Sigma^+$	$\oplus$	$\ominus$	$\oplus$	$\ominus$
(bottom)	$J =$	0	1	2	3

5.4.5.1 The case $\vec{\mathbf{J}} = \vec{\mathbf{N}}$ ($\vec{\mathbf{S}} = 0$)

The top line of the diagram gives the quantum number for each rotational level, R (or N in this case since $\Lambda = 0$). The middle line gives the symmetry of the rotational levels, positive = $\oplus$ or negative = $\ominus$. The bottom line gives the total electronic quantum number, J, for each rotational level. When $S = 1$ the diagram takes the form

$N = R =$	0	1	2	3
${}^3\Sigma^-$	$\ominus$	$\oplus\oplus\oplus$	$\ominus\ominus\ominus$	$\oplus\oplus\oplus$
$J =$	1	0 1 2	1 2 3	2 3 4

5.4.5.2 The case $\vec{\mathbf{J}} = \vec{\mathbf{N}} + 1$ ($\vec{\mathbf{S}} = 1$)

For the triplet states, there are three different values for J associated with each rotational level all of which have the same symmetry. There is only one value for the lowest level since $N = 0$.

Tab. 5.7 The symmetry of rotational levels of electronic states for homonuclear molecules.

g:	$\psi_{el} \rightarrow \psi_{el}$ upon exchange of nuclei
u:	$\psi_{el} \rightarrow -\psi_{el}$ upon exchange of nuclei
s:	after inversion and exchange of nuclei, $\psi \rightarrow \psi$
a:	after inversion and exchange of nuclei, $\psi \rightarrow -\psi$

Now what about homonuclear systems? In this case we have an additional symmetry we need to worry about, the exchange of the nuclei leading to *gerade* (g) and *ungerade* (u) states. Each level will now carry an additional label – symmetric (s) or antisymmetric (a). Levels labeled by s will have overall positive symmetry. We summarize the symmetries in Table 5.7 and 5.8. As an example, consider a ${}^3\Sigma_g^-$ state, again in case (b):

$$
\begin{array}{rcccc}
N = R = & 0 & 1 & 2 & 3 \\
{}^3\Sigma_g^- & \ominus & \underbrace{\oplus\,\oplus\,\oplus}_{s} & \underbrace{\ominus\,\ominus\,\ominus}_{a} & \underbrace{\oplus\,\oplus\,\oplus}_{b} \\
 & a & & & \\
J = & 1 & 0\ 1\ 2 & 1\ 2\ 3 & 2\ 3\ 4
\end{array}
$$

Tab. 5.8 The symmetry of rotational levels of Σ states for homonuclear molecules.

Σ_g^+:	$\oplus$ levels are symmetric
Σ_u^+:	$\ominus$ levels are symmetric
Σ_g^-:	$\oplus$ levels are symmetric
Σ_u^-:	$\ominus$ levels are symmetric

Now we will turn our attention to $\Lambda \neq 0$. We will look at two examples: one in case (b), a singlet ($S = 0$) state, and one in case (c), a triplet ($S = 1$) state. Since each electronic state with $\Lambda \neq 0$ is doubly degenerate, each rotational level will have both $\oplus$ and $\ominus$ symmetries. Among other things, the two-fold degeneracy means that we no longer have to carry the g and u specification for homonuclear molecules. The Herzberg diagram for the ${}^1\Pi$ state takes the form

$$
\begin{array}{rcccc}
N = & 1 & 2 & 3 & 4 \\
{}^1\Pi_1 & \oplus\ominus & \ominus\oplus & \oplus\ominus & \ominus\oplus \\
J = & 1 & 2 & 3 & 4
\end{array}
$$

Now in case (c) $\vec{J}_a = \vec{L} + \vec{S}$ (Ω is the projection of J_a onto the internuclear axis) and $J = \Omega + R$. Consider a ${}^3\Pi$ state. We will have three different situations, one for each Ω value, ${}^3\Pi_{0,1,2}$. Specifically, for $\Omega = 0$ ($J = R$) we have

$$
\begin{array}{rcccc}
R = & 0 & 1 & 2 & 3 \\
{}^3\Pi_0 & \oplus\ominus & \ominus\oplus & \oplus\ominus & \ominus\oplus \\
J = & 0 & 1 & 2 & 3
\end{array}
$$

For $\Omega = 1$ ($J = R + 1$), we have

$$
\begin{array}{rcccc}
R = & 0 & 1 & 2 & 3 \\
{}^3\Pi_1 & \oplus\ominus & \ominus\oplus & \oplus\ominus & \ominus\oplus \\
J = & 1 & 2 & 3 & 4
\end{array}
$$

Finally, for $\Omega = 2$ $(J = R + 2)$ we have

$$\begin{array}{ccccc} R = & 0 & 1 & 2 & 3 \\ {}^3\Pi_2 & \oplus\ominus & \ominus\oplus & \oplus\ominus & \ominus\oplus \\ J = & 2 & 3 & 4 & 5 \end{array}$$

5.4.6
Molecular Symmetries: Nuclear Spin

Thus far we have not mentioned the nuclear spin and how it affects the wavefunction. For a diatomic system, each nucleus has spin I that combine to give a total nuclear spin T, which takes on values

$$T = I_a + I_b, \ldots, |I_a - I_b|.$$

There are a total of $(2I_a + 1)(2I_b + 1)$ possible states. For example, for $I_a = 1$ and $I_b = 1/2$ there will be $3 \times 2 = 6$ possible states and

$$T = \underbrace{3/2,}_{4} + \underbrace{1/2}_{2} = 6 \text{ states}$$

For the homonuclear molecule there are $(2I + 1)^2$ states and

$$T = 2I, \ldots, 0.$$

However, we also need to consider what happens when the two nuclei are identical. Depending on whether I is an integer or half-integer, the nuclei behave as fermions or bosons.

Consider H_2 where $I = 1/2$. The overall wavefunction will take the form

$$\Psi = \psi_{el}\psi_{vib}\psi_{rot}\Phi_{spin},$$

and will transform like a fermion under the exchange of nuclei,

$$\widehat{R}_{exch}\{\Psi\} = -\Psi.$$

That is, Ψ must be antisymmetric. In this particular case,

$$\Psi = \psi_{el,rot,vib} \left\{ \Sigma_g^+ \begin{array}{c} \oplus \\ s \end{array} \right\} \Phi_{as}.$$

To make the overall wavefunction antisymmetric, the even rotational levels must combine with the antisymmetric spin states and the odd rotational levels must combine with the symmetric spin states.

5.4.6.1 Example

Question: How many of the possible $(2I+1)^2$ states for H_2 are symmetric and how many are antisymmetric?

Solution: To solve this problem we recognize that since H_2 is made of up of two protons, each with $I = 1/2$, Ψ_{tot} must be antisymmetric with respect to exchange of nuclei. We have two possible orientations for T,

$$\begin{array}{rll} T = & 0, & 1 \\ \text{No. of states} = & 1, & 3 \quad \Rightarrow \text{Four states all together} \end{array}$$

We will write them out in the form $\phi_1(m_1)\phi_2(m_2)$, where $m_{1,2}$ is m_I for each nucleus. The four possible combinations are then

$$\overset{1}{\phi_1(+\tfrac{1}{2})\phi_2(+\tfrac{1}{2})} \quad \overset{2}{\phi_1(-\tfrac{1}{2})\phi_2(-\tfrac{1}{2})}$$

$$\overset{3}{\phi_1(+\tfrac{1}{2})\phi_2(-\tfrac{1}{2})} \quad \overset{4}{\phi_1(-\tfrac{1}{2})\phi_2(+\tfrac{1}{2})}$$

The first two are obviously symmetric with respect to the exchange of nuclei. This leads to $m_T = m_1 + m_2 = \pm 1$, which is associated with $T = 1$. Evidently, $T = 1$, then, must correspond to symmetric levels. The third and fourth combinations are degenerate pairs that can be combined into a symmetric and an antisymmetric grouping. Specifically,

$$\phi_1(+\tfrac{1}{2})\phi_2(-\tfrac{1}{2}) + \phi_1(-\tfrac{1}{2})\phi_2(+\tfrac{1}{2}) \qquad \text{(symmetric)}$$

$$\phi_1(+\tfrac{1}{2})\phi_2(-\tfrac{1}{2}) - \phi_1(-\tfrac{1}{2})\phi_2(+\tfrac{1}{2}) \quad \text{(antisymmetric)}$$

Both of these produce $M_T = 0$ with one being associated with the singlet combination, $T = 0$, and the other with the triplet, $T = 1$. Clearly, $T = 0$ gives rise to the antisymmetric terms. In summary, there will be three symmetric terms and one antisymmetric term. Table 5.9 summarizes the general case. Special note: when $I = 0$, there are no antisymmetric terms and every other rotational level will be missing!

Tab. 5.9 Statistical weights for rotational levels.

I = half-integer	T = odd, associated with symmetric levels
I = integer	T = even, associated with symmetric levels
$(2I+1)(I+1)$	Symmetric statistical weight
$(2I+1)(I+1$	Antisymmetric statistical weight

5.4.7 Molecular State Labeling Convention

5.4.7.1 Rule 1: Ground State

The ground state is traditionally given the label X. Thus, the ground state for H_2 and all the alkalis is

$$X^1\Sigma_g^+.$$

5.4.7.2 Rule 2: Excited States with Ground-State Multiplicity

The first excited state of the same multiplicity as the ground state carries the label A. As an example, for Na_2 we have

$$A^1\Sigma_u^+.$$

The next state is labeled with B. This would be

$$B^1\Pi_u,$$

for Na_2.

5.4.7.3 Rule 3: Excited States with Different Multiplicity

The excited states with a multiplicity different from the ground state are labeled with lower case letters. For example, the first triplet state might be

$$a^3\Pi_u.$$

In general, the energy increases with the alphabet so that

$$\begin{array}{lllllll} \text{Energy} \Rightarrow & & & & & & \\ \quad A & < & B & < & C & < \ldots & \text{Ground-state multiplicity} \\ \quad a & < & b & < & c & < \ldots & \text{First with different multiplicity} \end{array}$$

As with most rules, there are a few notable exceptions, which stem primarily from history. This energy ordering is not followed in H_2 and He_2, for example. The upper case/lower case convention is not followed in N_2 where singlet states are labeled $a, b, \ldots$ and the triplets $A, B, \ldots$, even though the ground state is a singlet.

5.4.8 Molecular Orbital Theory

To complete this introduction to electronic states, we will look at how electrons congregate into configurations. While pure configurations rarely describe the electrons' motion exactly, they do provide a framework with which to organize our thoughts, characterize symmetries and determine which states contribute to bonding. The relationship between these configurations (labeled with lower case Greek letters) and the electronic states (upper case Greek letters) will be very similar to what we have for atoms. A single electron σ or π, for example, produces a $^2\Sigma$ or a $^2\Pi$ potential, respectively. In general, as with atoms, a molecular configuration will be composed of several states. Since the molecular symmetry is cylindrical (as opposed to spherical), the electronic state (potential) composed of more than one electron is a bit more complicated to construct than for atoms. It is the goal of this section to look at how to build the molecular state. In Section 1.2.3 we introduced the atomic orbital (AO) to represent the one-electron wavefunction. The corresponding one-electron wavefunction in the molecular case is called a *molecular orbital* (MO). In atoms, see Chapter 2, a state was represented by a collection of atomic orbitals. In a similar fashion, we will represent a molecular state by a collection of molecular orbitals.

The electrons, and hence the MOs, are characterized by one good quantum number, λ, the projection of the orbital angular momentum onto the internuclear axis. This quantum number assumes integer values with a naming convention similar to that for atomic states,

$$\begin{array}{lcccc} \lambda = & 0 & 1 & 2 & 3 \\ & \sigma & \pi & \delta & \phi \end{array}$$

Molecular orbitals sometimes carry two additional labels, the atomic n and l, which are not constants of the motion in general. These labels serve primarily as indicators of the atomic limit from which the MO was derived. There are two possible limits from which to build the MO known as the *united atom* and *separated atom* limits. To appreciate these limits, let us build some MOs for H_2.

In the united atom limit, we would start with an idealized He atom (one with just two protons and no neutrons) in a well-defined nl configuration (i.e., $1s^2$, $1s2s$, $1s2p$, etc.) and imagine separating the nuclei until the equilibrium internuclear separation, R_e, for H_2 is reached. If this separation is done adiabatically, the electrons will constantly adjust themselves until finally, at R_e, they will settle into a perturbed, cylindrically symmetric potential created by two charge centers.[16] While there are two distinct charge centers, there is only one well at the equilibrium separation. However, the perturbation is just the Stark effect. Recognizing this allows us to apply what we learned from the Stark effect (see Section 3.1.1.1) to determine how the degeneracy among the nl components of the H atom is removed. The $2p$ atomic state, for example,

16) Of course for H_2 the electrostatic interaction between the electrons contributes to the potential as well.

splits into $2p_{m=0}$ ($\equiv 2p\sigma$) and $2p_{m=\pm1}$ ($\equiv 2p\pi$) components. Thus, we designate the new state or states by $nl\lambda$ in this limit, where $\lambda = |m|$. Since we split He into two identical H atoms, we must account for the the g/u symmetry of the wavefunction upon inversion or exchange of nuclei. To determine how the symmetry of the atomic state maps to the molecular state, we use the results reported long ago by Wigner and Witmer [58], who demonstrated that the molecular orbital has the same symmetry as the atomic configuration from which it was derived. Specifically, even-parity (odd) states have g (u) symmetry [58] (see also Ref. [44]). Written in descending order, the first four configurations in the united atom limit with at most one excited electron are

$$\begin{array}{lcl} 1s2p & \rightarrow & \left\{ \begin{array}{l} (1s\sigma_g)(2p\pi_u) \\ (1s\sigma_g)(2p\sigma_u) \end{array} \right. \\ 1s2s & \rightarrow & (1s\sigma_g)(2s\sigma_g) \\ 1s^2 & \rightarrow & (1s\sigma_g)^2 \end{array}$$

The g/u symmetry does not exist for heteronuclear systems and, thus, these labels are omitted from the MOs that describe them. Again, since the electron is a spin-1/2 fermion, no two can have identical quantum numbers. Thus, σ orbitals can hold up to two electrons (two different orientations of the spin) and π (and higher) orbitals can hold up to four electrons. The $\lambda > 0$ orbitals hold four electrons because $\lambda \equiv |m|$ and m can be positive or negative. Consequently, there are two spin states for $+m$ and two spin states for $-m$ values.

In the opposite extreme, the separated atom limit, two H atoms are brought together adiabatically from infinity until their nl configurations, well defined at infinite separation, are again perturbed at R_e. The molecular states generated in this approach are distinguished from the united atom approach by placing the nl label after λ, λnl. Since the two H atoms are again identical, when brought together to form molecular states, a pair of states will always result, one being symmetric (g) and the other antisymmetric (u) depending on how the AOs are combined. The first few molecular orbitals in descending order with at most one excited electron are

$$\begin{array}{lcl} 1s + 2p & \rightarrow & \left\{ \begin{array}{l} (\sigma_g 1s)(\sigma_u 2p) \\ (\sigma_g 1s)(\pi_g 2p) \\ (\sigma_g 1s)(\pi_u 2p) \\ (\sigma_g 1s)(\sigma_g 2p) \end{array} \right. \\ 1s + 2s & \rightarrow & \left\{ \begin{array}{l} (\sigma_g 1s)(\sigma_u 2s) \\ (\sigma_g 1s)(\sigma_g 2s) \end{array} \right. \\ 1s + 1s & \rightarrow & \left\{ \begin{array}{l} (\sigma_g 1s)(\sigma_u 1s) \\ (\sigma_g 1s)^2. \end{array} \right. \end{array}$$

Again, the g/u labels are dropped for the heteronuclear MOs and replaced with labels on the nl to indicate which atom belongs to which atomic state;

$(\sigma 1s_A)(\pi 2p_B)$, for example. The ordering of MOs above is not sacred, but depends on the system.

Figure 5.12 compares the MOs constructed in the two limits. It is clear that the orderings on the left and right do not match up. Consider, for example, the first few states on the left – Σ_g ($= \sigma_g\sigma_g$), Σ_g ($= \sigma_g\sigma_g$), Σ_u ($= \sigma_g\sigma_u$), etc. However, on the right we have Σ_g ($= \sigma_g\sigma_g$), Σ_u ($= \sigma_g\sigma_u$) and Σ_g ($= \sigma_g\sigma_g$), etc. While the lowest level for both limits is a σ_g orbital occupied by two electrons (i.e., σ_g^2), the similarity stops there. Since symmetry and λ are sacred, a specific $nl\lambda_g$ ($nl\lambda_u$) state of the united atom construction must connect adiabatically to a specific $\lambda_g nl$ ($\lambda_u nl$) state of the separated atom construction. Assuming that the states are not degenerate, connecting states of the same symmetry and λ implies that the energy of a level changes as the internuclear axis changes. To connect the two limits adiabatically, we forbid two curves representing orbitals of the same symmetry from crossing. The correlation between the two constructions is indicated by dashed lines in Fig. 5.12.

To see that these correlations makes sense, we need to say a bit more about how an orbital is constructed mathematically. This will allow us to compare the charge distributions for specific orbitals generated from the two constructions.

5.4.8.1 United Atom Construction

To lowest order, an $nl\lambda_g$ orbital for a molecule X_2 is given by separating a configuration of an atom Y, where Y contains twice as many protons and electrons as atom X. For our H_2 example above (X = H and Y = He, albeit with just two protons), we generate the $1s\sigma_g$ orbital by starting with He in the $1s^2$ configuration. Separating the two protons leads to the $1s\sigma_g$ orbital occupied by two electrons. The charge distribution for the orbital occupied by one electron (i.e., the square of the wavefunction describing the electron in this orbital) is shown in the bottom row of Fig. 5.13, second from the left. The orbital has two distinct regions of charge centered on the protons, with charge spilling between them. The negative charge between the two positive centers is the glue holding the system together. Thus, this orbital contributes to bonding of the molecule and, hence, is called a *bonding orbital*. It is interesting to note that if one views the charge distribution along the internuclear axis (the z-axis) the distribution looks like a $1s$ state.

The first and second rows from the top in Fig. 5.13 show the construction of the $3d\pi_g$ and $2p\pi_u$ orbitals respectively from atomic $3d_{m=\pm1}$ and $2p_{m=\pm1}$ states. The charge distributions show that the $3d\pi_g$ orbital does not contribute to bonding (an antibonding orbital) while the $2p\pi_u$ orbital does (a bonding orbital). When either of these orbitals is viewed along the internuclear axis, they resemble $2p$ states.

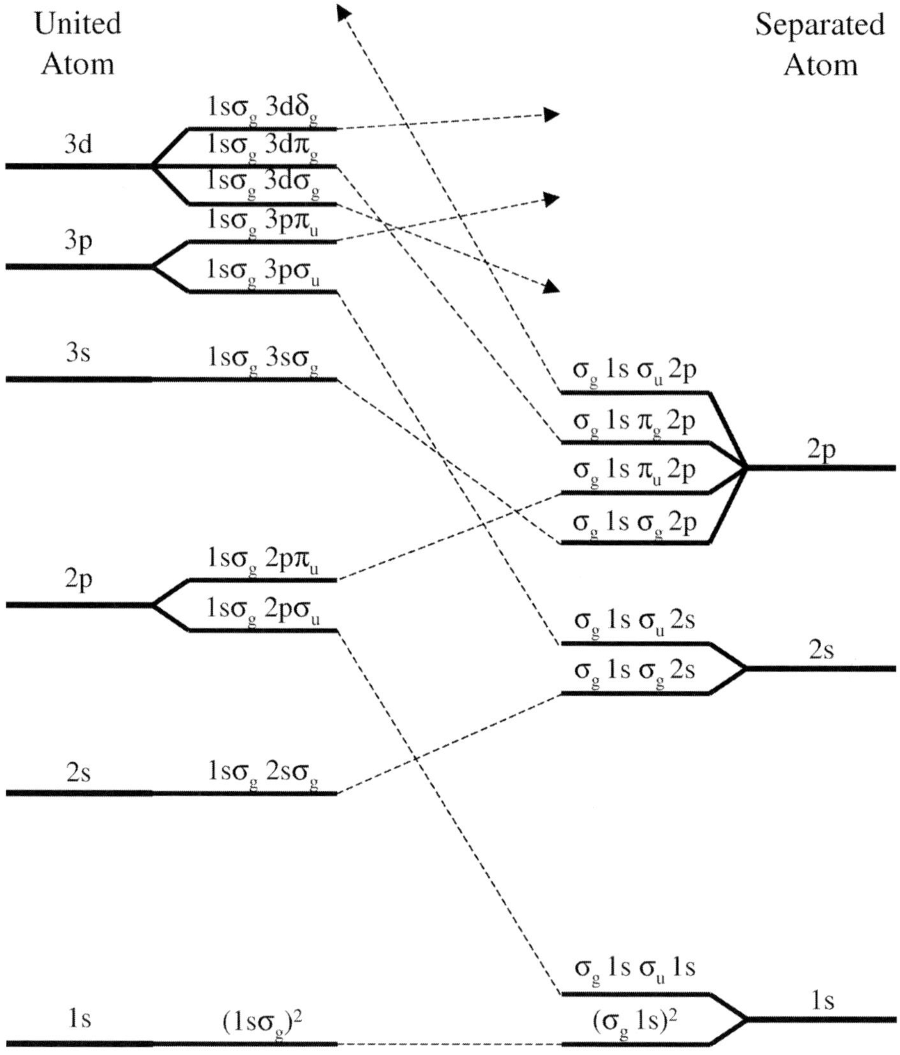

Fig. 5.12 Correlation between molecular orbitals derived from united and separated atom limits with one excited electron. Only the excited electron is indicated; each level has an additional $1s$ electron such that the ground-state limit is $1s^2$ and $1s + 1s$ for united and separated atom constructions. The dashed lines are to guide the eye only and do not represent the energy of the orbitals in going from one limit to the other.

5.4.8.2 **Separated Atom Construction**

The $\sigma_g 1s$ orbital is constructed, to lowest order, by bringing two H atoms in their $1s$ state together. Since the two H atoms are identical, the two degenerate states will perturb each other leading to two orbitals that will be linear combinations of the atomic states. One orbital will be symmetric with respect

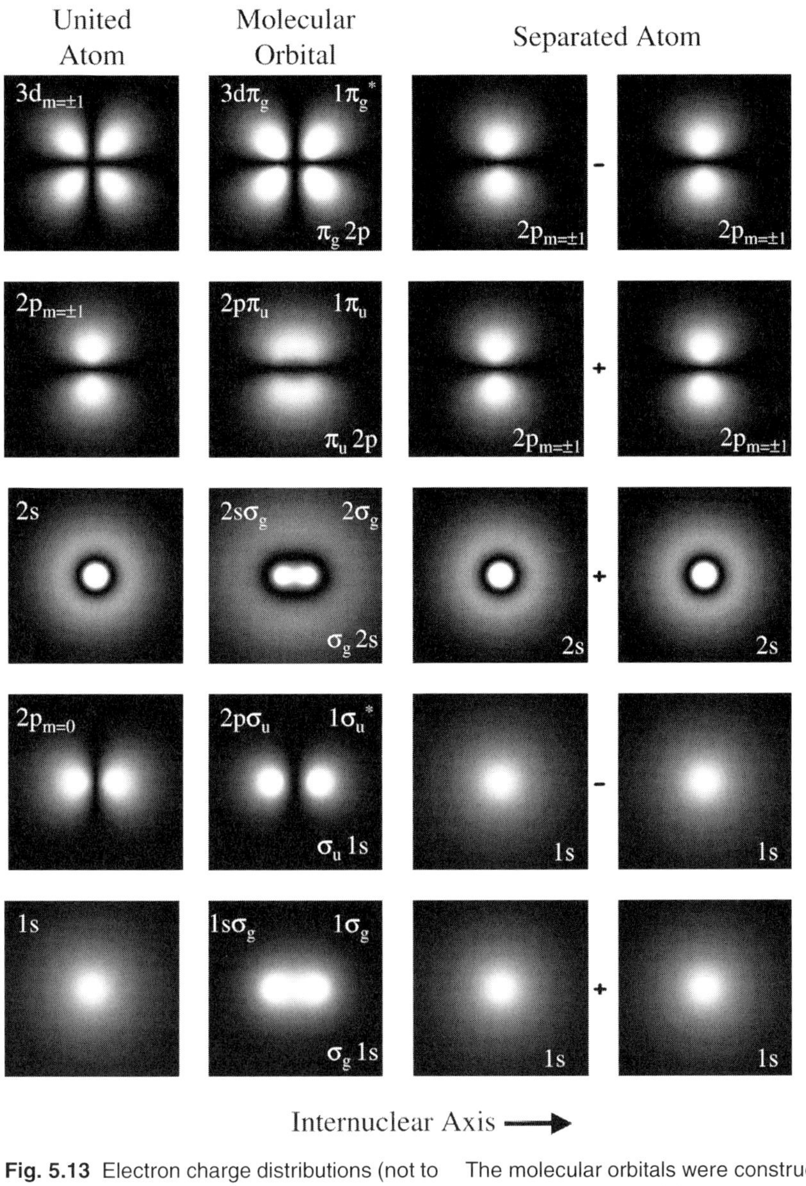

Fig. 5.13 Electron charge distributions (not to scale) for a few molecular orbitals along with the associated united atom (left-hand column) and separated atom (two right-hand columns) limits. The z-axis in these images is horizontal and in the plane of the page. The united and separated atom designations appear respectively in the upper left and lower right hand corners while the generic molecular label is given in the upper right hand corner. The molecular orbitals were constructed by squaring symmetric (+) and antisymmetric (–) sums of the atomic amplitudes used to create the separated atom distributions on the right but displaced from each other by R_e. The +/– signs between the two separated atom distributions indicate how the amplitudes were added. Each orbital, second column, is labeled by the three labeling schemes, where * indicates an antibonding orbital.

to exchange of protons (positive inversion symmetry) while the other will be antisymmetric (negative inversion symmetry). The symmetric combination, indicated by the + sign between the two $1s$ atomic charge distributions in the bottom row of Fig. 5.13, leads to a $\sigma_g 1s$ orbital occupied by two electrons. To lowest order, the resulting charge distribution for one electron in this orbital is the same as that generated with the united atom construction (see the charge distribution shown second from the left in the bottom row in Fig. 5.13).

The antisymmetric combination of the two $1s$ states, indicated by the "−" sign in the second row from the bottom in Fig. 5.13, produces the $\sigma_u 1s$ orbital. Since the wavefunction describing an electron in this orbital changes sign upon inversion, there is a conspicuous absence of charge over a wide region between the two protons since the charge distribution must vanish at the origin. This exclusion of charge leads to a stronger Coulomb repulsion between the protons. As a consequence, this orbital does not contribute to bonding and is called an *antibonding* orbital. The united atom construction leading to nearly the same charge distribution originates from a $2p_{m=0}$ atomic state as shown in the second from the bottom row in Fig. 5.13. Thus, the $2p\sigma_u$ and the $\sigma_u 1s$ orbitals are correlated. Again, looking along the z-axis, the charge distribution resembles a $1s$ state. Figure 5.13 shows MO constructions originating from $2s$ and $2p_{m=\pm 1}$ states as well, along with how they correlate to the united constructions.

We have just seen that to first order, the MOs can be constructed from either limit. The united atom label, however, is most appropriate for describing Rydberg states, where one electron spends most of its time far from the two nuclear centers and the nucleus looks a lot like an atom. The separated atom label, on the other hand, is more appropriate near dissociation limits, where the two nuclei are far apart and leading to two distinct positive charge centers. It turns out that the separated atom construction forms the basis for the construction method called the *linear combination of atomic orbitals* (LCAO). Consequently, one often finds the MOs labeled according to the separated atom scheme. Typically, however, the orbitals are labeled by a more neutral scheme, $m\lambda_{g/u}$, where m is a positive integer that simply indicates which state of a particular λ with a specific g/u symmetry is being discussed. While m may appear to be a molecular principal quantum number, it must be remembered that it is only a label. The configuration for O_2, which follows the ordering in Fig. 5.12, in this neutral scheme is

$$(1\sigma_g)^2(1\sigma_u^*)^2(2\sigma_g)^2(1\sigma_u^*)^2(3\sigma_g)^2(1\pi_u)^4(1\pi_g^*)^2. \tag{5.56}$$

In Eq. (5.56) we added a * to the antibonding orbitals to distinguish them from the bonding orbitals and indicate their *antibonding* nature. Figure 5.14 is an example of the two lowest potential curves for H_2 generated by two hydrogen atoms in their $1s$ state. This example shows the general property

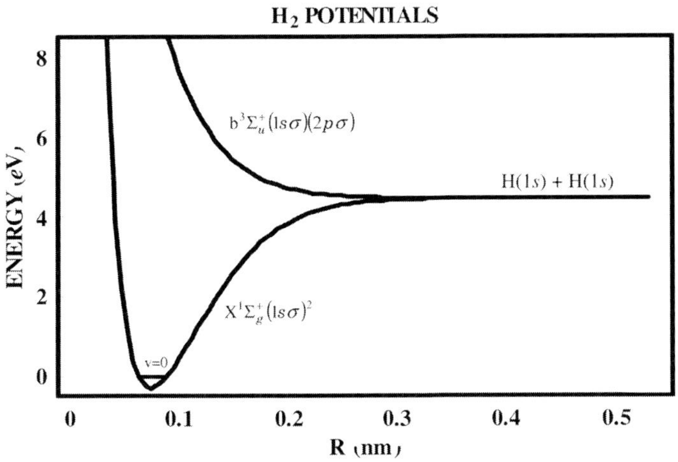

Fig. 5.14 The two molecular potentials derived from two H ($1s$) atoms. The lower, bound state is a singlet where the spins of the two electrons are antiparallel while the upper, unbound state is a triplet where the electron spins are aligned. These calculated potentials are taken from Ref. [48].

of molecular states; there are the same number of molecular states as there are pairs of atomic states. In this particular case, there are only two distinct combinations possible – one with the spins parallel (triplet state) and the other with them antiparallel (singlet state). The bonding and antibonding nature is also shown as well.

Further Reading

1 G. Herzberg, *Molecular Spectra and Molecular Structure I: Spectra of Diatomic Molecules*, Van Nostrand Reinhold, New York, NY, 1950.

2 C. S. Johnson, Jr and L. G. Pedersen, *Problems and Solutions in Quantum Chemistry and Physics*, Dover, New York, NY, 1986.

3 J. T. Hougen, *The Calculation of Rotational Energy Levels and Rotational Line Intensities in Diatomic Molecules*, NBS Monograph 115 [http://physics.nist.gov/Pubs/Mono115/contents.html].

4 I. Kovács, *Rotational Structure in the Spectra of Diatomic Molecules*, American Elsevier, New York, 1969.

5 H. Lefebvre-Brion and R. Field, *The Spectra and Dynamics of Diatomic Molecules*, Academic, New York, NY, 2004.

6 For a few additional problems with worked solutions see D. Budker, D. F. Kimball and D. P. DeMille, *Atomic Physics: An Exploration Through Problems and Solutions*, Oxford University Press, New York, NY, 2004.

Problems

5.1 Show that the spring constant for vibration is approximately given by

$$k_s = \frac{e^2}{4\pi\varepsilon_0}\frac{2}{R_e^3}. \tag{5.57}$$

5.2 Draw the Morse potential for the ground state of H_2 using molecular constants from Herzberg or Herzberg and Huber.

5.3 Show explicitly that for N_2, with nuclear spin 1, $T = 2, 0$ correspond to symmetric terms and occupy six of the possible terms. How many antisymmetric terms are there? What value(s) of T are relevant?

5.4 Write the configuration for the ground state of N_2, assuming that the electrons fill the orbitals in order according to the separated atom sequence of Fig. 5.12. To what united atom state do $\sigma_g 1s \sigma_g 2p$ correspond?

5.5 Use Fig. 5.13 to justify the correlation between the $2s\sigma_g - \sigma_g 2s$, $2p\pi_u - \pi_u 2p$ and $3d\pi_g - \pi_g 2p$ orbitals.

5.6 Use your favorite program to generate lowest-order charge distributions for the four MOs in the separated atom limit starting from $2p$ states of H. Your answers should look something like the images in Fig. 5.15. Label the charge distributions using the united and neutral schemes. Indicate which orbitals are bonding and which are antibonding.

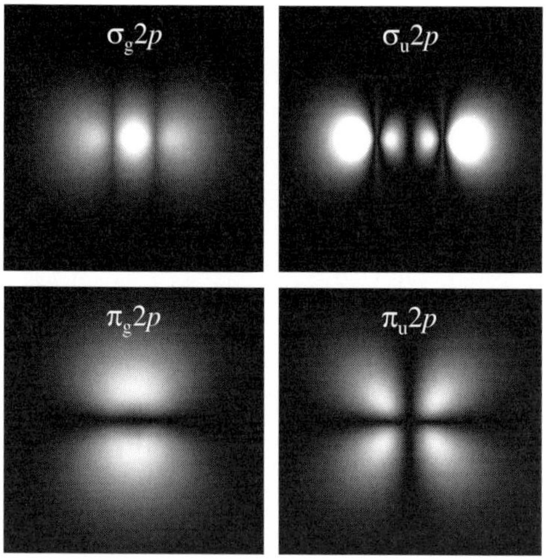

Fig. 5.15 Electron charge distributions for the four MOs generated in the separated atom construction from $2p$ atomic states. Each label corresponds to the separated atom labels.

6
Molecules in External Fields

6.1
Introduction

Much of what we know about diatomic molecules comes from spectroscopy, exposing systems to external fields to induce transitions or dynamics. As with atoms, the changes the system undergoes depend on the Hamiltonian perturbing the system. We will consider only two types of perturbations – those leading to radiative transitions and to field ionization. Most of the work required was done earlier for atoms so, here, we will be able to quote most of what we need.

This chapter is divided into two parts. In the first part, we will look at the conditions and selection rules for radiative transitions. We will then do a simple example to introduce Herzberg's method for bookkeeping. In the second part, we will look at field ionization of a diatomic molecule induced by an intense laser field.

6.2
Electronic Transitions

Spectroscopy involves transitions between levels. With sufficient resolution, one is able to identify individual rotational and vibrational levels. The wavelength of the transition between states will be given by $\Delta\nu = \nu' - \nu''$, where the prime indicates the upper level and the double prime indicates the lower level.[1] With the aid of Eq. (5.39) this can be written as

$$\Delta\nu = (\nu_e' - \nu_e'') + (G'(v') - G''(v'') + (F_v'(J') - F_v''(J'')), \tag{6.1}$$

where J is the total angular momentum apart from nuclear spin. As with atoms, not all values of J will be allowed and only a subset of the values of v will contribute. To determine which values are relevant, we must examine

1) The use of primes in this way corresponds to the convention employed by molecular spectroscopists.

Light-Matter Interaction: Atoms and Molecules in External Fields and Nonlinear Optics.
W. T. Hill and C. H. Lee

ISBN: 978-3-527-40661-6

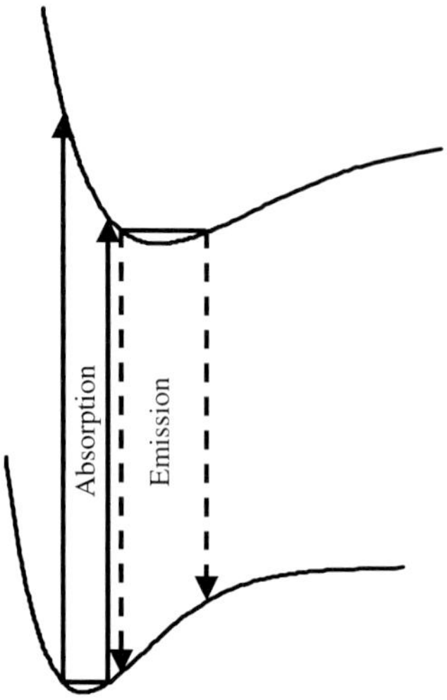

Fig. 6.1 Franck–Condon principle. Transitions are strongest for absorption and emission between the solid and dashed arrows, respectively.

the transition matrix element. As diatomic molecules have more degrees of freedom than atoms, it is not surprising that this exercise is more involved than it was with atoms. In this chapter, we will focus our attention on electric dipole transitions. In this case, the general matrix element takes the form

$$\langle \psi_e' \psi'(v') \psi'(J') | \, r \, | \psi_e'' \psi''(v'') \psi''(J'') \rangle \,, \tag{6.2a}$$

which can be written as

$$\langle \psi_e' \psi'(J') | \, r \, | \psi_e'' \psi''(J'') \rangle \, \langle \psi'(v') \, | \psi''(v'') \rangle \,. \tag{6.2b}$$

The second factor in Eq. (6.2b) will only be appreciable when there is significant overlap between the vibrational wavefunctions – the Franck–Condon principle. This leads to a so-called vertical transition as shown in Fig. 6.1. For the example shown, absorption would begin near the bottom of the lower well and terminate at the inner turning point of the upper well. Emission, on the other hand, would begin near the bottom of the upper well and terminate near the outer turning point of the lower well. The transitions originate where the electron probability density is highest.

The allowed transitions that render the first factor in Eq. (6.2b) nonzero depend on certain selection rules. Selection rules can be determined as was done for atoms by applying the Wigner–Eckart theorem. The rules fall into two categories – general rules that apply in all cases and case-specific rules. In the next three sections we will discuss these rules in the electric dipole approximation for the first three Hund's cases.

6.2.1 General Selection Rules

We begin with the general rules for electric dipole transitions. Since J is a good quantum number, its selection rules are the same as for atoms given in Eq. (4.54):

$$\Delta J = 0, \pm 1, \tag{6.3a}$$

with the restriction

$$J' = 0 \nleftrightarrow J'' = 0. \tag{6.3b}$$

Note that ΔM_J is not included here. It will come in, masquerading as different parameters under the specific selection rules. In Section 5.4.5 we discussed the symmetry of individual rotational levels. For both heteronuclear and homonuclear systems, each rotational level has either positive ($\oplus$) or negative ($\ominus$) symmetry. These levels obey the selection rules

$$\oplus \leftrightarrow \ominus, \oplus \nleftrightarrow \oplus, \ominus \nleftrightarrow \ominus. \tag{6.4}$$

Homonuclear systems carry the additional g/u symmetry, which when combined with the $\oplus/\ominus$ symmetry leads to levels being either symmetric (s) or antisymmetric (a). The selection rules that these obey are

$$s \longrightarrow s, a \longrightarrow a, s \nleftrightarrow a. \tag{6.5}$$

When identical nuclei have equal charge,

$$g \leftrightarrow u, g \nleftrightarrow g, u \nleftrightarrow u. \tag{6.6}$$

Given these few rules, we can organize spectral lines that link certain rotational levels into groups. The restriction on ΔJ limits the allowed transitions to three types, which go by the names P, Q and R branches. By convention,

Branch	J''	J'
P	J	$J-1$
Q	J	J
R	J	$J+1$

It should be understood that in absorption, the P branch means that J decreases by 1 while in emission, the P branch means that J increases by 1. That is, $J'' > J'$ for the P branch, $J'' < J'$ for the R branch and $J'' = J'$ for the Q branch.

6.2.2 Case-Specific Selection Rules

In this section we will give the rules for the first four cases. It is important to understand that the rules apply rigorously only when both the initial and final electronic cases are the same. When two different cases apply, only the common rules apply. With regard to quantum numbers and operators, in this section we are using the convention of Table 5.2.

6.2.2.1 **Hund's case (a)**

$$\Delta\Lambda = 0, \pm 1, \tag{6.7a}$$

with the restriction

$$\Sigma^+ \nleftrightarrow \Sigma^-, \tag{6.7b}$$

$$\Delta S = 0, \tag{6.7c}$$

$$\Delta\Sigma = 0, \tag{6.7d}$$

$$\Delta\Omega = 0, \pm 1, \tag{6.7e}$$

with the restriction

$$\Delta J \neq 0 \text{ when } \Omega = 0 \rightarrow \Omega = 0. \tag{6.7f}$$

The reader should note that this last restriction is the same restriction we have on ΔJ for atoms in Eq. (4.54).

6.2.2.2 **Hund's case (b)**

$$\Delta\Lambda = 0, \pm 1, \tag{6.8a}$$

with the restriction

$$\Sigma^+ \nleftrightarrow \Sigma^-, \tag{6.8b}$$

$$\Delta S = 0, \tag{6.8c}$$

$$\Delta N = 0, \pm 1, \tag{6.8d}$$

with the restriction

$$\Delta N \neq 0 \text{ when } \Sigma = 0 \rightarrow \Sigma = 0. \tag{6.8e}$$

6.2.2.3 **Hund's case (c)**

$$\Delta\Omega = 0, \pm 1, \tag{6.9a}$$

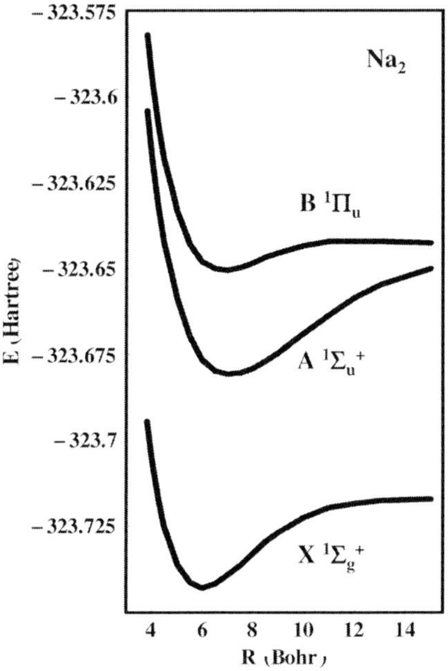

Fig. 6.2 *Ab initio* calculation (values taken from Ref. [59]) of three Na_2 potential curves.

with the restriction

$$\Omega = 0^+ \nleftrightarrow \Omega = 0^-. \tag{6.9b}$$

6.2.2.4 **Hund's case (d)**

$$\Delta N = 0, \pm 1, \tag{6.10a}$$

$$\Delta L = 0, \pm 1, \tag{6.10b}$$

$$\Delta R = 0. \tag{6.10c}$$

6.2.3 **Examples**

We are now ready to look at a couple of examples in a real system. Consider the potential curves for Na_2 shown in Fig. 6.2. The sodium dimer is well described by Hund's case (a) for transitions from the ground state to low-lying excited states. According to the selection rules above, transitions are possible to both the A and B states. We will first consider transitions to the A state. In this case we have a $^1\Sigma_u^+ \leftarrow ^1 \Sigma_g^+$ transition.[2] This is a homonuclear system, so the $g \leftrightarrow u$ restriction is obeyed as is $\Sigma^+ \leftrightarrow \Sigma^+$. Since $\Delta\Omega = 0$, there will be no $\Delta J = 0$ transitions. This means that only the P and R branches will exist; the Q branch is not allowed to this state. Consequently, the only transitions allowed will be $\Delta J = \pm 1$.

While this is a rather simple case, it is instructive to use Herzberg's bookkeeping method to keep track of the allowed transitions, which is very helpful in more complicated cases. We partially introduced the method back in Section 5.4.5. Now we must simply put two diagrams together and draw lines between the states that are allowed as we do in Fig. 6.3a. Note that both the $\oplus \nleftrightarrow \ominus$ and the $s \nleftrightarrow a$ restrictions are obeyed.

Now we will consider transitions to the B state. The Herzberg diagram for Hund's case (a) is shown in Fig. 6.3b. One should note two aspects of this diagram. First, each J level for the upper state has both a $\oplus$ (a) and a $\ominus$ (s) state due to Λ doubling. Second, all three branches are active.

6.3
AC Tunneling Ionization

In Section 3.1.1, we discussed AC tunneling ionization in atoms induced by strong laser fields. Tunneling, which is related to static-field ionization, occurs when the Keldysh parameter (γ, Eq. (3.6)) is small. A similar ionization process occurs in molecules as well. However, things are a bit more interesting with molecules for several reasons. For example, the electronic and nuclear motions are both important – we cannot assume that the nuclei are static. At the same time, after a sufficient number of electrons are removed from the parent molecule, the highly charged nuclei tend to dissociate, and rather violently at that. This strong-field-induced dissociative ionization often goes by the name of *Coulomb explosion* or laser-induced Coulomb explosion. While the genesis of Coulomb-explosion experiments can be traced to beam foil experiments [60],[3] laser-based experiments provide increased flexibility and a means for preparing the initial state and/or inducing exotic dynamics prior to the explosion. We will end our discussion of molecules with a brief discussion of this laser-induced Coulomb explosion.

2) The backwards arrow is the custom when describing absorption in molecular physics. The lower state is always on the right. Emission, for example, would flow from left to right.

3) Beam foil experiments involve stripping several electrons from a high-energy, low-charge molecular beam traversing a thin foil. After the foil, the highly stripped molecules undergo a Coulomb explosion. The atomic ions are then detected with position-sensitive detectors.

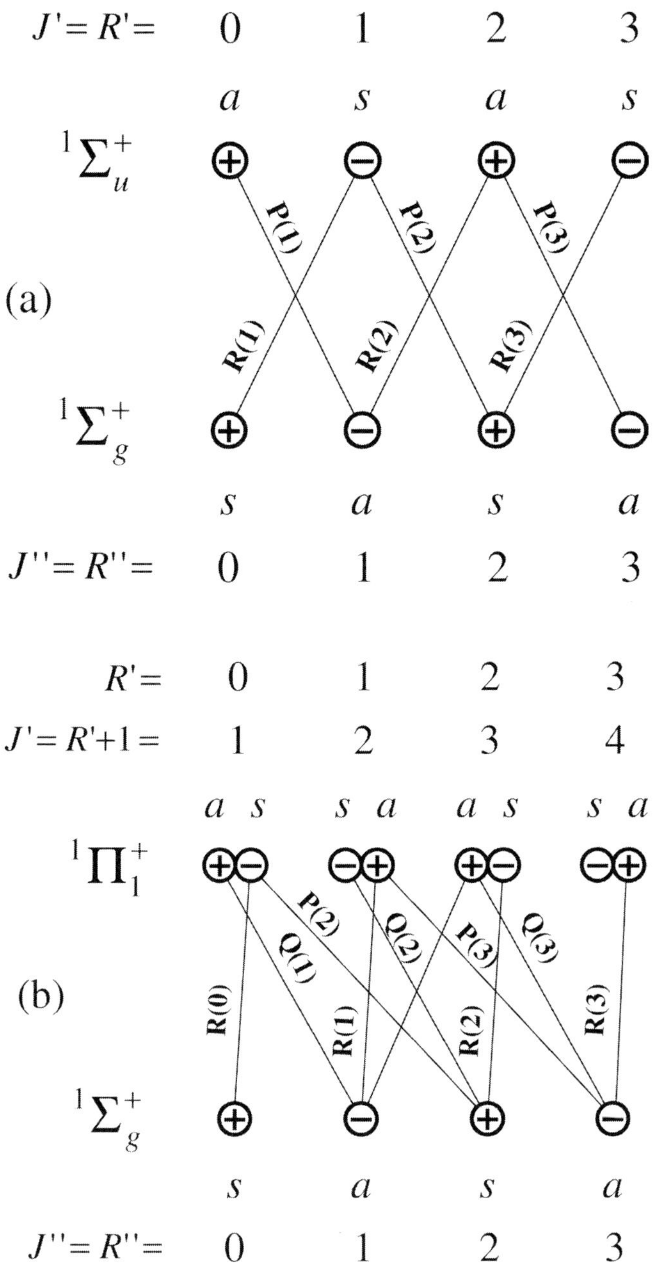

Fig. 6.3 Herzberg's bookkeeping diagram for (a) $^1\Sigma_u^+ \leftrightarrow ^1\Sigma_g^+$ transitions and (b) $^1\Pi_u \leftrightarrow ^1\Sigma_g^+$ transitions. The allowed transitions are indicated by the notation $\Delta J(J'')$. The subscript of the upper state in (b) is $\Omega = \Lambda + \Sigma$. Note that $J = N = R + 1$ in these diagrams because $S = 0$.

We begin by modifying the classical equations used to study field ionization of atoms (Eqs. (3.1b) and (3.2)) with a smoothed, *Eberlonium* potential. In this case, the potential energy resulting from the combined Coulomb and laser fields takes the form

$$V(z) = -\frac{e^2}{4\pi\varepsilon_0}\left(\frac{Z_1}{\sqrt{(z-R/2)^2+a^2}} + \frac{Z_2}{\sqrt{(z+R/2)^2+a^2}}\right) - eFz, \quad (6.11)$$

where $Z_{1,2}$ is the nuclear charge and R is the internuclear separation. The valence electron, apart from its dissociation energy, is bound by

$$E_0 = -\frac{1}{2}\left[\left(I_1^p + \frac{1}{4\pi\varepsilon_0}\frac{Z_1e^2}{R}\right) + \left(I_2^p + \frac{1}{4\pi\varepsilon_0}\frac{Z_2e^2}{R}\right)\right], \quad (6.12)$$

where $I_{1,2}^p$ is the ionization potential for each atom. At the equilibrium separation, R_e, the threshold intensity, I_{th}, required for over-the-barrier ionization can be determined as was done in Section 3.1.1.[4] For a model system, $I_1^p = I_2^p = 1$ a.u., $R = R_e = 1$ a.u. and $Z_1 = Z_2 = 1$,[5] Fig. 6.4a shows that $I_{th} \sim 1.4 \times 10^{15}$ W/cm^2. Ionization at R_e, in the absence of any residual bonding, would lead to an R_e explosion energy, potential energy turned into kinetic energy, of 14.4 eV, if the nuclei have the charge and mass of protons.[6] For sufficiently massive nuclei that do not move during the pulse, the explosion energy observed is nearly that predicted by R_e. More commonly, however, explosion energies significantly less than this are observed. They are typically consistent with separations $\simeq$ 2–3 $\times R_e$.

Explosion energy deficits can be understood in terms of a simple model based on peak tunneling rates occurring at extended separations [61–63]. The essence of the model, often called *enhanced ionization at* R_c, is captured in Fig. 6.4 for a model system. In the presence of the laser field, electrons are induced to move back and forth from one side of the molecule to the other. The rates for such transitions are sustained by field-enhanced charge-resonance states [64] that shift with the strength of the laser field and internuclear separation according to [65]

$$E_{\pm} = I_{\pm}^p \pm \frac{FR}{2}. \quad (6.13)$$

Above, the valence level is shifted upward (+) in the uphill potential well and downward (−) in the downhill potential well. It is important to recognize

4) Tunneling will occur at a slightly lower intensity, but I_{th} is more convenient to visualize. In this molecular case, a molecular ADK rate can be used to estimate the ionization rate [20].

5) Note that our model system does not correspond to the parameters of either H_2 or H_2^+.

6) The explosion energy, or so-called kinetic energy released, is the sum of the final energies of the two nuclei, protons in this model, at infinite separation subsequent to the explosion.

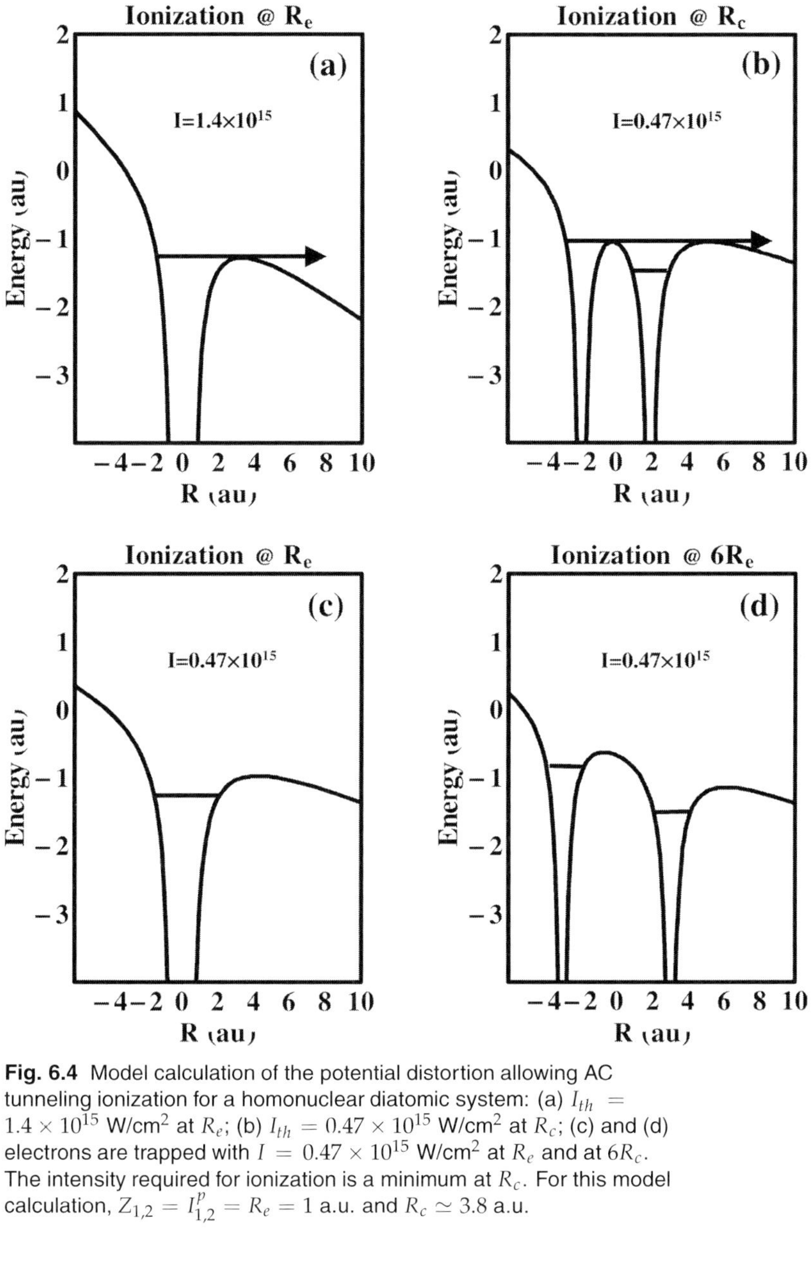

Fig. 6.4 Model calculation of the potential distortion allowing AC tunneling ionization for a homonuclear diatomic system: (a) $I_{th} = 1.4 \times 10^{15}$ W/cm^2 at R_e; (b) $I_{th} = 0.47 \times 10^{15}$ W/cm^2 at R_c; (c) and (d) electrons are trapped with $I = 0.47 \times 10^{15}$ W/cm^2 at R_e and at $6R_c$. The intensity required for ionization is a minimum at R_c. For this model calculation, $Z_{1,2} = I^p_{1,2} = R_e = 1$ a.u. and $R_c \simeq 3.8$ a.u.

that the shift increases with R. Eventually the uphill level will shift over the barrier allowing the electron to escape. Thus, it is easier to ionize the system at a larger R than at R_e. One must not lose sight of the fact that as the nuclei separate, an inner barrier develops that can trap the electron, which was free to move between the two nuclei, in one well or the other.[7] Once this occurs, it is possible to define I_{th} and R_c as the point where E_+ equals the heights of both inner and outer barriers. In the model system, for example Fig. 6.4b, I_{th} at R_c ($> R_e$) is less than half that required at R_e. Note as well that though the electron is in the downhill well it is still bound. Theoretically [62,65],

$$R_c \simeq \frac{4}{I^p}, \tag{6.14}$$

where atomic units are used. As R continues to increase, Fig. 6.4 shows that I_{th} also increases – R_c is unique in that I_{th} increases when $R > R_c$ or $R < R_c$. Even though we have presented these ideas in classical terms, they have been substantiated quantum mechanically [62,65].

Further Reading

1 G. Herzberg, *Molecular Spectra and Molecular Structure I: Spectra of Diatomic Molecules*, Van Nostrand Reinhold, New York, NY, 1950.

2 J. Posthumus (ed.), *Molecules and Clusters in Intense Laser Fields*, Cambridge University Press, New York, NY, 2001.

3 J. I. Steinfeld, *Molecules and Radiation: An Introduction to Modern Molecular Spectroscopy*, The MIT Press, Cambridge, MA, 1978.

Problems

6.1 Describe the spectrum for excitation between the ground state $X\,^1\Sigma_g^1$ and $A\,^1\Sigma_u^-$ of Na_2.

a) Write out the Herzberg diagrams to show the allowed transitions between the two states.

b) Estimate the wavelength for the lowest P, Q and R branch transitions in absorption to first order (i.e., only keep the lowest relevant constants) for the first $v = 0$ to $v = 0$ transition. Specify the J values for the lower and upper states. Also, if one of the branches does not exist, so state.

6.2 The selection rules for a magnetic dipole are

$$\oplus \not\leftrightarrow \ominus, \tag{6.15a}$$

$$\Delta J = 0, \pm 1, \tag{6.15b}$$

7) Clearly this must be the case, since at infinite separation an electron attached to either nucleus will simply behave like an atom or atomic ion having a single potential well.

with the restriction

$$J' = 0 \nleftrightarrow J'' = 0, \tag{6.15c}$$

$$\Delta\Lambda = 0, \pm 1. \tag{6.15d}$$

The restriction on the latter is that when $\Lambda = 0$ only intersystem combinations (single $\leftrightarrow$ triplet, etc.) are allowed.

Justify these selection rules using the Wigner–Eckhart theorem. Describe the spectrum for excitation between the ground state, $X^3\Sigma_g^-$, and the excited state, $b^1\Sigma_g^+$, of O_2, which approximately obeys Hund's case (b). Estimate the wavelengths of the lowest P, Q and R branch transitions to first order (i.e., only keep the lowest relevant constants). Specify the J values for the lower and upper states. If one or more of the branches does not exist, so state. For this problem, you will need to define these branches for ΔK and ΔJ transitions separately.

6.3 Derive Eq. (6.14). Hint, see Ref. [65].

6.4 Verify Eq. (6.14) for O_2^+ numerically by determining $I_{th}(R_c)$. Plot $I_{th}(R)$ for $R = R_e$ to $10R_e$.

Part 2
Light-Matter Interaction: Nonlinear Optics

Light-Matter Interaction: Atoms and Molecules in External Fields and Nonlinear Optics.
W. T. Hill and C. H. Lee

ISBN: 978-3-527-40661-6

7
Nonlinear Optics

7.1
Introduction

The linear relationship between the polarization and electric field is strictly true only if the susceptibility is not a function of the field. With the advent of the laser one can easily focus the laser beam in a medium, creating a very high field. In this case the susceptibility itself becomes dependent on the electric field. The relation between P and E will be nonlinear:

$$P(t) = \chi^{(1)}E(t) + \chi^{(2)}E^2 + \chi^{(3)}E^3 + \ldots \tag{7.1}$$

$$= P^{(1)}(t) + P^{(2)}(t) + P^{(3)}(t) + \ldots, \tag{7.2}$$

where the first term is the usual linear polarization and the others are referred to as the nonlinear polarization. The quantities $\chi^{(2)}$ and $\chi^{(3)}$ are the second- and third-order nonlinear optical susceptibilities, respectively. In general the magnitude of $\chi^{(n+1)}$ is approximately equal to $\chi^{(n)}/E_{atm}$, where E_{atm} is the atomic electrical field. From $E_{atm} \sim e/a_0^2$, where $-e$ is the charge of the electron and a_o is the Bohr radius of the hydrogen atom, one can calculate $E_{atm} \sim 5 \times 10^9$ V/cm (Eq. (A.11)). Thus, the effect of $\chi^{(2)}$ is a factor of E/E_{atm} smaller than $\chi^{(1)}$. This explains why the discovery of the optical nonlinear effect had to wait for the invention of the laser. Only by focusing a laser beam is one able to generate an E field with high enough strength such that it becomes possible to observe the nonlinear optical effect. In an extreme case, such as a table-top high-power femtosecond laser system, the electric field strength in the focus of the laser beam can exceed that of the atomic field, making higher-order nonlinear effects more probable than their lower-order counterparts and leading to the breakdown of perturbation theory, upon which Eq. (7.1) is based, altogether.

Historically, the first nonlinear optical experiment was optical second-harmonic generation reported by Franken et al. [66]. A Q-switched ruby laser was used as the fundamental beam at a wavelength of 694.3 nm. When the laser was focused onto a quartz crystal, a very small signal at half the original wavelength was observed amid the strong background radiation of the fun-

Light-Matter Interaction: Atoms and Molecules in External Fields and Nonlinear Optics.
W. T. Hill and C. H. Lee

ISBN: 978-3-527-40661-6

damental. This engendered the new field of nonlinear optics. Theoretically speaking, the nonlinear effect should be a fundamental property of materials. With either a strong enough fundamental field and/or a sensitive enough harmonic detector, one may proceed to look for new nonlinear effects. Starting as a scientific curiosity, the field of nonlinear optics has now grown into a vast technology empire with many applications.

In optical parametric generation, it is used to produce a coherent tunable output. Optical second- and third-harmonic generation provide good probes for the material properties. A second-harmonic generator, also referred to as a frequency doubler, is now a standard optical component one can purchase off-the-shelf. Even surface optical second harmonic generation can be applied to study the microscopic properties at the material surface with submicrometer spatial resolution by near-field harmonic imaging. In the extreme high field area, optical harmonic generation up to hundreds of orders provides a coherent beam in the extreme ultraviolet or soft X-ray region. In other words, nonlinear optics has become an inseparable part of everyday scientific phenomena and is used in many high-technology industries, such as optical communications.

There are many books written on the subject of nonlinear optics already. The purpose of this book is to provide readers with a sufficient background of the essence of nonlinear optics. We will first provide a phenomenological description of nonlinear optics. Wave propagation in a nonlinear medium will be discussed next. Some nonlinear phenomena will be discussed in detail. The quantum theory of nonlinear optical susceptibility will be presented using a density matrix formalism. Applications of nonlinear optics will be presented at the end.

7.2 Phenomenological Description of Nonlinear Optics

In Eqs. (7.1) and (7.2), $\chi^{(n)}$ may be regarded as the response function of the material. The polarization, when expressed in the frequency domain, can be written as

$$\begin{aligned} P_l(\omega_i) = {} & \chi_{lm}^{(1)}(\omega_i)\, E_m(\omega_i) \\ & + 2\chi_{lmn}^{(2)}(\omega_i;\omega_j,\omega_k)\, E_m(\omega_j)\, E_n(\omega_k)\, \delta(\omega_i;\omega_j+\omega_k) \\ & + 6\chi_{lmno}^{(3)}(\omega_i;\omega_j,\omega_k,\omega_h)\, E_m(\omega_j)\, E_n(\omega_k)\, E_n(\omega_k)\, E_o(\omega_h) \\ & \times \delta(\omega_i;\omega_j,\omega_k,\omega_h) \\ & + \text{higher terms in } E \text{ and other terms involving the} \\ & \quad \text{magnetic field } B \\ & + \ldots, \end{aligned} \tag{7.3}$$

where the subscripts l, m, n, o indicate Cartesian components and ω_i, ω_j, ω_k, etc., represent the frequencies of interest. The factor "2" in Eq. (7.3) is from our definition of the electric field:

$$\tilde{E}(t) = Ee^{-i\omega t} + c.c.$$

The notation used here follows that of Boyd [67]. The delta function assumes the value of unity when the frequencies in its argument satisfy the relation[1]

$$\omega_i = \omega_j + \omega_k + \omega_h.$$

For example, from the term

$$P_l^{(2)}(\omega_i) = 2\chi_{lmn}^{(2)}(\omega_i;\omega_j,\omega_k)\, E_m(\omega_j)\, E_n(\omega_k)\, \delta(\omega_i;\omega_j+\omega_k) \tag{7.4}$$

we may have one of the following nonlinear effects:

- optical three-wave mixing $\chi^{(2)}(\omega_i;\omega_j,\omega_k,)$, which includes sum- and difference-frequency generation,
- $\chi^{(2)}(2\omega;\omega,\omega)$, optical second-harmonic generation (SHG) or frequency doubling,
- $\chi^{(2)}(\omega;0,\omega)$, linear electrooptic or Pockels effects,
- $\chi^{(2)}(0;\omega,-\omega)$, optical rectification,
- $\chi^{(2)}(\omega_j+\omega_k;\omega_j,\omega_k)$, frequency upconversion, and
- $\chi^{(2)}(\omega_j-\omega_k;\omega_j,-\omega_k)$, parametric oscillation.

From the term

$$P_l = 6\chi_{lmno}^{(3)}(\omega_i;\omega_j,\omega_k,\omega_h)\, E_m(\omega_j)\, E_n(\omega_k)\, E_0(\omega_h)\, \delta(\omega_i;\omega_j+\omega_k+\omega_h) \tag{7.5}$$

we have

- $\chi_{lmno}^{(3)}(\omega_i;\omega_j,\omega_k,\omega_h)$, optical four-wave mixing,
- $\chi_{lmno}^{(3)}(3\omega;\omega,\omega,\omega)$, optical third-harmonic generation (THG),

1) See Appendix B.1:

$$\delta(\omega_i;\omega_j+\omega_k) = \begin{cases} 1, \omega_i = \omega_j+\omega_k \\ 0, \omega_i \neq \omega_j+\omega_k \end{cases}$$

$$\delta(\omega_i;\omega_j+\omega_k+\omega_h) = \begin{cases} 1, \omega_i = \omega_j+\omega_k+\omega_h \\ 0, \omega_i \neq \omega_j+\omega_k+\omega_h \end{cases}$$

- $\mathrm{Re}\chi^{(3)}_{lmno}(\omega;\omega,\omega,-\omega)$, a nonlinear refractive index,
- $\mathrm{Im}\chi^{(3)}_{lmno}(\omega;\omega,\omega,-\omega)$, two-photon absorption, and
- $\mathrm{Im}\chi^{(3)}_{lmno}(\omega_s;\omega_l,-\omega_l,+\omega_s)$, stimulated Raman scattering.

χ generally reflects the symmetry of the material system. For example, $\chi^{(2)}_{lmno}(\omega_i;\omega_j,\omega_k)$ vanishes in a medium with inversion symmetry.

7.2.1 Optical Second-Harmonic Generation (SHG) $\chi^{(2)}(2\omega;\omega,\omega)$

Historically, the most important nonlinear optical effect is SHG. The first experiment by Franken et al. [66] led to the birth of the field of nonlinear optics.

In a typical SHG experiment one sends into a nonlinear crystal a laser beam at frequency ω. In the output, one can detect a new optical field at frequency 2ω. Figure 7.1 shows the experimental arrangement. In a medium that lacks a center of inversion, this process is allowed. Physically, this nonlinear process can be described as a two-step process. First, the medium responds to the incident electric field of the laser beam in a nonlinear fashion, generating nonlinear polarization, $P^{(2)}(2\omega)$. This polarization serves as the driving force in the wave equation. The medium is driven by an ensemble of coherent oscillating dipoles, which is also referred to as P^{NLS} [68]. We have

$$-\nabla^2 E + \mu_0\,\sigma_c\,\frac{\partial E}{\partial t} + \frac{1}{c^2}\frac{\partial^2 E}{\partial t^2} = -\mu_0\,\frac{\partial^2 P^{NLS}}{\partial t^2}. \tag{7.6}$$

Here $P^{NLS}(2\omega)$ is the source term that propagates through the medium, creating a radiating field $E(2\omega, r_1)$ at r_1, an upstream position in the medium. The radiating field propagates in the medium as soon as it is generated. In the meantime, the laser beam with $E(\omega)$ also propagates through the medium,

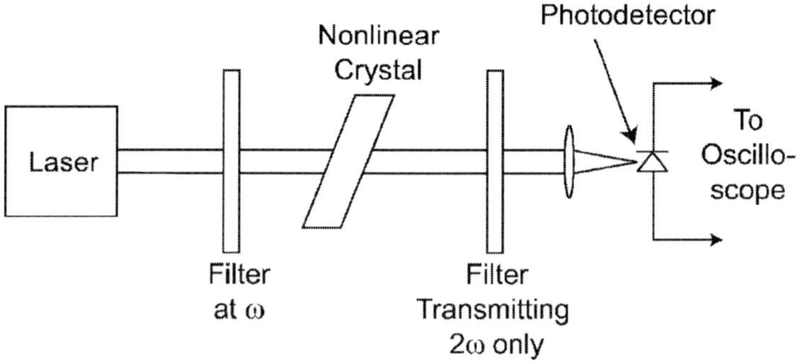

Fig. 7.1 Experimental setup for the detection of SHG light.

creating a new $P^{NLS}(2\omega, r)$ wave downstream ($r > r_1$). This propagating P^{NLS} will radiate $E(2\omega)$ as it propagates along. $E(2\omega, r)$, generated downstream will interfere with the $E(2\omega, r_1)$ generated earlier but which has propagated to the same downstream position r. Since all these are coherent processes, there is a definite phase relationship between the P^{NLS} wave with propagation vector $2k_1$ and the reradiating field $E(2\omega, r)$ with propagation vector k_2. Normally, because of dispersion, the fundamental wave which creates the $P^{NLS}(2\omega)$ will be out of step (out of phase) with the $E(2\omega)$. This results in an oscillatory amplitude of the $E(2\omega, r)$ as a function of position inside the crystal, causing periodic destructive and constructive interference between the $E(2\omega)$ generated at r and the $E(2\omega)_{r1}$ generated at an upstream position (r_1) but which has propagated to the same location r. This is the so-called non-phase-matched condition. It always limits the SHG conversion efficiency. A configuration can be devised to make the $P^{NLS}(2\omega)$ and $E(2\omega)$ waves propagate with the same phase velocity so that they always propagate in phase, leading only to constructive interference. This is the phase-matching configuration in SHG. Nowadays, vendors can supply crystals that have been properly cut and oriented for the phase-matching condition. In principle, with the most favored condition, SHG conversion efficiency can approach 100%. However, as a researcher or student in nonlinear optics, one still needs to understand the principle of phase matching in order to use the nonlinear crystal properly. We will discuss this in detail later. The interplay between $P^{NLS}(2\omega)$ and $E(2\omega)$ waves can be analyzed by the formalism developed for the wave propagation in a nonlinear medium [68].

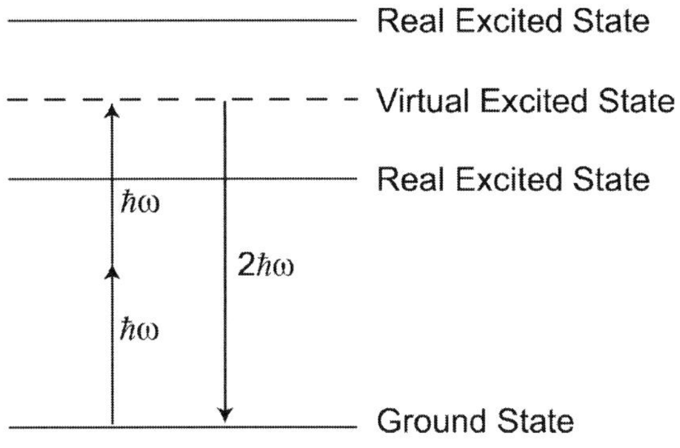

Fig. 7.2 Quantum mechanical description of optical second-harmonic generation.

In the wave–particle duality picture photons are quanta. The SHG process can be regarded as particle annihilation and generation: two fundamental photons are annihilated but at the same time a new particle at double the frequency is created as shown in Fig. 7.2. In this process energy is conserved:

$$\hbar\omega_2 = \hbar\omega_1 + \hbar\omega_1,$$
$$\omega_2 = 2\omega_1.$$

The process shown in Fig. 7.2 also does not disturb the population of the initial and final states. It is referred to as the parametric process. Since, under normal conditions, the conversion efficiency of SHG is always small, one usually needs to focus the laser beam very tightly into the nonlinear medium, increasing the intensity of the laser beam and hence the electrical field strength. Despite this focusing effort, in many instances the second-harmonic signals generated are still very weak as compared to the signal at the fundamental frequency. One needs good isolation to block out the intense fundamental beam before sending the second-harmonic beam into a photodetector.

7.2.2
Electrooptic Effect, $\chi^{(2)}(\omega; 0, \omega)$

The linear electrooptic effect may be regarded as a special $\chi^{(2)}$ process when one of the electric fields is a DC field, e.g., $\chi^{(2)}(\omega; 0, \omega)$. The refractive index of the extraordinary ray (e-ray) can be controlled by applying a DC field to an electrooptic crystal. For a wave propagating in a birefringent crystal, the equation of the index ellipsoid in the presence of a DC field is

$$\left(\frac{1}{n^2}\right)_1 x^2 + \left(\frac{1}{n^2}\right)_2 y^2 + \left(\frac{1}{n^2}\right)_3 z^2 + \left(\frac{1}{n^2}\right)_4 yz + \left(\frac{1}{n^2}\right)_5 xz + \left(\frac{1}{n^2}\right)_6 xy = 1. \quad (7.7)$$

The linear change in the coefficients $\left(\frac{1}{n^2}\right)'_i$, due to the presence of an arbitrary DC field $E\left(E_x, E_y, E_z\right)$, is expressed as

$$\triangle\left(\frac{1}{n^2}\right)_i = \sum_{j=1}^{3} r_{ij} E_j, \quad (7.8)$$

where $i = 1, 2, \ldots, 6$ and r_{ij} is the electrooptic coefficient, which is a third-rank tensor. The contraction of a third-rank tensor with the vector E results in a second-rank tensor similar to $\chi^{(2)}$. The linear electrooptic effect is the principle utilized in electrooptic modulators (Mach–Zehnder type) and Pockels cells. A recent but challenging application is using the modulator for a broadband RF fiber link for transporting broadband microwave or millimeter-wave signals over optical fibers, instead of conventional copper transmission lines [69]. The key component of this type of system is a modulator which is used to impress a RF subcarrier onto optical carrier waves [70]. A high-performance modulator requires high spurious free dynamic range, low third-order intermodulation distortion and a high degree of linearization. To achieve broadband operation, the modulator electrodes must be part of the transmission line.

7.2.3 Optical Rectification $\chi^{(2)}(0;\omega,-\omega)$

Optical rectification is represented by $\chi^{(2)}(0;\omega,-\omega)$. Two fundamental waves beat together to generate a zero-frequency (DC or carrierless) signal. Optical rectification usually refers to the generation of a DC polarization component, which is not of particular interest. However, with the advent of femtosecond lasers, this effect becomes responsible for terahertz (THz) pulse generation from dielectrics. The output from the process of

$$P_i^2(0) = \sum_{j,k} \chi^{(2)}(0;\omega,-\omega)E_j(\omega)E_k^*(\omega) \tag{7.9}$$

is an envelope of the E field with its carrier stripped off. This effect becomes particularly interesting when the envelope function of the optical wave lasts less than 1 picosecond (ps). The picosecond electrical pulse radiates a short THz impulse, consisting of only one and one-half cycles (a Mexican hat shape pulse). Pulsed THz sources have been used for time-resolved far-infrared (FIR) imaging of a concealed object [71].

In order to generate THz pulses by optical rectification processes, subpicosecond optical pulses are required to excite the nonlinear crystal. Materials capable of generating THz pulses include semiconductors (GaAs, ZnSe, ZnTe, InP, CdTe, etc.), electrooptic materials ($LiNbO_3$, $LiTaO_3$, $BaTiO_3$, KTP, ZnO, etc.), and organic crystals (dimethyl amino 4-N-methylstilbazolium tosylate or DAST, etc.). Detecting the THz pulses requires special techniques. To achieve the time resolution required, a physical effect with a quick response must be employed. The electrooptic effect is such an effect. When a THz pulse impinges onto an electrooptic crystal it will induce an instantaneous electrooptic effect, causing the polarization of the probe pulse to rotate. By carefully measuring the amount of rotation as a function of time delay between the pump, which generates the THz pulse, and the probe pulse, one can map out the temporal shape of the THz pulse. This technique turns out to be the only way to measure the waveform of the THz pulses, since no other detecting system can provide the necessary bandwidth to resolve the waveform [72].

7.2.4
Parametric Generation $\chi^{(2)}\left(\omega_s;\omega_p,-\omega_i\right)$

The second-order nonlinear optical susceptibility, $\chi^{(2)}\left(\omega_s;\omega_p,-\omega_i\right)$, can be used to generate a coherent output by a strong pump beam at frequency ω_p such that

$$\hbar\omega_p \Rightarrow \hbar\omega_s + \hbar\omega_i, \tag{7.10}$$

i.e., the pump photon breaks up into two photons, one at a signal frequency, ω_s, and the other at an idler frequency ω_i, respectively.[2] Signal and idler fields are coupled together through the pump field. The theory of parametric generation will be presented in more detail in the next chapter.

Any parametric interaction satisfies energy and momentum conservation conditions

$$\hbar k_p \Rightarrow \hbar k_s + \hbar k_i \tag{7.11}$$

and will emerge as the dominant process, generating a coherent output at ω_s and ω_i, respectively [73]. From Eq. (7.10) it is clear that as long as the condition $\hbar\omega_p \Rightarrow \hbar\omega_s + \hbar\omega_i$ is met, there will be an infinite number of values of ω_s satisfying the energy-conservation condition, leading to a tunable coherent output. Two commercial optical devices, the optical parametric oscillator (OPO) and the optical parametric amplifier (OPA), have been built according to the principle of parametric generation discussed here. Using intense second-harmonic pulses from a commercial Ti:sapphire regenerative amplifier as the pump pulses, one can generate a tunable pulse in the infrared.

7.2.5
Third-Order Nonlinear Effect

With the third-order $\chi^{(3)}$ nonlinear effect the most important process is four-wave mixing $\chi^{(3)}_{lmno}\left(\omega_i;\omega_j,\omega_k,\omega_h\right)$.

Energy conservation gives

$$\hbar\omega_i = \hbar(\omega_j + \omega_k + \omega_h), \tag{7.12}$$

2) In the parametric generation process the energy taken away from the pump beam is fed into two beams of lower frequency. Consequently, the process can be used as an amplifier: a weak signal beam is made to interact with a strong, high-frequency pump beam, and both the original signal, known as the "signal", and the generated difference frequency, known as the "idler", are amplified. Note that the denominations "signal" and "idler" have specific meanings only for the parametric amplifier. In the oscillator, either of the two lower frequencies can be called the "signal" or the "idler".

where the j, k, h indices stand for 1, 2 or 3 respectively and the frequencies may take positive or negative values. For example, $\omega_i = \omega_1 + \omega_1 + \omega_1 = 3\omega_1$ represents third-harmonic generation, $\omega_4 = \omega_1 + \omega_2 - \omega_3$ represents four-wave mixing and $\omega_1 = \omega_1 + \omega_1 - \omega_1$ represents a nonlinear refractive-index change.

Let us consider

$$\tilde{E}(t) = E_1\,\mathrm{e}^{-i\omega_1 t} + E_2\,\mathrm{e}^{-i\omega_2 t} + E_3\,\mathrm{e}^{-i\omega_3 t} + c.c.$$

The $\chi^{(3)}$ nonlinear term gives rise to the nonlinear polarization

$$P^{NL}(\omega_1) = \chi^{(3)}\left(3E_1E_1^* + 6E_2E_2^* + 6E_3E_3^*\right)E_1, \tag{7.13}$$

$$P^{NL}(\omega_1 + \omega_2 - \omega_3) = 6\chi^{(3)}E_1E_2E_3^* \text{ and} \tag{7.14}$$

$$P^{NL}(2\omega_1 - \omega_2) = 3\chi^{(3)}E_1^2E_2^*, \text{ etc.} \tag{7.15}$$

Of particular interest is the degenerate case, where $\omega_1 = \omega_2 = \omega_3 = \omega$ and

$$P^{NL}(\omega_1) = 3\chi^{(3)} \mid E \mid^2 E. \tag{7.16}$$

This is the nonlinear polarization at the same frequency of the fundamental field. This term will give rise to the intensity-dependent refractive index

$$n = n_0 + n_2 I(r, t), \tag{7.17}$$

where $I(r, t) = (n_0 c/2\pi) \mid E(\omega, r, t) \mid^2$ is the intensity of the optical beam. For a short-pulse laser, the intensity of the laser is a function of position (spatial Gaussian distribution) and time (short-pulse envelope function) at the carrier frequency ω. The intensity-dependent refractive-index nonlinearity is responsible for phenomena such as self-focusing [74], self-phase modulation and soliton formation [75].

7.2.6
Nonlinear d coefficient

We have seen previously with Eq. (7.4) that the nonlinear polarization for a $\chi^{(2)}$ effect is given by

$$P^{NL}(\omega_n + \omega_m, \omega_n, \omega_m) = \sum_{jk}\sum_{nm}\chi_{ijk}^{(2)}(\omega_n + \omega_m, \omega_n, \omega_m) \times E_j(\omega_n)\,E_k(\omega_m)\,\mathrm{e}^{-i(\omega_n+\omega_m)t},$$

where the indices i, j, k are referred to as the Cartesian components of the analytical E field. Therefore, i, j, k may represent x, y or z while n and m represent the specific frequency of the component field.

Experimentally, one can measure the electric field, which is a real quantity:

$$E = \varepsilon \cos(\omega t - \phi), \tag{7.18}$$

where ε is a real amplitude of the electric field with an angular frequency ω and phase angle ϕ. Analytically, this electric field can be expressed as

$$E = E(\omega)e^{-i\omega t} + E^*(\omega)e^{+i\omega t}. \tag{7.19}$$

The reality of the electric field requires that

$$E(\omega) = E^*(\omega) = E(-\omega) = \frac{1}{2}\varepsilon,$$

and thus

$$\begin{aligned} E &= \frac{1}{2}\,\varepsilon\, e^{i\phi}e^{-i\omega t} + \frac{1}{2}\,\varepsilon\, e^{-i\phi}e^{i\omega t} \\ &= \varepsilon \cos(\omega t - \phi) = 2E(\omega)\cos(\omega t - \phi). \end{aligned} \tag{7.20}$$

We see that the analytical amplitude of the electric field is

$$E(w) = \frac{1}{2}\,\varepsilon. \tag{7.21}$$

We have used a script letter to denote a real quantity. With the usual second-order ($\chi^{(2)}$ process) effect the real nonlinear polarization may be expressed as

$$p = 2d\varepsilon^2, \tag{7.22}$$

where d is a nonlinear coefficient used by experimentalists; d is in general a third-rank tensor. The d tensors for various nonlinear materials can be found in table form in many nonlinear optics books. The factor "2" in Eq. (7.22) is used so that the d coefficients conform with those listed in the other books. The application of Eq. (7.22) to a nonlinear crystal with two fields present is now illustrated below:

$$\varepsilon = \varepsilon_1(z,t)\cos(\omega_1 t - k_1 z) + \varepsilon_2(z,t)\cos(\omega_2 t - k_2 z);$$

for simplicity we assume that these two fields at ω_1 and ω_2 are propagating collinearly,

$$\begin{aligned} p &= 2d\left[\varepsilon_1\cos(\omega_1 t - k_1 z) + \varepsilon_2\cos(\omega_2 t - k_2 z)\right]^2 \\ &= d\left(\varepsilon_1^2 + \varepsilon_2^2\right) + d\,\varepsilon_1^2\cos\left[2(\omega_1 t - k_1 z)\right] \\ &\quad + d\,\varepsilon_2^2\cos\left[2(\omega_2 t - k_2 z)\right] \\ &\quad + 2d\,\varepsilon_1\varepsilon_2\cos\left[(\omega_1 t + \omega_2)t - (k_1 + k_2)z\right] \\ &\quad + 2d\,\varepsilon_1\varepsilon_2\cos\left[(\omega_1 - \omega_2)t - (k_1 - k_2)z\right] \\ &= p_{dc} + p(2\omega_1) + p(2\omega_2) + p(\omega_1 + \omega_2) + p(\omega_1 - \omega_2). \end{aligned} \tag{7.23}$$

Equation (7.23) defines the real nonlinear polarization,

$$p_{dc} = d\left(\varepsilon_1^2 + \varepsilon_2^2\right),$$

$$\begin{aligned} p(2\omega_1) &= d\,\varepsilon_1^2 \cos\left[2(\omega_1 t - k_1 z)\right] \\ &= d\,\varepsilon_1^2\,\frac{1}{2}\left[e^{-i2(\omega_1 t - k_1 z)} + e^{-i2(\omega_1 t - k_1 z)}\right] \\ &= P(2\omega_1) + P^*(2\omega_1), \end{aligned}$$

where $P(2\omega_1)$ and $P^*(2\omega_1)$ are the analytical amplitudes of the nonlinear polarization. Note that

$$P^*(2\omega_1) = P(-2\omega_1).$$

On the other hand, we have

$$P(2\omega_1) = \chi^{(2)}(2\omega_1, \omega_1, \omega_1)\,E(\omega_1)\,E(\omega_1)\mathrm{e}^{-i2(\omega_1 t - k_1 z)},$$

since $E(\omega_1) = \frac{1}{2}\,\varepsilon_1$.
By comparing these two expressions, one has

$$\frac{1}{2} d\,\varepsilon_1^2 = \chi^{(2)}\,E(\omega_1)\,E(\omega_1),$$

$$d = \frac{1}{2}\,\chi^{(2)}. \tag{7.24}$$

Equation (7.24) enables one to write the nonlinear polarization either in real-amplitude form or as an analytical expression. For the nondegenerative three-wave-mixing case,

$$\begin{aligned} P_i(\omega_1 + \omega_2) = \sum_{j,k} \Big[&\chi_{ijk}^{(2)}(\omega_1 + \omega_2, \omega_1, \omega_2) E_j(\omega_1) E_k(\omega_2) \mathrm{e}^{-i(\omega_1+\omega_2)t} \\ &+ \chi_{ijk}^{(2)}(\omega_2 + \omega_1, \omega_2, \omega_1) E_j(\omega_2) E_k(\omega_1) \mathrm{e}^{-i(\omega_2+\omega_1)t} \Big]. \end{aligned} \tag{7.25}$$

When the summation is carried out for all j and k values, there are a total of 18 terms in Eq. (7.25). However, the difference between the first and second terms is merely an effect of interchanging the sequence of fields in the equation. The sequence of the fields only affects the appearance of the mathematical expressions. It does not give any physically discernible difference. Mathematically, for a given index i (x, y or z) there are only six distinct terms in Eq. (7.25).

Passing to the d notation, one should realize that there are only six independent d coefficients corresponding to a given index, i. One can devise a contracted notation with a single running index ℓ to represent the double indices jk as follows:

$$\begin{array}{cccc} d\,\underbrace{ixx}_{(1)} & d\,\underbrace{ixy}_{(6)} & \longleftarrow & d\,\underbrace{ixz}_{(5)} \\ & \searrow & & \uparrow \\ & d\,\underbrace{iyy}_{(2)} & & d\,\underbrace{iyz}_{(4)} \\ & & \searrow & \uparrow \\ & & & d\,\underbrace{izz}_{(3)} \end{array}$$

$$\begin{array}{lcl} \text{or } jk & : & xx\ yy\ zz\ yz\ xz\ xy \\ \ell & : & 1\ \ 2\ \ 3\ \ 4\ \ 5\ \ 6 \end{array} \tag{7.26}$$

For example,
$$d_{yyz} \rightarrow d_{24}, \text{ etc.,}$$
$$d_{ixx} \rightarrow d_{i1}.$$

In general, $d_{i\ell}$ has 18 independent terms with index "i" representing x, y or z and "ℓ" representing $1, 2, \ldots, 6$.

With the contracted notation given in Eq. (7.26), one can use matrix multiplication to represent the second-harmonic generation of Eq. (7.25):

$$P_i(2\omega_1) = 2\, d_{i\ell} F_\ell \;=\; \sum_\ell 2 d_{i\ell} F_\ell,$$

$$\begin{aligned} \text{where } F_\ell &= \left(1 - \frac{1}{2}\delta_{jk}\right)\ [E_j(\omega_1)E_k(\omega_2) + E_k(\omega_1)E_j(\omega_2),] \\ \text{or}\quad F_1 &= \left(1 - 1\frac{1}{2}\delta_{xx}\right)\ [E_x(\omega_1)E_x(\omega_2) + E_x(\omega_1)E_x(\omega_2)] \\ &= \frac{1}{2}\, 2E_x(\omega_1)E_x(\omega_2) = E_x^2(\omega_1), \\ F_4 &= \left(1 - \frac{1}{2}\delta_{yz}\right)\ [E_y(\omega_1)E_z(\omega_2) + E_z(\omega_1)E_y(\omega_2)] \\ &= E_y(\omega_1)E_z(\omega_2) + E_z(\omega_1)E_y(\omega_2). \end{aligned}$$

When $\omega_1 = \omega_2$,

$$F_4 = 2\, E_y(\omega_1)\, E_z(\omega_1), \text{ etc.}$$

$$\text{or} \begin{pmatrix} P_x(2\omega) \\ P_y(2\omega) \\ P_z(2\omega) \end{pmatrix} = 2 \begin{pmatrix} d_{11} & d_{12} & d_{13} & d_{14} & d_{15} & d_{16} \\ d_{21} & \cdot & \cdot & \cdot & \cdot & d_{26} \\ d_{31} & \cdot & \cdot & \cdot & \cdot & d_{36} \end{pmatrix} \begin{pmatrix} E_x^2(\omega) \\ E_y^2(\omega) \\ E_z^2(\omega) \\ 2E_y(\omega)E_z(\omega) \\ 2E_x(\omega)E_z(\omega) \\ 2E_x(\omega)E_y(\omega) \end{pmatrix}, \tag{7.27}$$

$$\begin{pmatrix} P_x(\omega_1+\omega_2) \\ p_y(\omega_1+\omega_2) \\ P_z(\omega_1+\omega_2) \end{pmatrix} = 4 \begin{pmatrix} d_{11} & \dots & d_{16} \\ & \dots & \\ d_{31} & \dots & d_{36} \end{pmatrix} \begin{pmatrix} E_x(\omega_1)E_x(\omega_2) \\ E_y(\omega_1)E_y(\omega_2) \\ E_z(\omega_1)E_z(\omega_2) \\ E_y(\omega_1)E_z(\omega_2) + E_z(\omega_1)E_y(\omega_2) \\ E_x(\omega_1)E_z(\omega_2) + E_z(\omega_1)E_x(\omega_2) \\ E_x(\omega_1)E_y(\omega_2) + E_y(\omega_1)E_x(\omega_2) \end{pmatrix}. \tag{7.28}$$

The extra factor of two results when one sums over ω_1 and ω_2 sequentially. It should be noted that, depending on the symmetry of the nonlinear crystal, many d coefficients vanish and some may be equal to other elements. For example, all d coefficients vanish if the crystal has inversion symmetry (an inversion center). This can be illustrated as follows:

$$\vec{P}(2\omega, r) = 2d\vec{E}(r)\vec{E}(r).$$

For a medium with inversion symmetry, as shown in Fig. 7.3,

$$\vec{E}(-r) = -\vec{E}(r),$$

$$\vec{P}(2\omega, -r) = -\vec{P}(2\omega, r).$$

Applying an inversion operation on the equation

$$-\vec{P}(2\omega, +r) = 2d(-\vec{E})(-\vec{E}) = 2d\vec{E}\vec{E},$$

$$\text{or } \vec{P}(2\omega, r) = -2d\vec{E}\vec{E} = 2d\vec{E}\vec{E}, \quad \text{for arbitrary } \vec{E},$$

we have

$$d \equiv 0.$$

Similarly for a crystal with a 222 symmetry class, by repeatedly applying a two-fold rotation symmetry operation about x-, y- and z-axes, it can be shown that the only nonzero terms in the d matrix are $d_{xyz}(d_{14})$, $d_{yxz}(d_{25})$ and $d_{zxy}(d_{36})$, i.e. when all three indices are different. The form of the d_{il} matrix for all 21 crystal classes can be found in a number of books. For example,

$$d = \begin{pmatrix} \cdots & d_{14} & \cdot & \cdot \\ \cdots & \cdot & d_{14} & \cdot \\ \cdots & \cdot & \cdot & d_{36} \end{pmatrix},$$

showing only two nonzero independent elements, d_{14} and d_{36}. The values of these d coefficients can also be found in many books, for example, $KTiOPO_4$ (KTP) belongs to the point group mm2, with

$$d_{31} = d_{15} = 3.7\ \text{pm/V and} \quad d_{33} = 14.6\ \text{pm/V},$$

while d_{36} for $AgGaSe_2$ is 49.3 pm/V.[3]

It should be pointed out that different books use different units for the d coefficient; when comparing the d coefficients listed from different sources one needs to make sure that they are in the same unit. Since the reader of this book only needs to understand some of the principles of nonlinear optics, we need not concern ourselves with how to derive the entire d matrix.

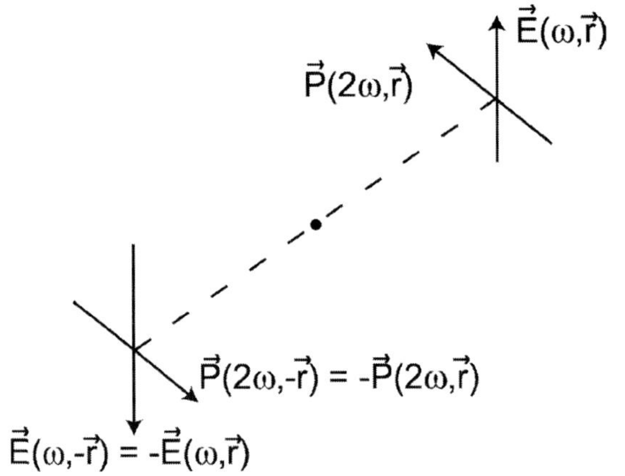

Fig. 7.3 $E_\omega(\vec{r})$ and $\vec{P}_{2\omega}(\vec{r})$ vectors in a medium with inversion symmetry. $d = 0$ in this case.

3) Data are from Table 9.3 of J. M. Liu, *Photonic Devices*, Cambridge University Press, 2005.

Further Reading

1 R. W. Boyd, *Nonlinear Optics*, Academic, San Diego, CA, 1992

2 N. Bloembergen, *Nonlinear Optics*, 4th edn., World Scientific, Singapore, 1996

3 F. Zernike and J. E. Midwinter, *Applied Nonlinear Optics*, Wiley, New York, NY, 1973

4 R. H. Pantell and H. E. Puthoff, *Fundamentals of Quantum Electronics*, Wiley, New York, NY, 1969

5 Y. R. Shen, *The Principles of Nonlinear Optics*, Wiley, New York, NY, 1984

6 J. M. Liu, *Photonic Devices*, Cambridge University Press, Cambridge, UK, 2005

Problems

7.1 The electric flux density (displacement vector) $\vec{D}$ is expressed as

$$\vec{D} = \epsilon_0 \vec{E} + \vec{P} = \epsilon_0 (1 + \chi) \vec{E}$$

in the MKS system of units, and

$$\vec{D} = \vec{E} + 4\pi \vec{P} = (1 + 4\pi\chi) \vec{E}$$

in the CGS Gaussian system of units.

If one includes nonlinear effects, the polarization $\vec{P}$ can be expressed as

$$\vec{P}(t) = \chi^{(1)} \vec{E}(t) + \chi^{(2)} \vec{E}^2(t) + \chi^{(3)} \vec{E}^3(t) + \ldots$$

in the Gaussian system of units, and

$$\vec{P}(t) = \epsilon_0 [\chi^{(1)} \vec{E}(t) + \chi^{(2)} \vec{E}^2(t) + \chi^{(3)} \vec{E}^3(t) + \ldots]$$

in the MKS system, where $\epsilon_0 = 8.85 \times 10^{-12}$ F/m denotes the permittivity of free space.

In some of the literature, the second-order nonlinear term is also expressed as

$$\vec{P}^{(2)}(t) = \chi^{(2)} \vec{E}^2(t) = 2d\vec{(E)}^2(t)$$

in CGS and

$$\vec{P}^2(t) = \epsilon_0 \chi^{(2)} \vec{E}^2(t) = 2d\vec{E}^2(t)$$

in MKS. The d coefficient in CGS is in the unit of "esu" and in MKS, C/V^2.

a) Show that $\chi_{MKS} = 4\pi\chi_{CGS}$.

b) Derive the ratio d_{MKS}/d_{CGS}.

c) d_{MKS}/ϵ_0 is also referred to as the nonlinear coefficient in m/V.

The d_{il}'s for the following materials are given as

$$\begin{array}{lll} \text{Quartz,} & d_{11} & = 0.96 \times 10^{-9}\ \text{esu,} \\ \text{Proustite,} & d_{22} & = 68 \times 10^{-9}\ \text{esu,} \\ \text{KDP,} & d_{36} & = 1.1 \times 10^{-9}\ \text{esu.} \end{array}$$

Calculate the corresponding nonlinear coefficient d_{il}/ϵ_0 for these materials in m/V.

d) The following materials have the d_{il} coefficient given in the unit of $(1/9) \times 10^{-22}\ \text{C/V}^2$.

$$\begin{array}{ll} \text{LiIO}_3, & d_{15} = 4.4, \\ \text{GaAs,} & d_{14} = 72, \\ \text{CdGeAs}_2, & d_{36} = 360. \end{array}$$

Calculate the corresponding d_{il} in esu and d_{il}/ϵ_0 in m/V.

7.2 Find the d matrix for a crystal with $\bar{4}2m$, $\bar{4}3\text{m}$ and 3m symmetry.

8
Wave Propagation in Nonlinear Media

8.1
Nonlinear Wave Equation

For simplicity, let us use optical second-harmonic generation as an example. We have seen from the preceding chapter that when a nonlinear optical medium is illuminated by a strong laser beam, an induced polarization at a frequency twice the laser frequency is created:

$$\bar{P}^{NL}(2\omega,\bar{r},t) = \chi^{(2)}(2\omega)\,\bar{E}(\omega,r)\,\bar{E}(\omega,r)e^{i2\omega t} + c.c. \tag{8.1}$$

Since $\bar{E}(\omega,r,t) = E(\omega)e^{i(\omega t-\bar{k}.\bar{r})}$ is a propagating wave, the $\bar{P}^{NL}$ term induced by this nonlinear interaction is also a propagating wave. It propagates collinearly with the fundamental beam with propagation constant $2k$. This $\bar{P}^{NL}$ serves as the driving term (source) for generating second-harmonic waves according to the nonlinear wave equation

$$\begin{aligned}\nabla\times(\nabla\times\bar{E}(2\omega,\bar{r},t)) &+ \frac{\eta A}{c}\frac{\partial\bar{E}(2\omega,\bar{r},t)}{\partial t}\\ &+\frac{\eta^2}{c^2}\frac{\partial^2\bar{E}(2\omega,r,t)}{\partial t^2} = -\mu_0\frac{\partial^2\bar{P}^{NL}}{\partial t^2},\end{aligned} \tag{8.2}$$

where

$$\bar{E}(2\omega,r,t) = \frac{1}{2}\bar{E}(2\omega)e^{i(2\omega t-\bar{k}_2\cdot\bar{r})} + c.c. \tag{8.3}$$

is the second-harmonic field generated by this nonlinear process. For simplicity, let us assume that the light wave is propagating along the z direction, and further assume that the medium is transparent to ω and 2ω waves, e.g., $A = 0$. Figure 8.1 shows the collinear second-harmonic generation process.

For a uniform plane wave approximation, we have

$$\nabla\times(\nabla\times\bar{E}) = -\nabla^2\bar{E} = -\frac{d^2}{dz^2}\,\bar{E}(2\omega,z),$$

where

$$\bar{E}(2\omega,z) = \frac{1}{2}E_{2\omega(z)}e^{i(2\omega t-k_2z)} + c.c.$$

Light-Matter Interaction: Atoms and Molecules in External Fields and Nonlinear Optics.
W. T. Hill and C. H. Lee

ISBN: 978-3-527-40661-6

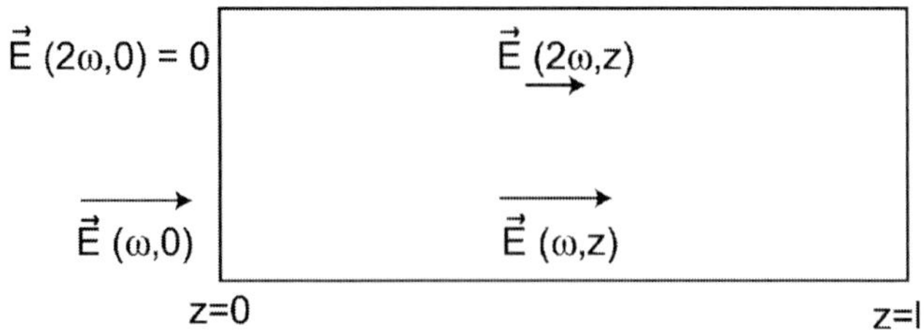

Fig. 8.1 Second-harmonic wave generated by laser beam in a nonlinear crystal.

In the small-signal approximation, the fundamental field does not depreciate, but the second-harmonic field will grow; we have

$$-\frac{d^2}{dz^2}\bar{E}(2\omega,z) = \frac{-1}{2}\left[\left(\frac{d^2\bar{E}_{2\omega}(z)}{dz^2} - ik_2\frac{d\bar{E}_{2\omega}(z)}{dz}\right) e^{-ik_2z} \right. \\ \left. -ik_2\left(\frac{d\bar{E}_{2\omega}(z)}{dz} - ik_2\,\bar{E}_{2\omega}(z)\right) e^{-ik_2z}\right]. \tag{8.4}$$

For a slowly varying amplitude approximation, we have

$$\frac{d^2E_{2w}(z)}{dz^2} \ll k_2\,\frac{dE_{2w}(z)}{dz}.$$

The nonlinear wave equation becomes

$$\frac{-1}{2}\left[-i2k_2\frac{dE_{2\omega}(z)}{dz} - k_2^2\,E_{2\omega}(z) + \frac{n_{2\omega}^2}{c^2}\,4\omega^2\,E_{2\omega}(z)\right] e^{i(2\omega t-k_2z)} \\ = \frac{1}{2}\left(4\mu_0\omega^2\,\tilde{P}(2\omega)\right)\,e^{i(2\omega t-2kz)}. \tag{8.5}$$

Using the relations

$$\omega = \frac{kc}{n_\omega},\; 2\omega = \frac{k_2c}{n_{2\omega}},$$

we can substitute

$$k_2^2 = \frac{n_{2\omega}^2}{c^2} \times 4\omega^2$$

and realize that the second and third terms on the left-hand side of the equation cancel each other. We finally have

$$\frac{dE_{2\omega}(z)}{dz} = \frac{4\mu_0\omega^2\tilde{P}(2\omega)}{i2k_2}\,e^{i(k_2-2k)z}, \tag{8.6}$$

which governs the growth of the $\vec{E}_{2\omega}$ field as a function of z. Introducing

$$\Delta k = k_2 - 2k,$$

the growth of the second-harmonic field can be obtained by integrating Eq. (8.6) with respect to z over the length of the nonlinear medium,

$$\begin{aligned} E_{2\omega}(z = \ell) &= \int_0^\ell \left[-i \frac{2\mu_0 \omega^2 \tilde{P}(2\omega)}{k_2} \right] \mathrm{e}^{i\Delta k z} dz \\ &= \frac{2\mu_0 \omega^2 \tilde{P}(2\omega)}{k_2 \Delta k} \left(1 - \mathrm{e}^{i\Delta k \ell)}\right) \\ &= A \left(1 - \mathrm{e}^{i\Delta k \ell}\right), \end{aligned} \tag{8.7}$$

$$\text{where } A = \frac{2\mu_0 w^2 \tilde{P}(2w)}{k_2 \Delta k}. \tag{8.8}$$

The intensity of the second-harmonic wave at the exit of the nonlinear crystal is given by

$$\begin{aligned} I_{2\omega}(\ell) &= \frac{1}{2} n_{2\omega}\, c\, \epsilon_0 \mid E_{2\omega}(\ell) \mid^2 \\ &= \frac{1}{2} n_{2\omega} c \epsilon_0 \mid A \mid^2 \left(1 - \mathrm{e}^{i\Delta k \ell}\right) \left(1 - \mathrm{e}^{-i\Delta k \ell}\right) \\ &= \frac{2 n_{2\omega} c \epsilon_0 \mu_0^2 \omega^4 \chi^2(2\omega) \ell^2 E_0^4(\omega)}{k_0^2} \frac{\sin^2 \frac{\Delta k \ell}{2}}{(\frac{\Delta k \ell}{2})^2}. \end{aligned} \tag{8.9}$$

In general, one can define a coherence length, ℓ_c, for second-harmonic generation as

$$\sin \frac{\Delta k \ell_c}{2} = 1,$$

$$\begin{aligned} \frac{\Delta k \ell_c}{2} &= \frac{\pi}{2} \text{ or } \ell_c = \frac{\pi}{\Delta k} = \frac{\pi}{k_2 - 2k} \\ \ell_c &= \frac{\lambda}{4\,(n_{2\omega} - n_\omega)}. \end{aligned} \tag{8.10}$$

In a nonlinear crystal with normal dispersion one usually has $n_{2\omega} \neq n_\omega$. The maximum second-harmonic signal one can expect to obtain from such a crystal is when the nonlinear crystal length ℓ is such that

$$\sin \frac{\Delta k \ell}{2} = \sin(2m + 1)\frac{\pi}{2},$$

or $\ell = (2m+1)\ell_c$. In this situation the maximum harmonic intensity one can expect is

$$I_{max}(2\omega, \ell) = \frac{2n_{2\omega}c\epsilon_0\mu_0^2\omega^4\chi^2(2\omega)E_0^4(\omega)}{k_0^2\left(\frac{\Delta k}{2}\right)^2}. \tag{8.11}$$

The harmonic intensity is limited by the k-vector mismatch. Since $p = \hbar k$ represents the photon momentum, the condition $\Delta k\hbar = \hbar k_2 - 2\hbar k$ is also referred to as the momentum mismatch. If one is clever, one may choose a nonlinear medium in which one has $n_{2\omega} = n_\omega$. In this case $\ell_c = \infty$ or $\Delta k = 0$. This is the situation of momentum matching or phase matching. Since with phase matching the indices of the fundamental and second-harmonic waves must be equal, it is also referred to as index matching in the literature. Under this condition the phase velocity of the fundamental and second harmonic are equal; the two light waves always propagate in phase. Under the phase-matching condition

$$\frac{\sin^2\frac{\Delta k\ell}{2}}{\left(\frac{\Delta k\ell}{2}\right)^2} \to 1,$$

$I_{2\omega}(\ell) \propto \ell^2$. The longer the nonlinear crystal, the stronger the second-harmonic intensity. In practice, the second-harmonic intensity will always be less than the intensity of the fundamental wave. As the second-harmonic intensity grows stronger, the small-signal approximation breaks down. In reality, when the second-harmonic intensity becomes comparable to the fundamental, it can act as the pump field, generating a new wave at the fundamental frequency by the nonlinear process

$$P^{NL}(\omega, 2\omega, -\omega) = \chi^{(2)}(\omega, 2\omega, -\omega)E(2\omega)E^*(\omega). \tag{8.12}$$

In this case, full coupled-wave equations between the ω and 2ω waves must be solved. If a proper phasing condition is maintained, it is possible to convert all fundamental energy into second harmonics [68].

8.2 Phase Matching in Second-Harmonic Generation

In a crystal with normal dispersion, the indices of refraction of the fundamental wave and the second harmonic are always different, and therefore phase matching is not possible. However, in a birefringent nonlinear crystal the index of refraction of the ordinary ray (o-ray) and that of the extraordinary ray (e-ray) are different at a given wavelength. With a clever arrangement it may

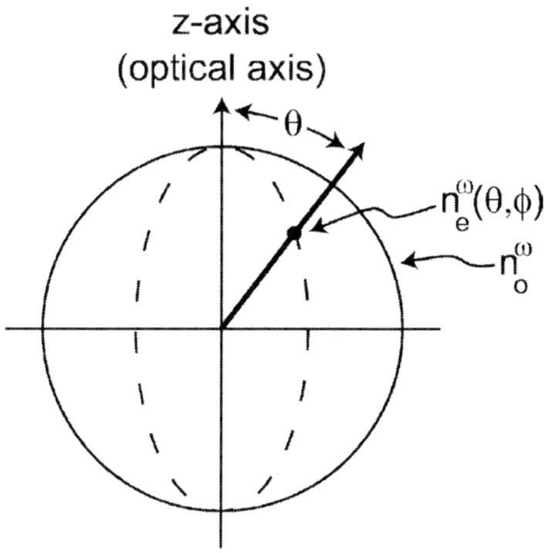

Fig. 8.2 Normal (index) surfaces for the ordinary and extraordinary rays in a negative $(n_e < n_o)$ uniaxial crystal.

be possible to obtain $n_{2\omega} = n_\omega$ by forcing the fundamental wave to propagate as an o-ray and the second harmonic as the e-ray or vice versa, depending on whether the crystal is positive uniaxial or negative uniaxial. This is the usual trick nonlinear optical engineers utilize to achieve index matching. In the early days researchers in nonlinear optics had to know the crystal optics so that they could figure out how to manipulate the crystal orientation, laser polarization, etc., to achieve index matching. Nowadays, vendors of nonlinear optical materials can usually provide a crystal that is already cut for the index-matching configuration. A researcher after receiving the nonlinear crystal is merely required to follow the instructions of the crystal supplier to perform the phase-matched frequency-doubling process. In this text we only show one example of how to achieve phase matching in potassium dihydrogen phosphate (KDP). Many new nonlinear crystals have been grown with various properties for a variety of different applications, and these crystals and their properties can readily be found via various vendors' online resources.[1] Researchers or graduate students, however, should know the fundamental principle of phase matching as illustrated in the following example.

8.2.1
Phase Matching of Second-Harmonic Generation in KDP

Potassium dihydrogen phosphate (KDP) is a negative uniaxial crystal; the normal (index) surface (note that it is not the index ellipsoidal surface) is shown in Fig. 8.2. A normal (index) surface reads the refractive-index value along the

1) See for example, Cleveland Crystal Inc., www.clevelandcrystals.com.

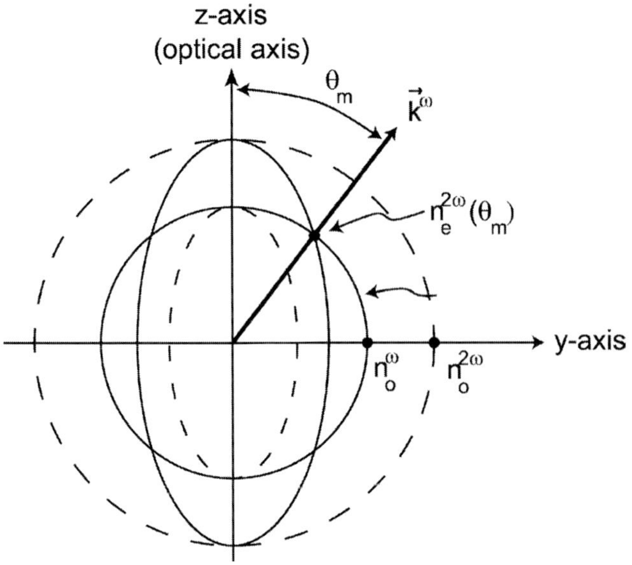

Fig. 8.3 Using the normal surface to illustrate the phase-matching condition for a negative uniaxial crystal, such as KDP. The fundamental wave is propagating as an o-ray and the second harmonic as an e-ray. At the phase-matching angle θ_m, $n_e^{2w}(\theta_m) = n_o^w$.

direction of the $\bar{k}$ vector for the e- and o-rays. This is much more convenient to use in illustrating the index-matching process. Figure 8.3 shows the index of the e- and o-rays of KDP as a function of frequency. According to normal dispersion, the index of the higher-frequency light wave is higher than that of the lower frequency. The index surface at 2ω is larger than that at ω. For a negative uniaxial crystal, the e-ray surface is within the o-ray surface. Therefore, if the 2ω wave propagates as an e-ray and the ω wave as an o-ray their normal (index) surfaces will intersect at an angle θ_m. This means that the fundamental laser beam should propagate along the θ_m direction, the phase-matching direction, as an o-ray, i.e., with its electric field polarized perpendicular to both the optical axis (z-axis) and the $\bar{k}$ vector. The generated second-harmonic wave will propagate along exactly the same direction, $\bar{k}^{2\omega}$, but polarized as an e-ray. This configuration is called collinear phase matching.

In order to find the phase-matching angle θ_m, one needs to use the equation for the index ellipsoid of KDP (negative uniaxial crystal) and thus go back to the index of an ellipsoidal surface

$$\frac{x^2}{n_0^2} + \frac{y^2}{n_0^2} + \frac{z^2}{n_e^2} = 1. \tag{8.13}$$

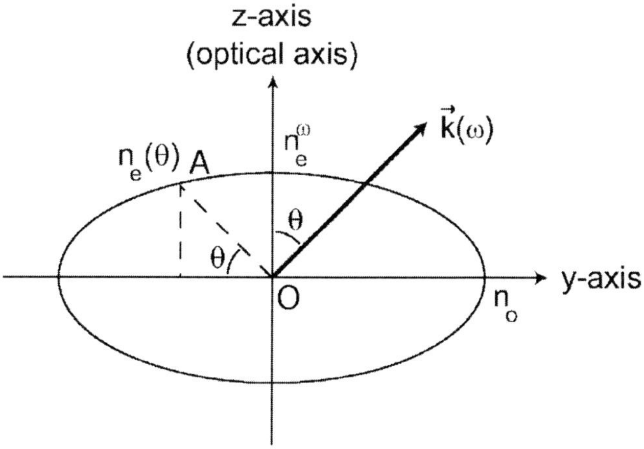

Fig. 8.4 Index ellipsoidal surface of KDP, $n_e < n_o$, at the $x = 0$ plane.

Figure 8.4 represents the index ellipsoid of KDP at the $x = 0$ plane. We have

$$\begin{aligned} | 0A | = n_e(\theta), \quad z &= | 0A | \sin\theta = n_e(\theta)\sin\theta, \\ y &= | 0A | \cos\theta = n_e(\theta)\cos\theta, \end{aligned}$$

$$\frac{n_e^2(\theta)\cos^2\theta}{n_0^2} + \frac{n_e^2(\theta)\sin^2\theta}{n_e^2} = 1, \tag{8.14}$$

$$\frac{1}{n_e^2(\theta)} = \frac{\cos^2\theta}{n_0^2} + \frac{\sin^2\theta}{n_e^2}.$$

At the phase-matching angle θ_m, we have

$$n_0^{\omega} = n_e^{2\omega}(\theta_m)$$

$$\frac{1}{\left(n_0^{\omega}\right)^2} = \frac{1}{\left[n_e^{2\omega}(\theta_m)\right]^2} = \frac{\cos^2\theta_m}{\left(n_0^{2\omega}\right)^2} + \frac{\sin^2\theta_m}{\left(n_e^{2\omega}\right)^2}$$

and, solving for θ_m,

$$\sin^2\theta_m = \frac{\dfrac{1}{\left(n_0^{\omega}\right)^2} - \dfrac{1}{\left(n_0^{2\omega}\right)^2}}{\dfrac{1}{\left(n_e^{2\omega}\right)^2} - \dfrac{1}{\left(n_0^{2\omega}\right)^2}}. \tag{8.15}$$

The calculated phase-matching angle, θ_m, means that the frequency-doubling process will have $\Delta k = 0$ when the laser is propagating along this direction. From Eq. (8.9),

$$I_{2\omega}(\ell) = \frac{2n_{2\omega}c\,\epsilon_0\mu_0^2\omega^4\chi^2\ell^2\,E_0^4(\omega)}{k_0^2} \propto \chi^2 E_0^4(\omega)\ell^2. \tag{8.16}$$

Under this condition the coherence length approaches infinity. $I_{2\omega}$ is limited, among other things, by the length of the nonlinear crystal. Actually, the above equation is an oversimplified case since we have neglected the crystal orientation and the polarization of the optical field. To analyze the frequency-doubling process properly one needs to take these into account. We have for KDP

$$\begin{pmatrix} P_x^{NL}(2\omega) \\ p_y^{NL}(2\omega) \\ P_z^{NL}(2\omega) \end{pmatrix} = \begin{pmatrix} 0 & 0 & 0 & d_{14} & 0 & 0 \\ 0 & 0 & 0 & 0 & d_{14} & 0 \\ 0 & 0 & 0 & 0 & 0 & d_{36} \end{pmatrix} \begin{pmatrix} E_x^2(\omega) \\ E_y^2(\omega) \\ E_z^2(\omega) \\ 2E_y(\omega)E_z(\omega) \\ 2E_x(\omega)E_z(\omega) \\ 2E_x(\omega)E_y(\omega) \end{pmatrix}, \quad (8.17)$$

where the d coefficient instead of χ is used. In general,

$$d_{ijk} = \frac{1}{2}\chi_{ijk}^{(2)}$$

and, furthermore, using the contracted notation for the third-rank tensor, we have $d_{14} = d_{123} = d_{132}, d_{36} = d_{321} = d_{312}$, etc. The optical fields are expressed as the component fields along the crystallographic axes, x, y and z:

$$\begin{aligned} P_x^{NL}(2\omega) &= 2d_{14}E_y(\omega)E_z(\omega) \\ P_y^{NL}(2\omega) &= 2d_{14}E_x(\omega)E_z(\omega) \\ P_z^{NL}(2\omega) &= 2d_{36}E_x(\omega)E_y(\omega). \end{aligned} \quad (8.18)$$

The phase matching for this negative uniaxial crystal requires that the fundamental laser beam must propagate as an o-ray, i.e., $\bar{E}_L \perp$ z-axis, or $E_z(\omega) = 0$. The only nonzero $P^{NL}(2\omega)$ is along the z-axis, or

$$P_z^{NL}(2\omega) = 2d_{36}\, E_x(\omega)\, E_y(\omega).$$

For $\vec{k}(\omega)$ in the (θ, ϕ) direction, $\bar{E}_L(\omega)$, which must be perpendicular to both the z-axis and $\bar{k}(\omega)$, lies in the x–y plane only; from Fig. 8.5

$$\begin{aligned} E_x(\omega) &= E_L \sin\phi \\ E_y(\omega) &= -E_L \cos\phi, \end{aligned}$$

$$P_z^{NL}(2\omega) = -2d_{36}E_L^2 \sin\phi\cos\phi. \quad (8.19)$$

The second-harmonic polarization has a z component; thus, it will create an optical field that propagates as an e-ray. When the second-harmonic field radiates, only the component perpendicular to the direction of propagation matters, i.e., the effective $\bar{P}^{NL}$ is that perpendicular to $\vec{k}\ (2w)$, which is collinear

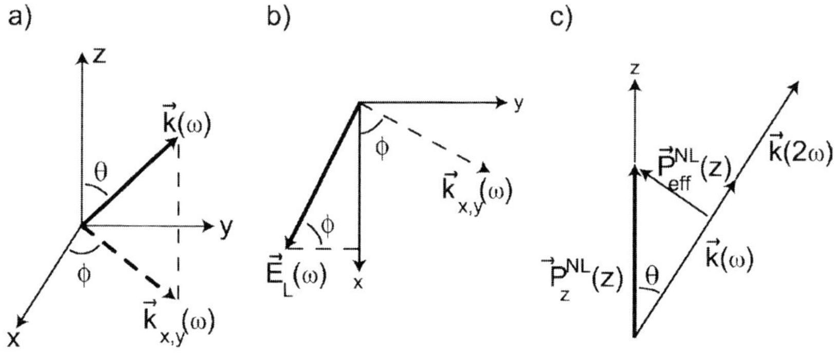

Fig. 8.5 Relationship of (a) $\vec{k}\ (\omega)$, (b) $\vec{E}_L\ (\omega)$ and (c) $p^{NL}_{eff}(2\omega)$ and $\vec{k}\ (2\omega)$ with both the fundamental and the second-harmonic waves propagating along the (θ, ϕ) direction.

with $\vec{k}\ (w)$:

$$\begin{aligned} P^{NL}_{\perp}(2w) &= P^{NL}_{eff}(2w) \\ &= P^{NL}_{z}(2w)\sin\theta \\ &= -2d_{36}\, E^2_L(w)\sin\theta\sin\phi\cos\phi \\ &= d_{eff}E^2_L = -d_{36}\sin\theta\sin 2\phi E^2_L, \end{aligned}$$

$$d_{eff} = -d_{36}\sin\theta\sin 2\phi. \tag{8.20}$$

For the phase-matching case $\theta = \theta_m$ one still needs to optimize d_{eff}, leading to $\phi = 45^\circ$ for this simple example. Alternatively, the d_{eff} formula suggests that the polarization of the E_L field not only should lie in the x–y plane to qualify as an o-ray, it should not be polarized along the x- or y-axis. In other words, $P^{NL}_z(2\omega)$ will be maximum when $\phi = 45^\circ$ or when $|E_x| = |E_y| = (1/\sqrt{2})E_L$. This simple fact suggests that one still needs to know what one is doing by orientating the crystal (by rotating it with respect to the crystal axis) and/or polarizing the laser beam correctly.

The phase-matching example just described is the manifestation of the energy and momentum conservation condition for the fundamental and second-harmonic photons:

$$\begin{aligned} \hbar\omega + \hbar\omega &= \hbar(2\omega) \\ \hbar\,\vec{k}_L + \hbar\,\vec{k}_L &= \hbar\,\vec{k}\ (2\omega). \end{aligned} \tag{8.21}$$

In this example, two fundamental photons are both polarized as an o-ray. This is called type I phase matching. In a different case, two fundamental photons are polarized orthogonal to each other, i.e., one o-ray and one e-ray. This is known as type II phase matching. The d_{eff}'s for various crystal classes for both type I and type II phase matching have been previously calculated and they can be found in Ref. [76].

8.2.2 Noncollinear Momentum Matching

The index-matching example described in the previous section is a special case of momentum matching when the fundamental and second-harmonic beams are propagating collinearly. To describe the general concept of momentum matching it is more instructive to use the $\vec{k}$ representation. One can modify the index-matching diagram using a normal (index) surface by a $\vec{k}$ vector diagram in the collinear case first:

$$k_\omega = \frac{\omega}{c} n_\omega$$
$$k_{2\omega} = \frac{2\omega}{c} n_{2\omega}.$$

Under the index-matching condition $n_\omega = n_{2\omega}$, we have

$$2 \mid k_\omega \mid \; = \; \mid k_{2\omega} \mid$$

$$\text{or} \quad \vec{k}(\omega) + \vec{k}(\omega) = \vec{k}(2\omega),$$

$$\Delta \vec{k} = \vec{k}(2\omega) - \vec{k}(\omega) - \vec{k}(\omega) = 0.$$

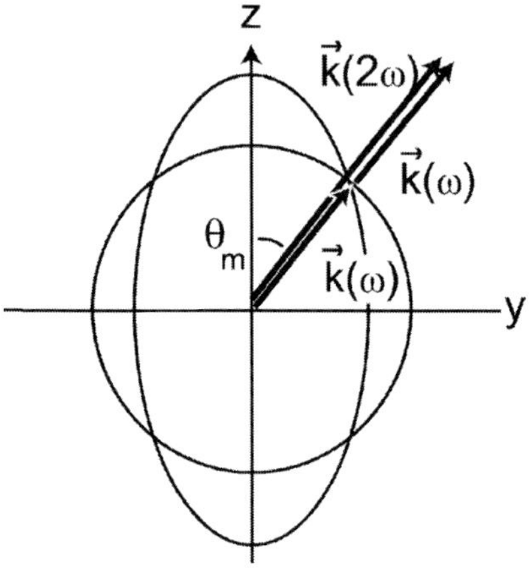

Fig. 8.6 Momentum matching in the collinear case.

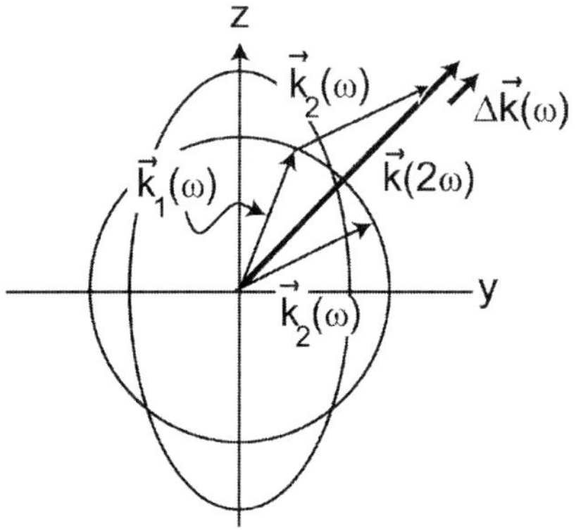

Fig. 8.7 Noncollinear propagating fundamental waves showing momentum mismatch with $\Delta \vec{k} \neq 0$.

This is depicted in Fig. 8.6. When the two $\vec{k}(\omega)$'s are not collinear but make a small angle with respect to each other (Fig. 8.7),

$$\vec{k}_1(\omega) + \vec{k}_2(\omega) + \Delta\vec{k} = \vec{k}(2\omega) \text{ with } \Delta\vec{k} \neq 0.$$

It may be possible, as shown in Fig. 8.8, that $\vec{k}_1(\omega) + \vec{k}_2(\omega) = \vec{k}(2\omega)$ as in a true noncollinear situation. As long as the triangle closes, one has a momentum-matching condition. Although the noncollinear momentum-matching condition satisfies mathematically the momentum-matching condition, the process in general is less efficient in generating second harmonics, particularly when focused beams are used. In this case, all beams cannot remain overlapped over a long propagating distance, resulting in the "walk-

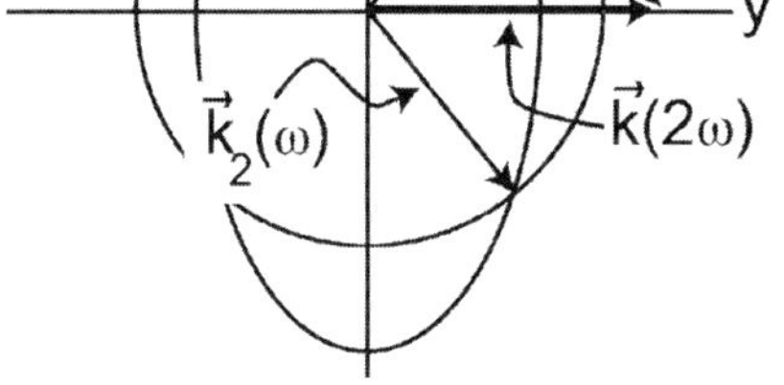

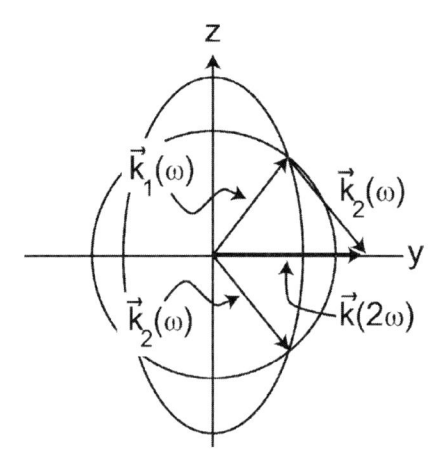

Fig. 8.8 One example of noncollinear momentum matching.

off" effect where the fundamental and the second harmonic beams walk off from each other. In other words, the interaction length between the various beams becomes a limiting factor, instead of the physical length of the nonlinear crystal. However, nonlinear momentum-matching configurations have been used for ultrashort optical pulse measurement, since in this configuration the second-harmonic and fundamental beams are spatially separated, resulting in a background-free second-harmonic intensity autocorrelator [77].

8.2.3
Experimental Arrangement for Phase-Matched Second-Harmonic Generation

From the discussion in the previous section, at phase matching, the intensity of the generated second harmonic $I_{2\omega}$ is proportional to

$$I_{2\omega} \propto d_{eff}^2 \, I_L^2(\omega)\ell^2. \tag{8.22}$$

The goal of phase-matching SHG is to convert maximum laser light to that of the SH. To achieve this, the SHG is done in a transmission arrangement by choosing an appropriate optically polished nonlinear crystal [78]. The crystal is cut and optically polished to ensure maximum transmission of light. The orientation of the crystal is such that the crystal axis (the direction perpendicular to the end faces) is along the phase-matching direction, and that the z-axis of the crystal makes an angle θ_m with respect to the direction perpendicular to the end faces. As mentioned earlier, nowadays these considerations are handled by the vendor. The user only needs to know what polarization the laser beam needs to be in order to satisfy the particular type of phase-matching condition (i.e., whether the laser should be polarized as an o-ray or an e-ray, etc.). Since the transmission arrangement is used, the crystal has to be transparent to both the fundamental and second-harmonic wavelengths. One then focuses the laser beam into the crystal as shown in Fig. 8.9.

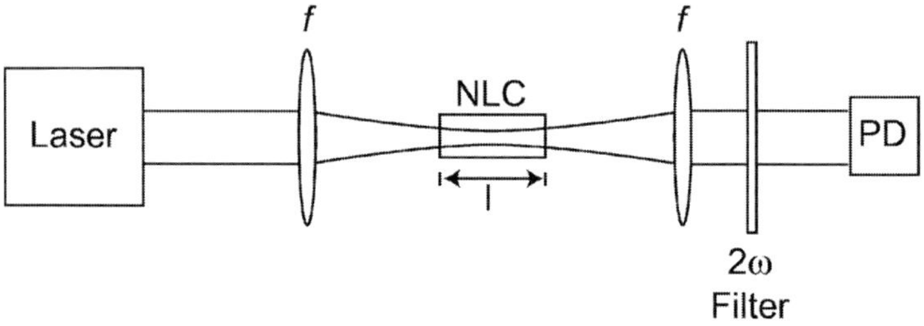

Fig. 8.9 Experimental arrangement for detecting phase-matched SHG. The laser beam is focused into the nonlinear crystal, NLC, of length ℓ, and recollimated by another lens, f. A second-harmonic, 2ω, filter blocks the fundamental and passes the second-harmonic light, which is detected by the photodetector, PD.

In the focused arrangement with a Gaussian beam, it is not always optimum to focus as tight as possible since this will limit the effective crystal length to the Rayleigh range of the lens, which may be much shorter than the length of the nonlinear crystal. Too tight a focus may also cause damage. Usually, the optimal situation is to match the Rayleigh range to the length of the crystal. The filter shown in Fig. 8.9 is used to block the fundamental but let through the second harmonic. In practice, the filter used is a high-pass step function type filter which passes all light with a wavelength shorter than a certain value but blocks all light longer than this wavelength. Since the second harmonic is in the pass band and the fundamental in the stop band of the filter, the filter provides a very effective means of separating the second harmonic from the fundamental wave. This type of pass-band filter usually is not expensive. Some colored glass filters may be used. Another way to discriminate the second harmonic from the fundamental beam is by spatial dispersion. For example, in the noncollinear phase-matching case the second harmonic is in the direction bisecting the two noncollinear fundamental beams. Another simple method to spatially separate the second harmonic from the fundamental is to let it pass through a dispersive prism. The second harmonic and fundamental beam will propagate along different directions after emerging from the prism.

8.3 Parametric Interaction

8.3.1 Coupled Equations for Parametric Interaction

We have seen how two photons at the fundamental frequency ω can be combined to generate a second-harmonic photon at 2ω frequency. In a more general sense, two photons at frequencies ω_1 and ω_2 can be combined to generate a sum-frequency photon at $\omega_1 + \omega_2$:

$$\hbar\omega_3 + \hbar\omega_2 \Longrightarrow \hbar(\omega_1 + \omega_2) = \hbar\omega_3. \tag{8.23}$$

The arrow in the above equation is pointing from $\omega_1 + \omega_2$ toward ω_3. Nonlinear optical processes can occur in the reverse direction, namely, a higher-frequency photon at ω_3 can be broken up into photons at ω_1 and ω_2, respectively. In other words, the arrow in the above equation is pointing as follows:

$$\hbar\omega_3 \Longrightarrow \hbar\omega_1 + \hbar\omega_2. \tag{8.24}$$

This is called parametric interaction. The nonlinear polarization for the sum frequency generation process is

$$P_i^{NL}(\omega_3 = \omega_1 + \omega_2) = 2d_{ijk}(\omega_3 = \omega_1 + \omega_2)E_j(\omega_1)E_k(\omega_2), \tag{8.25}$$

where

$$\begin{aligned} E_j(\omega_1, z, t) &= \frac{1}{2}\left[E_{1j}(z)e^{i(\omega_1 t - k_1 z)} + c.c.\right], \\ E_k(\omega_2, z, t) &= \frac{1}{2}\left[E_{2k}(z)e^{i(\omega_2 t - k_2 z)} + c.c.\right], \\ E_i(\omega_3, z, t) &= \frac{1}{2}\left[E_{3_i}(z)e^{i(\omega_3 t - k_3 z)} + c.c.\right]. \end{aligned} \tag{8.26}$$

One can write down the equation for the parametric process, $P_j^{NL}(\omega_1 = \omega_3 - \omega_2)$ and $P_k^{NL}(\omega_2 = \omega_3 - \omega_1)$, in a similar manner:

$$P_k^{NL}(\omega_2 = \omega_3 - \omega_1) = 2d_{kij}E_i(\omega_3)E_j^*(\omega_1), \tag{8.27}$$

$$\text{where } E_j^*(\omega_1, z, t) = \frac{1}{2}\left[E_{1j}^* e^{-i(\omega_1 t - k_1 z)} + c.c.\right], \tag{8.28}$$

and where $\omega_3 = \omega_1 + \omega_2$ is assumed. The wave equation is

$$-\nabla^2 \vec{E} + \frac{nA}{c}\frac{\partial \vec{E}}{\partial t} + \frac{n^2}{c^2}\frac{\partial^2 \vec{E}}{\partial t^2} = -\mu_0 \frac{\partial^2 \vec{P}^{NL}}{\partial t^2}, \tag{8.29}$$

where n is the refractive index; $\frac{nA}{c} = \mu_0\sigma$, σ being the conductivity of the medium. We have chosen the light wave at various frequencies to propagate along the z direction and also assumed that there is no x and y variation in the field expression, i.e., we still deal with infinite plane waves for the purpose of keeping the mathematics simple. The wave equation under this special situation reduces to

$$-\frac{d^2\vec{E}}{dz^2} + \frac{nA}{c}\frac{\partial \vec{E}}{\partial t} + \frac{n^2}{c^2}\frac{\partial^2 \vec{E}}{\partial t^2} = -\mu_0 \frac{\partial^2 \vec{P}^{NL}}{\partial t^2}. \tag{8.30}$$

The field and polarization in the nonlinear wave equation can represent the total field or polarization

$$\vec{E} = \vec{E}_1(\omega_1, z, t) + \vec{E}_2(\omega_2, z, t) + \vec{E}_3(\omega_3, z, t),$$

$$\vec{P}^{NL} = \vec{P}_1^{NL}(\omega_1, z, t) + \vec{P}_2^{NL}(\omega_2, z, t) + \vec{P}_3^{NL}(\omega_3, z, t),$$

or the quantity at an individual frequency. Let us examine the ith component of the E field at ω_3. It satisfies the nonlinear Maxwell equation

$$-\frac{d^2E_i(\omega_3, z, t)}{dz^2} + \frac{nA}{c}\frac{\partial E_i(\omega_3, z, t)}{\partial t} + \frac{n^2}{c^2}\frac{\partial^2 E_i(\omega_3, z, t)}{\partial t^2} = -\mu_0 \frac{\partial^2 P_i^{NL}(\omega_3, z, t)}{\partial t^2}. \tag{8.31}$$

Differentiating the first term twice and assuming the slowly varying envelope approximation,

$$k_3 \frac{dE_{3i}(z)}{dz} = \frac{2\pi}{\lambda_3} \frac{dE_{3i}(z)}{dz} \gg \frac{d^2E_{3i}(z)}{dz^2}, \tag{8.32}$$

i.e., the change in the curvature of the amplitude as a function of z is much less than the derivative over a distance of a wavelength. One can drop the second-derivative term; the nonlinear Maxwell equation then becomes

$$\begin{aligned} 2ik_3 \frac{dE_{3i}(z)}{dz} &+ k_3^2 E_{3i}(z) + \frac{n_3 A}{c} i\omega_3 E_{3i}(z) - \frac{\omega_3^2 n_3^2}{c^2} E_{3i}(z) \\ &= -\mu_0 d_{ijk}(\omega_3 = \omega_1 + \omega_2)(-\omega_3^2) E_{1j}(z) E_{2k}(z) \mathrm{e}^{i[k_3-(k_1+k_2)]z}. \end{aligned} \tag{8.33}$$

Dividing the equation by $2ik_3$ and realizing that

$$\frac{\omega_3}{k_3} = \frac{c}{n_3},$$

the equation becomes

$$\frac{dE_{3i}(z)}{dz} = -\frac{1}{2} A E_{3i}(z) - \frac{\omega_3 i c \mu_0}{n_3} \frac{1}{2} d_{ijk}\, E_{1j}(z) E_{2k}(z)\, \mathrm{e}^{i[k_3-(k_1+k_2)]z},$$

since

$$\frac{nA}{c} = \mu_0 \sigma,$$

or

$$A = \frac{c\mu_0}{n}\sigma = \sqrt{\frac{\epsilon_0}{\epsilon}}\sqrt{\frac{\mu_0}{\epsilon_0}} = \sqrt{\frac{\mu_0}{\epsilon}}\,\sigma.$$

Finally, we can express the equation as

$$\begin{aligned} \frac{dE_{3i}(z)}{dz} &= -\frac{1}{2}\sqrt{\frac{\mu_0}{\epsilon}}\,\sigma_3 E_{3i}(z) - \frac{i\omega_3}{2}\sqrt{\frac{\mu_0}{\epsilon_3}}\, d_{ijk}(\omega_3 = \omega_1 + \omega_2) \\ &\times E_{1j}(z) E_{2k}(z)\, \mathrm{e}^{i[k_3-(k_1+k_2)]z}. \end{aligned} \tag{8.34}$$

Similarly, one can obtain

$$\begin{aligned} \frac{dE_{1j}(z)}{dz} &= -\frac{1}{2}\sqrt{\frac{\mu_0}{\epsilon_1}}\,\sigma_1 E_{1j}(z) - \frac{i\omega_1}{2}\sqrt{\frac{\mu_0}{\epsilon_1}}\, d_{jik}(\omega_1 = \omega_3 - \omega_2) \\ &\times E_{3i}(z) E_{2k}^*(z)\, \mathrm{e}^{i[k_1-(k_3-k_2)]z}, \end{aligned} \tag{8.35}$$

$$\begin{aligned} \frac{dE_{2k}(z)}{dz} &= -\frac{1}{2}\sqrt{\frac{\mu_0}{\epsilon_1}}\,\sigma_2 E_{2k}(z) - \frac{i\omega_2}{2}\sqrt{\frac{\mu_0}{\epsilon_2}} d_{kij}(\omega_2 = \omega_3 - \omega_1) \\ &\times E_{3i}(z) E_{1j}^*(z)\, \mathrm{e}^{i[k_2-(k_3-k_1)]z} \end{aligned} \tag{8.36}$$

or, if one takes the complex conjugate of the above equation,

$$\frac{dE_{2k}^*(z)}{dz} = -\frac{1}{2}\sqrt{\frac{\mu_0}{\epsilon_2}}\,\sigma_2 E_{2k}^*(z) + \frac{i\omega_2}{2}\sqrt{\frac{\mu_0}{\epsilon_2}}\,d_{kij}(\omega_2 = \omega_3 - \omega_1) \times E_{3i}^*(z)E_{1j}(z)\ \mathrm{e}^{i[-k_2+(k_3-k_1)]z}. \tag{8.37}$$

Defining $\Delta k = k_3 - (k_1 + k_2)$ and also noting that the symmetry of the d coefficient satisfies

$$\begin{aligned} d_{ijk}(\omega_3, \omega_1, \omega_2) &= d_{jik}(\omega_1, \omega_3 - \omega_2) \\ &= d, \end{aligned}$$

etc., we finally have a set of coupled wave equations:

$$\frac{dE_{3i}(z)}{dz} = -\frac{1}{2}\sqrt{\frac{\mu_0}{\epsilon_3}}\,\sigma_3 E_{3i}(z) - \frac{i\omega_3}{2}\sqrt{\frac{\mu_0}{\epsilon_3}}\,dE_{1j}(z)E_{2k}(z)\,\mathrm{e}^{i\Delta kz}, \tag{8.38}$$

$$\frac{dE_{1j}(z)}{dz} = -\frac{1}{2}\sqrt{\frac{\mu_0}{\epsilon_1}}\,\sigma_1 E_{1j}(z) - \frac{i\omega_1}{2}\sqrt{\frac{\mu_0}{\epsilon_1}}\,dE_{3i}(z)E_{2k}^*(z)\,\mathrm{e}^{-i\Delta kz}, \tag{8.39}$$

$$\frac{dE_{2k}^*(z)}{dz} = -\frac{1}{2}\sqrt{\frac{\mu_0}{\epsilon_2}}\,\sigma_2 E_{2k}^*(z) + \frac{i\omega_2}{2}\sqrt{\frac{\mu_0}{\epsilon_2}}\,dE_{3i}^*(z)E_{1j}(z)\,\mathrm{e}^{i\Delta kz}. \tag{8.40}$$

This set of coupled wave equations governs the growth of the amplitude of the specific optical field. The first equation describes the sum-frequency generation. If we denote $\omega_3 = \omega_p$, the frequency of the pump wave, $\omega_1 = \omega_s$, the signal wave and $\omega_2 = \omega_i$, the idler wave, then the second and third equations describe the parametric process. It is instructive to express the intensity in terms of photon density from the relation

$$I(\omega) = \frac{1}{2} n_\omega\, c_0 \epsilon_0 \mid E(\omega) \mid^2 = \frac{1}{2} n_\omega \sqrt{\frac{\epsilon_0}{\mu_0}} \mid E(\omega) \mid^2 . \tag{8.41}$$

Let us introduce

$$A_\ell = \sqrt{\frac{\mu_\ell}{\omega_\ell}}\,E_\ell,$$

$$I_\ell = \frac{1}{2} n_\ell \sqrt{\frac{\epsilon_0}{\mu_0}\frac{\omega_\ell}{n_\ell}} \mid A_\ell \mid^2 = \frac{1}{2}\sqrt{\frac{\epsilon_0}{\mu_0}}\,\omega_\ell \mid A_\ell \mid^2 . \tag{8.42}$$

On the other hand, the intensity can also be expressed as the product of the photon density q_ℓ, $(\#/cm^3) \times \hbar\omega_\ell$ (energy per photon) $\times c$:

$$I_\ell = q_\ell\, \hbar\omega_\ell\, c = \frac{1}{2}\sqrt{\frac{\epsilon_0}{\mu_0}}\,\omega_\ell \mid A_\ell \mid^2$$

or

$$q_\ell = \frac{1}{\hbar c}\,\frac{1}{2}\sqrt{\frac{\epsilon_0}{\mu_0}}\mid A_\ell \mid^2 . \tag{8.43}$$

Substituting E's everywhere in the three coupled equations by A_ℓ's, the equation can be written as

$$\begin{aligned}\frac{dA_{3i}(z)}{dz} &= \frac{-1}{2}\sqrt{\frac{\mu_0}{\epsilon_3}}\sigma_3\sqrt{\frac{\omega_3}{n_3}}\,A_{3i}(z)\\ &\quad -\frac{i\omega_3}{2}\sqrt{\frac{\mu_0}{\epsilon_3}}d\sqrt{\frac{\omega_1}{n_1}}\sqrt{\frac{\omega_2}{n_2}}\,A_{1j}(z)A_{2k}(z)\;e^{i\Delta kz}.\end{aligned}$$

Substituting $\sqrt{\frac{1}{\epsilon_i}} = \frac{1}{n_i}\frac{1}{\sqrt{\epsilon_0}}$, $i = 1, 2$ or 3, one can write the equation as

$$\frac{dA_{3i}(z)}{dz} = \frac{-1}{2}\sqrt{\frac{\mu_0}{\epsilon_3}}\,\sigma_3 A_{3i}(z) - \frac{i}{2}\sqrt{\frac{\mu_0}{\epsilon_0}}d\sqrt{\frac{\omega_1\omega_2\omega_3}{n_1n_2n_3}}A_{1j}(z)A_{2k}(z)e^{i\Delta kz}. \tag{8.44}$$

Since the loss is

$$\alpha_i = \sqrt{\frac{\mu_0}{\epsilon_i}}\,\sigma_i, \tag{8.45}$$

and defining a coupling coefficient λ,

$$\lambda = \sqrt{\frac{\mu_0}{\epsilon_0}}d\sqrt{\frac{\omega_1\omega_2\omega_3}{n_1n_2n_3}}\,, \tag{8.46}$$

we finally arrive at

$$\frac{dA_{3i}(z)}{dz} = \frac{-1}{2}\alpha_3 A_{3i}(z) - \frac{i}{2}\lambda A_{1j}(z)A_{2k}(z)e^{i\Delta kz}. \tag{8.47}$$

Similarly,

$$\frac{dA_{1j}(z)}{dz} = -\frac{1}{2}\alpha_1 A_{1z}(z) - \frac{i}{2}\lambda A_{3i}(z)A^*_{2k}(z)e^{-i\Delta kz}, \tag{8.48}$$

$$\frac{dA^*_{2k}(z)}{dz} = -\frac{1}{2}\alpha_2 A^*_{2k}(z) + \frac{i}{2}\lambda A^*_{3i}(z)A_{1j}(z)e^{i\Delta kz}. \tag{8.49}$$

Equations (8.47)–(8.49) are referred to as the coupled wave equations (or simply as coupled equations). Note that there is only a single coupling coefficient λ involved in all equations. This set of three equations forms the basis for analyzing parametric interaction as will be discussed in the following section.

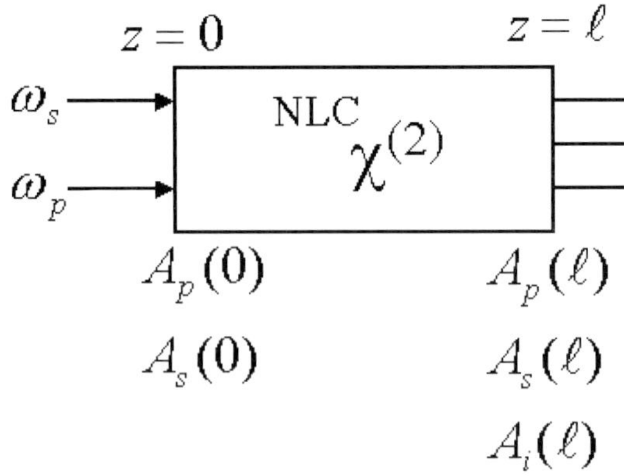

Fig. 8.10 Schematics of an optical parametric amplifier. $A_p(0) \approx A_p(\ell), A_s(\ell) \gg A_s(0)$.

8.3.2
Parametric Amplification

A single-pass parametric interaction may lead to the amplification of a signal wave when the nonlinear crystal is pumped by a strong pump beam. This is the basis for the optical parametric amplifier (OPA) [79]. The simple schematics of a parametric amplifier are shown in Fig. 8.10.

The nonlinear crystal cut correctly with proper orientation is pumped by a strong pump beam at $\omega_3 = \omega_p$, the signal wave is at frequency $\omega_1 = \omega_s$ and the idler is $\omega_2 = \omega_i$, where

$$\omega_p = \omega_s + \omega_i,$$

so that energy-conservation relation is satisfied. At the entrance face, we assume that $A_p(0), A_s(0)$ and $A_i(0)$ are known (in fact, either $A_s(0)$ or $A_i(0)$ may be zero). We further assume that the pump beam is not depleted. This means that $A_p(z) = A_p(0)$, independent of z. We thus only need to deal with two coupled equations. Let us further assume, for simplicity, that there is no loss at ω_s and ω_i, $\alpha_s = \alpha_i = 0$. We can also assume that the pump beam is a real field, $A_p(0) = A_p^*(0)$. Let us also assume that the polarizations of A_p, A_s and A_i are consistent with the crystallographic orientation and that we do not need to worry about them and that we can drop the i, j, k indices denoting Cartesian components. Then

$$\begin{aligned}\frac{dA_s(z)}{dz} &= -\frac{i}{2}\,\lambda\, A_p\, A_i^*(z)\mathrm{e}^{-i\Delta kz}\\ \frac{dA_i(z)}{dz} &= \frac{i}{2}\,\lambda\, A_p\, A_s(z)\mathrm{e}^{i\Delta kz}.\end{aligned} \tag{8.50}$$

Also, if we denote

$$\lambda A_p = g, \tag{8.51}$$

the equations become

$$\frac{dA_s(z)}{dz} = -\frac{i}{2}\, g\, A_i^*(z)\, \mathrm{e}^{-i\Delta kz}, \tag{8.52a}$$

$$\frac{dA_i^*(z)}{dz} = \frac{i}{2}\, g\ A_s\, \mathrm{e}^{i\Delta kz}. \tag{8.52b}$$

Differentiating Eq. (8.52a) with respect to z,

$$\frac{d^2A_s(z)}{dz^{(2)}} = -\frac{i}{2}\, g \left[\frac{dA_i^*(z)}{dz}\, \mathrm{e}^{-i\Delta kz} - i\Delta k A_i^*(z)\mathrm{e}^{-i\Delta kz}\right]$$

and, substituting $dA_i^*(z)/dz$ from Eq. (8.52b), we have

$$\frac{d^2A_s(z)}{dz^2} = \frac{1}{4}\, g^2\, A_s(z) - i\Delta k\left[\frac{-i}{2}\, g\, A_i^*(z)\mathrm{e}^{-i\Delta kz}\right].$$

Then, substituting the second term on the right-hand side of the above equation, we arrive at

$$\frac{d^2A_s(z)}{dz^2} + i\Delta k\, \frac{dA_s(z)}{dz} - \frac{g^2}{4}\, A_s(z) = 0. \tag{8.53}$$

The solution for $A_s(z)$ takes the following form:

$$A_s(z) = A\, \mathrm{e}^{\left[-\frac{i\Delta k}{2} + \frac{1}{2}\sqrt{g^2-\Delta k^2}\right]z} + B\, \mathrm{e}^{\left[-\frac{i\Delta k}{2} - \frac{1}{2}\sqrt{g^2-\Delta k^2}\right]z}. \tag{8.54}$$

Substituting the expression of $A_s(z)$ back into Eq. (8.52a), we can obtain the expression for $A_i^*(z)$:

$$\begin{aligned} A_i^*(z) = \frac{2i}{g} &\left[A\left(-\frac{i\Delta k}{2} + \frac{1}{2}\sqrt{g^2-\Delta k^2}\right)\mathrm{e}^{\left[\frac{i\Delta k}{2} + \frac{1}{2}\sqrt{g^2-\Delta k^2}\right]z}\right. \\ &\left. + B\left(-\frac{i\Delta k}{2} - \frac{1}{2}\sqrt{g^2-\Delta k^2}\right)\mathrm{e}^{\left(i\frac{i\Delta k}{2} - \frac{1}{2}\sqrt{g^2-\Delta k^2}\right)z}\right]. \end{aligned} \tag{8.55}$$

The undetermined coefficients A and B can be fixed by applying the boundary conditions.

A. Phase-Matching Case, $\Delta k = 0$

One of the important applications of the parametric nonlinear process is parametric amplification. A parametric amplifier is a nonlinear optical device that is designed to generate a maximum-intensity light wave at signal frequency.

Therefore, the nonlinear optical crystal is already cut and the experimental configuration has been set such that the phase-matching condition is satisfied:

$$\Delta k = k_p - (k_s - k_i) = 0.$$

For such a situation $\vec{k}_p, \vec{k}_s$ and $\vec{k}_i$ are collinear to have maximum interaction length. Therefore, the vector k relation becomes the scalar relation $k_p = k_s + k_i$. This leads to the solution of parametric amplification for the phase-matching case of $\Delta k = 0$:

$$A_s(0) = A + B,$$

$$A_i^*(0) - \frac{2i}{g}\left[A\frac{1}{2}g - B\frac{1}{2}g\right] = i(A - B),$$

when A and B are expressed as

$$A = \frac{1}{2}\left[A_s(0) - iA_i^*(0)\right],$$

$$B = \frac{1}{2}\left[A_s(0) + iA_i^*(0)\right],$$

and are substituted back into the expressions for $A_s(z)$ and $A_i^*(z)$ to give

$$\begin{aligned} A_s(z) &= A_s(0)\frac{1}{2}\left[e^{\frac{1}{2}gz} + e^{\frac{-1}{2}gz}\right] + iA_i^*(0)\frac{1}{2}\left[e^{\frac{-1}{2}gz} - e^{\frac{1}{2}gz}\right] \\ &= A_s(0)\cos h\frac{gz}{2} - iA_i^*(0)\sin h\frac{gz}{2}. \end{aligned} \tag{8.56}$$

Following a similar procedure, we can obtain

$$A_i^*(z) = iA_s(0)\sin h\frac{gz}{2} + A_i^*(0)\cos h\frac{gz}{2}. \tag{8.57}$$

B. Non-Phase-Matching Case $\Delta k \neq 0$

In the general case when there is no phase matching, i.e., $\Delta k \neq 0$, one can still obtain a mathematical expression for the solution of $A_s(z)$ and $A_i^*(z)$. Defining

$$b = \frac{1}{2}\sqrt{g^2 - \Delta k^2},$$

$$\begin{aligned} A_s(z)\, e^{\frac{i\Delta kz}{2}} &= A_s(0)\left[\cos h(bz) + \frac{i\Delta k}{2b}\sin h(bz)\right] \\ &\quad - i\frac{g}{2b}A_i^*(0)\sin h(bz) \end{aligned} \tag{8.58}$$

and

$$A_i^*(z)\,e^{\frac{-i\Delta kz}{2}} = A_i^*(0)\left[\cos h(bz) - \frac{i\Delta k}{2b}\sin h(bz)\right] + i\frac{g}{2b}A_s(0)\sin h(bz). \tag{8.59}$$

From the solution shown in Eqs. (8.56–8.57) ($\Delta k = 0$) and Eqs. (8.58–8.59) ($\Delta k \neq 0$), it is clear that initially only one wave (either signal or idler wave) needs to be present in order to generate a growing signal or idler wave. When the process is not phase matched, in order for the signal or idler wave to grow, it must have a real b value; b is a real quantity only if $g > \Delta k$. When this condition is satisfied the amplitude of the signal or idler wave will grow exponentially according to e^{bz}. Otherwise, b is imaginary and $\cos h(bz)$ becomes an oscillatory function of z. Therefore, the mismatch Δk must be smaller than

$$g = d\sqrt{\frac{\mu_0\omega_1\omega_2\omega_3}{\epsilon_0 n_1 n_2 n_3}}\left(\sqrt{\frac{\mu_3}{\omega_3}}\,E_3(0)\right).$$

In the case of phase matching, $\Delta k = 0$ and also assuming that $A_i(0) = 0$,

$$A_s(z) = A_s(0)\cos h\frac{gz}{2}.$$

$$\text{When } gz \gg 1, \quad \frac{|A_s(z)|^2}{|A_s(0)|^2} \to \frac{1}{4}\,e^{gz};$$

the signal wave grows exponentially as a function of z. This is the theoretical basis for the commercial parametric amplifier. To achieve $gz \gg 1$, one should try to use a long crystal. In this case, the crystal length L limits the interaction length.

$$gL = \sqrt{\frac{\mu_0\omega_1\omega_2\omega_3}{\epsilon_0\, n_1 n_2 n_3}}\sqrt{\frac{\mu_3}{\omega_3}}\,dE_3(0)L \gg 1.$$

It is obvious that we need: (1) a crystal with a large d coefficient, (2) a long crystal (provided that the phase-matchable interaction can be maintained over the entire length of the crystal) and (3) a strong pump field. Strong pump fields can always be accomplished by focusing. However, focusing too strongly will shorten the phase-matchable interaction length (this is generally the main limiting length factor in the parametric amplification process, not the physical length of the crystal).

Based on these guiding principles, a commercial parametric amplifier can be built. The wavelength tuning of the OPA is achieved by angular rotation of the OPA crystal. One essentially lets the phase-matching process select a pair of signal and idler wavelengths so that both energy and momentum are conserved.

8.4 Parametric Oscillation

The parametric amplification process is essentially a single-path process which does not require feedback or recycling of the signal wave through the pumped medium. In contrast to the single-path process, one can build a cavity which allows both the signal and idler waves to recirculate through the nonlinear crystal. With proper feedback control one can construct an oscillator based on the parametric process. The optical parametric oscillator (OPO) in every aspect looks like a laser. An OPO cavity is formed by two feedback reflectors in a Fabry–Perot cavity arrangement as shown in Fig. 8.11. To achieve the best result, the cavity should be designed to provide mode matching between the pump and signal waves.

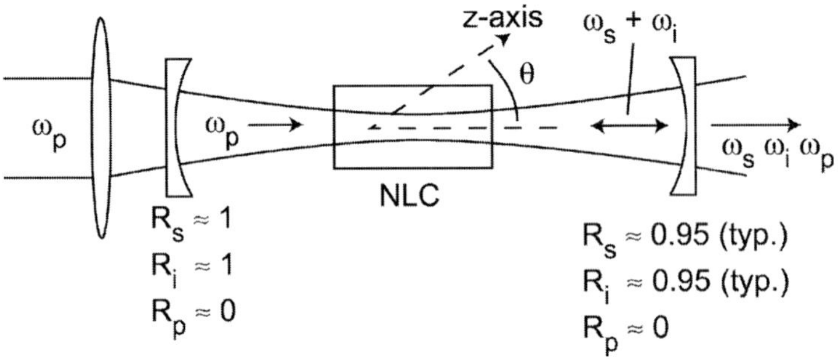

Fig. 8.11 Schematic diagram of optical parametric oscillator. The pump wave at ω_p gives rise to oscillations at ω_s and ω_i in an optical cavity that contains the nonlinear crystal, NLC, and resonates at ω_s and ω_i.

The signal and idler waves are expressed as

$$E_s = E_s(z,t)e^{i(\omega_s t - \vec{k}_s\cdot\vec{r})},$$

$$E_i = E_i(z,t)e^{i(\omega_i t - \vec{k}_i\cdot\vec{r})},$$

respectively. The associated "photon" field through the relation of $A_s = \sqrt{\frac{n_s}{\omega_\ell}}\,E_\ell$, etc., is given by

$$A_s = A_s(z,t)e^{i(\omega_s t - \vec{k}_s\cdot\vec{r})}$$

and

$$A_i = A_i(z,t)e^{i(\omega_i t - \vec{k}_i\cdot\vec{r})}. \tag{8.60}$$

In the OPA case, one is interested in investigating the amplitude growth as a function of position (inside the nonlinear crystal), i.e., $A_s(z)$ and $A_i(z)$. In the OPO case, one wants to examine how the amplitude of A_s and A_i grow as a function of time. It would be clearer if the spatially dependent function and differential equation are converted into temporally dependent ones. Let us recall the coupled equations for parametric interaction, Eqs. (8.47)–(8.49),

$$\begin{aligned}\frac{dA_s(z)}{dz} &= \frac{-1}{2}\alpha_s A_s(z) - \frac{i}{2}\lambda A_p A_i^*(z)e^{-i\Delta kz},\\ \frac{dA_i^*(z)}{dz} &= \frac{-1}{2}\alpha_i A_i^*(z) + \frac{i}{2}\lambda A_p A_s(z)e^{i\Delta kz},\end{aligned} \tag{8.61}$$

where we have taken $A_p^* = A_p$. To convert this set of equations from z dependence to time dependence, let us recall that

$$z = vt = \frac{c}{n}t,$$

$$\frac{d}{dz} = \frac{n}{c}\frac{d}{dt}.$$

Also, realizing that the only interesting case for an OPO is when $\Delta k = 0$, Eq. (8.61) becomes

$$\frac{n_s}{c}\frac{dA_s(t)}{dt} = -\frac{1}{2}\alpha_s A_s(t) - \frac{i}{2}\lambda A_p A_i^*(t), \tag{8.62}$$

$$\frac{n_i}{c}\frac{dA_i^*(t)}{dt} = \frac{-1}{2}\alpha_i A_i^*(t) + \frac{i}{2}\lambda A_p A_s(t). \tag{8.63}$$

We are interested in investigating how the amplitude function of $A_s(t)$ and $A_i^*(t)$ grows at a fixed point inside the nonlinear crystal as it is pumped by a cw pump $A_p\, e^{i(\omega_p t - \vec{k}_p \cdot \vec{r})}$. If the device leads to oscillation, $A_s(t)$ and $A_i^*(t)$ will reach a constant value. Therefore,

$$\frac{dA_s(t)}{dt} = 0, \quad \frac{dA_i^*(t)}{dt} = 0$$

at oscillation. These conditions convert the set of differential equations in Eqs. (8.62)–(8.63) into a set of simultaneous algebraic equations:

$$\begin{cases}\frac{-1}{2}\alpha_s A_s & -\frac{i}{2}\lambda A_p A_i^* = 0,\\ \frac{-1}{2}\alpha_i A_i^* & +\frac{i}{2}\lambda A_p A_s = 0.\end{cases}$$

The necessary condition for a nontrivial solution of A_s and A_i^* to exist is for the secular equation to be zero, or

$$\begin{vmatrix}\frac{-1}{2}\alpha_s & \frac{-i}{2}\lambda A_{p,th}\\ \frac{i}{2}\lambda A_{p,th} & \frac{-1}{2}\alpha_i\end{vmatrix} = 0,$$

where we have denoted the A_p that satisfies the secular equation as the threshold pump wave amplitude $A_{p,th}$ with a threshold value of

$$\lambda^2 \mid A_{p,th} \mid^2 = g^2 = \alpha_i \alpha_s. \tag{8.64}$$

We reach essentially the same conclusion with an OPO as in a laser, i.e., the gain equal to the loss sets the threshold condition for lasing.

Remember that $\lambda^2 = \frac{\mu_0}{\epsilon_0} d^2 \frac{\omega_p \omega_s \omega_i}{n_p n_s n_i}$.

We have the threshold pump beam intensity

$$\begin{aligned} I_{th}(\omega_p) &= \frac{1}{2}\sqrt{\frac{\epsilon_0}{\mu_0}}\,\omega_p \mid A_{p,th} \mid^2 \\ &= \frac{1}{2}\left(\frac{\epsilon_0}{\mu_0}\right)^{\frac{3}{2}} \frac{n_p n_s n_i}{d^2} \frac{\alpha_i \alpha_s}{\omega_i \omega_s}. \end{aligned} \tag{8.65}$$

Since the losses α_i and α_s are distributed loss parameters, if there is no such loss in the OPO, one may define the total loss from the OPO as

$$\alpha_i \alpha_s \, \ell_R^2 = (1 - R_i)(1 - R_s),$$

where ℓ_R is the length of the OPO cavity and R_i and R_s are the reflectivities of the idler and signal waves, respectively. The threshold pumping intensity is then

$$I_{th}(\omega_p) = \frac{1}{2}\left(\frac{\epsilon_0}{\mu_O}\right)^{\frac{3}{2}} \frac{n_p n_s n_i (1 - R_i)(1 - R_s)}{\omega_s \omega_i \, \ell_R^2 d^2}. \tag{8.66}$$

In a singly resonated OPO, one may have $R_i = 0$ to suppress the oscillation of the idler wave at the expense of high threshold pumping intensity.

8.4.1
Tuning of OPO

The output wavelength of the OPO is governed by the phase-matching condition (momentum matching)

$$\frac{\omega_p n_p}{c} = \frac{n_s \omega_s}{c} + \frac{n_i \omega_i}{c} \tag{8.67}$$

and the energy-conservation condition

$$\omega_p = \omega_s + \omega_i. \tag{8.68}$$

For the sake of discussion, let us assume that the nonlinear crystal is birefringent and the pump beam is propagating as an e-ray. Initially along the θ direction one has

$$\omega_p n_p^e(\theta) = \omega_s^0 n_s^0 + \omega_i^0 n_i^0, \tag{8.69}$$

where $n_p^e(\theta)$ is a function of θ. We may tune the angle from θ to $\theta + \Delta\theta$, in which case

$$n_p^e \to n_p^e + \Delta n_p$$
$$n_s^0 \to n_s^0 + \frac{\partial n_s}{\partial \omega_s} \mid \Delta\omega_s$$
$$n_i^0 \to n_i^0 + \frac{\partial n_i}{\partial \omega_i} \mid \Delta\omega_i.$$

The energy-conservation condition leads to

$$\Delta\omega_s = -\Delta\omega_i,$$

since ω_p is fixed. The differential relation of Eq. (8.68) is

$$\omega_p \Delta n_p^e + \Delta\omega_p n_p^e = \omega_s^0 \Delta n_s^0 + \Delta\omega_s^0 n_s^0 + \omega_i^0 \Delta n_i^0 + \Delta\omega_i^0 n_i^0,$$

$$\omega_p \left(n_p^e + \frac{\partial n_p^e}{\partial \theta} \Delta\theta \right) = \omega_s^0 \left(n_s^0 + \frac{\partial n_s^0}{\partial \omega} \Delta\omega_s \right) + \Delta\omega_s n_s^0 + \omega_i^0 \left(n_i^0 + \frac{\partial n_i}{\partial \omega} \Delta\omega_i \right) + \Delta\omega_i n_i^0,$$

from which one obtains

$$\frac{\Delta\omega_s}{\Delta\theta} = \frac{\omega_p \frac{\partial n_p^e}{\partial \theta}}{\left(n_s^0 - n_i^0\right) + \left[\omega_s^0 \frac{\partial n_s^0}{\partial \omega} - \omega_i^0 \frac{\partial n_i^0}{\partial \omega}\right]}, \tag{8.70}$$

since $n_p(\theta)$ satisfies

$$\frac{1}{n_p^2(\theta)} = \frac{\cos^2(\theta)}{n_0^2(\omega_p)} + \frac{\sin^2\theta}{n_e^2(\omega_p)}.$$

After some calculation, it is possible to show that

$$\frac{\partial n_p(\theta)}{\partial \theta} = \frac{n_p^3(\theta)\sin 2\theta}{2} \left[\frac{1}{n_0^2(\omega_p)} - \frac{1}{n_e^0(\omega_p)} \right], \tag{8.71}$$

or

$$\frac{\Delta\omega_s}{\Delta\theta} = \frac{\frac{1}{2}\omega_p n_p^2(\theta_0)\sin 2\theta \left[\frac{1}{n_0^2(\omega_p)} - \frac{1}{n_e^2(\omega_p)} \right]}{\left(n_s^0 - n_i^0\right) + \left[\omega_s^0 \frac{\partial n_s^0}{\partial \omega} - \omega_i^0 \frac{\partial n_i^0}{\partial \omega}\right]}, \tag{8.72}$$

as the angular tuning relationship for the signal wave. This relation shows that the birefringence, dispersion and initial angle all play important roles in frequency-tuning characteristics of the OPO [80].

8.5
The Manley–Rowe Relations

In the parametric interaction, photons at higher energy (pump photons) are converted to photons of lesser energy (signal and idler photons). The rules that govern the photon number density at various frequencies are the Manley–Rowe relations [81]. They can be derived from the coupled equations, Eqs. (8.47)–(8.49).

From the expression of

$$\frac{d}{dz} \mid A_p(z) \mid^2 = A_p(z) \frac{dA_p^*(z)}{dz} + \frac{dA_p(z)}{dz} A_p^*(z),$$

one substitutes from the coupled equation, Eq. (8.47),

$$\frac{dA_p(z)}{dz} = \frac{-i}{2} \lambda A_s(z) A_i(z) \mathrm{e}^{i\Delta kz}$$

and its complex conjugate to obtain

$$\frac{d \mid A_p(z) \mid^2}{dz} = \frac{-i}{2} \lambda A_s(z) A_i(z) A_p^*(z)\, \mathrm{e}^{i\Delta kz} + \frac{i}{2} \lambda A_s^* A_i^* A_p\, \mathrm{e}^{-i\Delta kz}. \tag{8.73}$$

Similarly, one can obtain

$$\frac{d \mid A_s(z) \mid^2}{dz} = \frac{i}{2} \lambda A_s A_p^* A_i\, e^{i\Delta kz} - \frac{i}{2} \lambda A_p A_i^* A_s^* \mathrm{e}^{-i\Delta kz} \tag{8.74}$$

and

$$\frac{d \mid A_i(z) \mid^2}{dz} = \frac{i}{2} \lambda A_i A_p^* A_s e^{i\Delta kz} - \frac{i}{2} \lambda A_i^* A_p A_s^* \mathrm{e}^{-i\Delta kz}. \tag{8.75}$$

We observe that the right-hand side of Eq. (8.74) is the same as that of Eq. (8.75) and has opposite sign to the right-hand side of Eq. (8.73); we then have

$$\frac{d \mid A_s(z) \mid^2}{dz} = \frac{d \mid A_i \mid^2}{dz} = -\frac{d \mid A_p \mid^2}{dz}.$$

Since

$$I_\ell = \frac{1}{2} \sqrt{\frac{\epsilon_0}{\mu}}\, \omega_\ell \mid A_\ell \mid^2 = \frac{P(\omega_\ell)}{S} = q_\ell \hbar \omega_\ell c,$$

where $P(\omega_\ell)$ is the power at ω_ℓ and s the cross-sectional area of the beam, we obtain

$$\frac{dq_s(\omega_s)}{dz} = \frac{dq_i(\omega_i)}{dz} = -\frac{dq_p(\omega_p)}{dz}. \tag{8.76}$$

We have

$$\frac{d}{dz}\frac{P_s(z)}{\omega_s} = \frac{d}{dz}\frac{P_i(z)}{\omega_i} = -\frac{d}{dz}\frac{P_p(z)}{\omega_p}. \tag{8.77}$$

Equation (8.76) states that the signal photon density increase per unit length in the medium is exactly the same as the increase in idler photon density and at the expense of the decreasing pump photon density. In other words, in this parametric process the one pump photon is split into one signal and one idler photon.

By integrating Eq. (8.77) and rearranging, we arrive at

$$\frac{(P_s)_{out} - (P_s)_{in}}{(P_p)_{in} - (P_p)_{out}} = \frac{\omega_s}{\omega_p}, \tag{8.78}$$

$$\frac{(P_i)_{out} - (P_i)_{in}}{(P_p)_{in} - (P_p)_{out}} = \frac{\omega_i}{\omega_p}, \tag{8.79}$$

where the frequency relation is shown in Fig. 8.12. Note that $\omega_s/\omega_p < 1$ and $\omega_i/\omega_p < 1$.

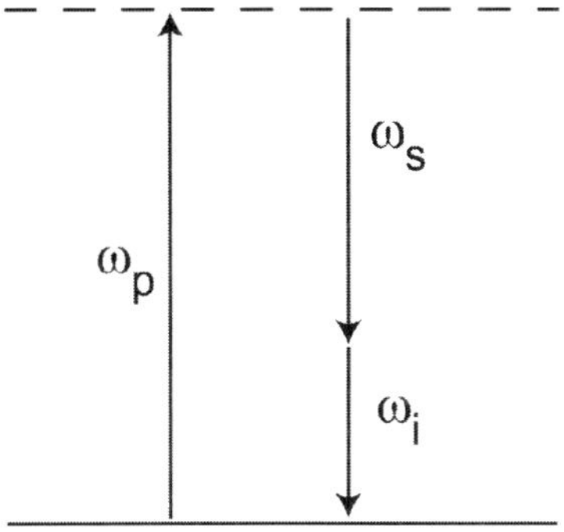

Fig. 8.12 Manley–Rowe relation showing that one pump photon generates one signal and one idler photon.

The equations (8.76–8.79) are referred to as the Manley–Rowe relations. They mean that the net power gain in signal power is always less than the power loss in pump photons according to the ratio in frequency. In order to have maximum power conversion from pump to signal wave, the frequency of the signal wave should be close to that of the pump wave.

8.6 Parametric Upconversion

An inverse process to parametric generation is parametric upconversion,[2] where

$$\omega_{in} + \omega_p \rightarrow \omega_{co}. \tag{8.80}$$

This relation indicates that an initial photon at frequency ω_{in} is upconverted to a photon at higher frequency ω_{co} with the aid of a strong pump beam at frequency ω_p. One application of the parametric upconversion process is to convert a photon in a spectral region where there is no sensitive photodetector available into a photon at higher frequency where a sensitive photodetector is readily available. Since this conversion process is coherent, the converted photon keeps all the characteristics of the original photon. Assuming that $\Delta k = 0$,

$$\begin{aligned} \frac{dA_{co}(z)}{dz} &= \frac{-i}{2}\,\lambda A_{in}(z) A_p(z) \\ \frac{dA_{in}}{dz} &= -\frac{i}{2}\,\lambda A_{co} A_p^*(z). \end{aligned} \tag{8.81}$$

This set of equations differs from the set describing parametric interaction in that the signs of the terms on the right-hand side are both negative in the current case, while the signs are opposite in the parametric interaction case. Let us assume that A_p is real and define the gain similar to Eq. (8.51),

$$g = \lambda\, A_p;$$

we then have

$$\frac{d^2 A_{co}(z)}{dz^2} = -\left(\frac{g}{2}\right)^2 A_{co},$$

which leads to the oscillatory solution

$$\begin{aligned} A_{co}(z) &= A_{co}(0)\,\cos\frac{gz}{2} - i\,A_{in}(0)\,\sin\frac{gz}{2} \\ A_{in}(z) &= A_{in}(0)\,\cos\frac{gz}{2} - i\,A_{co}\,\sin\frac{gz}{2}. \end{aligned} \tag{8.82}$$

The Manley–Rowe relation becomes

$$\frac{P_{in}(z) - P_{in}(0)}{\omega_{in}} = \frac{P_{co}(0) - P_{co}(z)}{\omega_{co}}.$$

2) Parametric amplification and oscillation are parametric downconversion processes, where a pump photon is split into two photons of lower frequencies. In parametric upconversion, the process is reversed, i.e., a lower-frequency photon is converted to a higher-frequency photon with the aid of a pump photon.

Since

$$P_{co}(0) = 0,$$

this can be written as

$$\frac{P_{in}(0)}{\omega_{in}} = \frac{P_{co}(z)}{\omega_{co}} + \frac{P_{in}(z)}{\omega_{in}} \tag{8.83}$$

or

$$n_{in}(0) = n_{co}(z) + n_{in}(z). \tag{8.84}$$

In other words,

$$n_{co}(z) \leq n_{in}(0).$$

In this upconversion process, the maximum number of upconverted photons is limited by the number of initial photons at the original frequency ω_{in}. The above relation can also be obtained by examining the expressions of $A_{in}(z)$ and $A_{co}(z)$, realizing that $A_{co}(0) = 0$,

$$| A_{in}(z) |^2 + | A_{co}(z) |^2 = | A_{in}(0) |^2$$

are equivalent to the expression seen in Eq. (8.84). The upconversion efficiency is defined as

$$\eta = \frac{P_{co}(\ell)}{P_{in}(0)} = \frac{\omega_{co}}{\omega_{in}} \sin^2\left(\frac{g\ell}{2}\right).$$

For $g\ell \ll 1$ and from Eqs. (8.46) and (8.51),

$$g = \lambda A_p = \sqrt{\frac{\mu_0}{\epsilon_0}}\, d \sqrt{\frac{\omega_{in}\omega_p\omega_{co}}{n_{in}n_p n_{co}}}\, A_p.$$

We have

$$\eta = \frac{w_{co}}{w_{in}} \frac{g^2\ell^2}{4} = \frac{w_{co}^2\, \ell^2 d^2}{2\eta_{in}\eta_p\eta_w} \left(\frac{\mu_0}{\epsilon_0}\right)^{\frac{3}{2}} I_p, \tag{8.85}$$

where I_p is the intensity of the pump. The upconversion process is sometimes used to obtain a visible image of an original infrared image [82]. If I_p is from an ultrashort laser pulse, it can be used as the gate pulse to gate an optical event taking place at the infrared spectral region. This effect is used to obtain time-resolved detection of infrared waves [83].

Problems

8.1 Derive d_{eff} for 622, 422 and 3m crystals.

8.2 The first nonlinear optical experiment was done in crystalline quartz, which belongs to the trigonal symmetry group of class 32.

a) Show that in order to generate optical second harmonics, the incident laser beam cannot propagate along the x direction.

b) In a standard optical harmonic generation experiment, a z-cut quartz crystal normal to the platelet (z-cut means that the z-axis of the quartz crystal is normal to the platelet) is used as a reference to monitor the square of the laser intensity, $[I(w)]^2$. The z-cut quartz crystal is chosen mainly because of the convenience in not having to worry about the orientation of the laser polarization with respect to the crystalline axis. Show explicitly with a formula that this is the case. The nonlinear d tensor of the 32 class is

$$\begin{pmatrix} d_{11} & -d_{11} & 0 & d_{14} & 0 & 0 \\ 0 & 0 & 0 & 0 & -d_{14} & -d_{14} \\ 0 & 0 & 0 & 0 & 0 & 0 \end{pmatrix}.$$

8.3 How to cut a rectangular $LiIO_3$ crystal for type I SHG ($\lambda_1 = 1.06$ µm)?

$n_0(\omega) = 1.8517, \quad n_0(2\omega) = 1.8976,$

$n_e(\omega) = 1.7168, \quad n_e(2\omega) = 1.7473.$

8.4 In a birefringent crystal, the direction of electromagnetic wave propagation is different from the direction of energy transport. Discuss the implication of this phenomenon on the phase matching of optical harmonic generation, assuming that the laser beam is a very thin pencil of light and the nonlinear crystal is very long.

8.5 a) The nonlinear polarization for the third-harmonic generation in an isotropic medium is given by

$P_i^{(3)}(3\omega) = 3CE_i(\omega)(\vec{E}(\omega) \cdot \vec{E}(\omega)),$

where C can be regarded as a constant. Show that a circularly polarized fundamental beam *cannot* generate a third-harmonic wave in this medium while a linearly polarized beam can.

b) Using the effect described in Part 8.5, devise a method to measure the pulse duration of a picosecond optical pulse. (Assume that you have a series of identical pulses emitted at a regular time interval.)

c) Is it possible to achieve phase matching for the process shown in Part 8.5? How?

8.6 Use a normal (index) surface to make a sketch to illustrate

a) Collinear type I phase matching.

b) A noncollinear type I phase-matching process so that the second-harmonic wave is propagating along the direction normal to the optical axis.

c) Phase matching for the sum-frequency generation process.

8.7 The $LiNbO_3$ crystal has a crystal symmetry of 3m. The electrooptic coefficients are of the form

$$\begin{bmatrix} 0 & -\gamma_{22} & \gamma_{13} \\ 0 & \gamma_{22} & \gamma_{13} \\ 0 & 0 & \gamma_{33} \\ 0 & \gamma_{51} & 0 \\ \gamma_{51} & 0 & 0 \\ -\gamma_{22} & 0 & 0 \end{bmatrix}.$$

The index of an ellipsoid of a $LiNbO_3$ crystal is of the form

$$\frac{x^2}{n_o^2} + \frac{y^2}{n_o^2} + \frac{z^2}{n_e^2} = 1.$$

If we now apply a DC electric field along the c-axis (or z-axis) of the crystal and send in a linearly polarized optical beam propagating along the a-axis (or x-axis), derive an expression for the phase retardation of the $LiNbO_3$ plate with thickness ℓ_1 assuming that the polarization of the light field is 45° with respect to the c-axis.

8.8 Assume that a nonlinear crystal can be fabricated in such a way that (see Fig. 8.13) it consists of alternating sections each with the thickness equal to the coherence length but the inversion image of the other. It is known that a single such section can convert the fundamental to the second harmonic with 1% conversion efficiency. Assuming that all the other experimental parameters remain the same, what would be the expected conversion efficiency if the crystal contains five (5) sections?

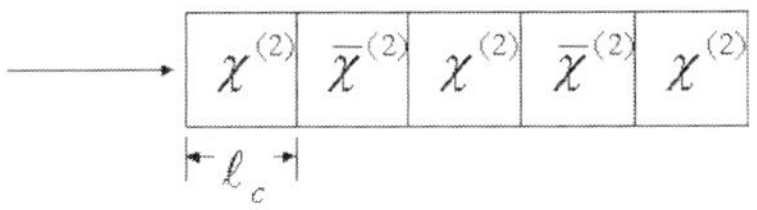

Fig. 8.13 Nonlinear crystal with alternating sections.

9
Quantum Theory of Nonlinear Processes

9.1
Introduction

Up to this point we have treated nonlinear interactions in optical media phenomenologically. We assumed that there was a nonlinear susceptibility associated with a particular nonlinear process. For readers who are interested in exploring the theoretical analysis from fundamental physical principles, it is pointed out here that nonlinear susceptibilities and multiphoton transition probabilities can be calculated using quantum mechanical analysis to obtain explicit expressions. The objective for the presentation in this chapter is to provide readers with an overall perspective of how nonlinear coefficients may be derived from basic quantum principles. Examples of calculations are limited to some specific, and usually simple, cases. Readers who are interested in a more advanced theoretical treatment should consult Refs. [67,84,85].

We begin the theoretical discussion by reviewing the density operator formalism and modifying it for nonlinear processes. For a nonparametric process only the two-photon absorption cross section is presented. For the parametric process only $\chi^{(2)}(2\omega)$ and $\chi^{(3)}$ will be calculated.

9.2
Density Matrix Formalism for Calculation of Nonlinear Coefficient

The equation of motion for the density operator ρ is [86]

$$\frac{\partial \rho}{\partial t} = \frac{1}{i\hbar}[H,\rho]. \tag{9.1}$$

The equation for the off-diagonal matrix elements of ρ is

$$\frac{\partial \rho_{ij}}{\partial t} = \frac{1}{i\hbar}[H,\rho]_{ij}, \tag{9.2}$$

where

$$\rho_{ij} = < u_i \mid \rho \mid u_j >$$

Light-Matter Interaction: Atoms and Molecules in External Fields and Nonlinear Optics.
W. T. Hill and C. H. Lee

ISBN: 978-3-527-40661-6

and where the $| \ u_i \ >$'s are the eigenstates of the unperturbed Hamiltonian H_0[1]:

$$H_0 \mid u_i > \ = \ E_i \mid u_i > \tag{9.3}$$

and

$$H = H_0 + H_1. \tag{9.4}$$

H_1 is the perturbing Hamiltonian; Eq. (9.1) becomes

$$i\hbar \frac{\partial \rho_{ij}}{\partial t} = (E_i - E_j)\rho_{ij} + [H_1, \rho]_{ij}, \tag{9.5}$$

where $[H_1, \rho]_{ij}$ is the off-diagonal matrix element of the commutator of the perturbing Hamiltonian H_1 and the density operator ρ. H_1 has two contributions: H', the external perturbation and H^r, the randomized internal perturbation which restores the system back to the equilibrium condition after it was perturbed:

$$H_1 = H' + H^r. \tag{9.6}$$

Substituting back, one has

$$i\hbar \frac{\partial \rho_{ij}}{\partial t} = \hbar\omega_{ij}\, \rho_{ij} + [H', \rho]_{ij} + [H^r, \rho]_{ij}, \tag{9.7}$$

where $\hbar\omega_{ij} = E_i - E_j$.

In the absence of an external perturbation, $H', \rho_{ij}(t \to \infty) \to 0$ for $i \neq j$. It is customary to introduce a phenomenological relaxation time for the off-diagonal density operator elements, such that

$$[H^r, \rho]_{ij} \ \rightarrow \ \frac{i\hbar}{\tau_{ij}}\, \rho_{ij} \tag{9.8}$$

and

$$\rho_{ij}(t) = \rho_{ij}(0)\ \mathrm{e}^{-t/\tau_{ij}}\ \mathrm{e}^{-iw_{ij}t}. \tag{9.9}$$

The diagonal elements ρ_{jj} approach a time-independent equilibrium value ρ^e_{jj} as $H' \to 0$. The equation of motion for the diagonal element ρ_{jj} is

$$i\hbar \frac{\partial \rho_{jj}}{\partial t} = [H', \rho]_{jj} + i\hbar\, \Sigma_k \left(\rho_{kk}\, W_{kj} - \rho_{jj}\, W_{jk}\right), \tag{9.10}$$

where the terms $\rho_{kk} W_{kj}$ in Eq. (9.10) represent the per-unit-time increase in the probability of occupation of the state j resulting from a transition from state

1) For a review of quantum mechanics, please see Appendix B.8.

k as a result of a randomized perturbation. The terms $-\rho_{jj}W_{jk}$ represent the per-unit-time decrease in the probability of occupation of the j state resulting from a transition from the j state to the k state. With no external perturbation, $H' = 0$, one demands that $\partial\rho_{jj}/\partial t = 0$ when the system is in equilibrium. Therefore,

$$\Sigma_k \left(\rho^e_{kk}W_{kj} - \rho^e_{jj}W_{jk}\right) = 0.$$

When the system does not have any net emission or absorption of energy at any of the transition frequencies, the principle of detailed balance requires that

$$\rho^e_{kk}\, W_{kj} = \rho^e_{jj}\, W_{jk} \tag{9.11}$$

or

$$\frac{\rho^e_{kk}}{W_{jk}} = \frac{\rho^e_{jj}}{W_{kj}}.$$

We define the relaxation times $T_{jk} = \rho^e_{kk}/W_{jk}$ and $T_{kj} = \rho^e_{jj}/W_{kj}$. From the principle of detailed balance it is obvious that $T_{kj} = T_{jk} = T_i$, the longitudinal relaxation time. We have

$$i\hbar\,\frac{\partial\rho_{jj}}{\partial t} = [H',\rho]_{jj} + i\hbar\Sigma_k\left(\rho_{kk}\,\frac{\rho^e_{jj}}{T_1} - \rho_{jj}\frac{\rho^e_{kk}}{T_1}\right),$$

$$i\hbar\,\frac{\partial\rho_{jj}}{\partial t} = [H',\rho]_{jj} + \frac{i\hbar}{T_1}\left(\rho^e_{jj} - \rho_{ij}\right). \tag{9.12}$$

We note that

$$\rho^e_{jj} = \frac{\mathrm{e}^{-E_j/kT}}{\Sigma_m\,\mathrm{e}^{-E_m/kT}} \tag{9.13}$$

follows a Boltzmann distribution.

For a two-level system, one has

$$i\hbar\,\frac{\partial\rho_{11}}{\partial t} = [H',\rho]_{11} + \frac{i\hbar}{T_1}\,(\rho^e_{11} - \rho_{11}), \tag{9.14}$$

$$i\hbar\,\frac{\partial\rho_{22}}{\partial t} = [H',\rho]_{22} + \frac{i\hbar}{T_1}\,(\rho^e_{22} - \rho_{22}), \tag{9.15}$$

$$i\hbar\,\frac{\partial\rho_{12}}{\partial t} = \hbar w_{12}\,\rho_{12} + [H',\rho]_{12} - \frac{i\hbar}{T_2}\,\rho_{12}. \tag{9.16}$$

9.3 Perturbation Method in Density Operator Formalism

We will now apply the perturbation method to the quantum system. Let us choose a perturbation parameter λ such that

$$H = H_0 + \lambda H', \tag{9.17}$$

where H_0 is the unperturbed Hamiltonian and H' represents the external perturbation. With $\lambda = 1$ the perturbation is fully turned on and, with $\lambda = 0$, there is no perturbation. One can also expand ρ in terms of powers of λ following the justification given below.

From

$$\frac{\partial \rho}{\partial t} = -i\hbar^{-1}[H,\rho],$$
$$\int_0^t d\rho(t) = -i\hbar^{-1}\int_0^t [H,\rho(t')]dt',$$
$$\rho(t) = \rho(0) - i\hbar^{-1}\int_0^t [H,\rho(0)]dt',$$

for the first-order approximation, where we simply substitute $\rho(t')$ by $\rho(0)$. To go to higher orders we need to use the recursion formula for $\rho(t)$ repeatedly. Carrying out this procedure to the second order in ρ, we have

$$\rho(t) - \rho(0) = \underbrace{-i\hbar^{-1}\int_0^t [H,\rho(0)]dt'}_{\rho(1)} \underbrace{- i\hbar^{-2}\int_0^t dt' \int_0^{t'} dt''\,[H,[H,\rho(0)]]}_{\rho(2)}, \quad (9.18)$$

where the first integral may be defined as $\rho^{(1)}$ and the second, $\rho^{(2)}$, etc. One may write

$$\rho = \Sigma_p\, \lambda^p \rho^{(p)} \text{ and}$$
$$\rho_{ij} = \Sigma_p\, \lambda^p \rho_{ij}^{(p)}.$$

Substituting into the equation

$$i\hbar\frac{\partial \rho_{jj}}{\partial t} = [\lambda H',\rho]_{jj},$$

we have

$$i\hbar\Sigma_p\lambda^p\,\frac{\partial \rho_{jj}^{(p)}}{\partial t} = \Sigma_p\lambda^{p+1}\,[H',\rho^{(p)}]_{jj}$$

and, equating λ^p terms on both sides, we have the recursion formula for $\rho^{(p)}$:

$$i\hbar\frac{\partial}{\partial t}\rho_{jj}^{(p)} = [H',\rho^{(p-1)}]_{jj},$$

for the case excluding a damping term. When the damping terms are included, the equation becomes

$$i\hbar\frac{\partial}{\partial t}\rho_{jj}^{(p)} = [H',\rho^{(p-1)}]_{jj} + i\hbar\,\Sigma_k\left(\rho_{kk}^{(p)}W_{kj} - \rho_{jj}^{(p)}W_{jk}\right). \quad (9.19)$$

Similarly, one can derive the recursion formula for the off-diagonal elements, ρ_{ij},

$$i\hbar\left(\frac{\partial}{\partial t}+i\omega_{ij}+\frac{1}{\tau_{ij}}\right)\rho_{ij}^{(p)} = [H',\rho^{(p-1)}]_{ij}. \tag{9.20}$$

9.4
Transition Probability in Electric Dipole Interaction

In a nonparametric process, the quantum system will be at a different state after it interacts with the photons. The physical picture of this interaction is shown in Fig. 9.1. The transition probability for single and/or multiple-photon absorption will be evaluated by the perturbation procedure introduced in the previous section. Assuming that the ground state is state $\mid 1 >$ and the system is in its ground state at $t = 0$, we have

$$\rho_{ij}(t=0) = \delta_{i1}\delta_{j1}. \tag{9.21}$$

The only nonzero matrix element for ρ is $\rho_{11} = 1$ at $t = 0$. The transition probability from the ground state $\mid 1 >$ to any $\mid k >$ is by definition

$$W = \overline{\frac{\partial\rho_{kk}}{\partial t}},$$

where the overbar indicates a time average over a time much longer than the period of the perturbation.

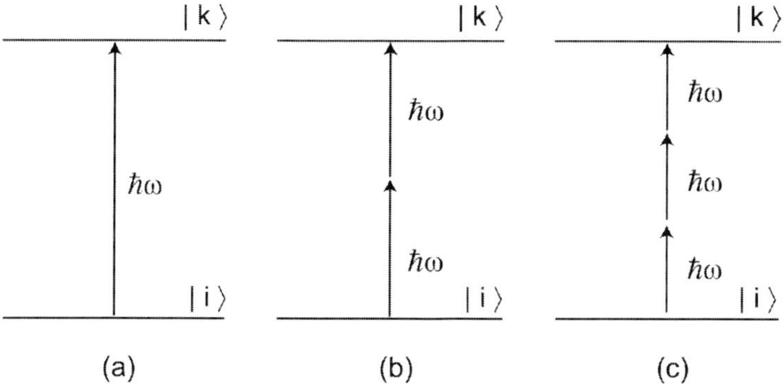

Fig. 9.1 Interaction of photons with quantum systems. (a) Single-photon, (b) two-photon and (c) three-photon interaction.

In the perturbation procedure outlined above,

$$
\begin{aligned}
W &= \overline{\frac{\partial}{\partial t}\left[\Sigma_p \rho_{kk}^{(p)}\right]} \\
&= \overline{\frac{\partial \rho_{kk}^{(0)}}{\partial t}} + \overline{\frac{\partial \rho_{kk}^{(1)}}{\partial t}} + \overline{\frac{\partial \rho_{kk}^{(2)}}{\partial t}} + \cdots .
\end{aligned}
\tag{9.22}
$$

If the time average for the lowest order $\overline{\partial \rho_{kk}^{(p)}/\partial t}$ is nonzero, one does not need to evaluate the higher-order quantity. In order to calculate the multiple-photon transition probability, one should begin by calculating the single-photon transition probability first. Assuming that the system is in thermal equilibrium with the reservoir, then $\rho_{11}^{(0)}, \rho_{22}^{(0)}, \rho_{33}^{(0)}, \ldots$, etc., will follow a Boltzmann distribution. If the states are electronic states, and $E_2 - E_1 \gg kT$, then we can assume that $\rho_{ij}(t = 0) = \delta_{i1}\delta_{j1}$ as in Eq. (9.21). $\rho_{kk}^{(0)} = \text{constant} = 0$ before the perturbation H' is turned on. To zeroth order $\overline{\partial \rho_{kk}^{(0)}/\partial t} = 0$. One needs to go the next order to calculate $\overline{\partial \rho_{kk}^{(1)}/\partial t}$:

$$
i\hbar \frac{\partial \rho_{kk}^{(1)}}{\partial t} = [H', \rho^{(0)}]_{kk} = \Sigma_j \left(H'_{kj}\rho_{jk}^{(0)} - \rho_{kj}^{(0)} H'_{jk} \right),
\tag{9.23}
$$

where

$$
\text{the } \rho_{jk}^{(0)\prime}\text{s are } \rho_{1k}^{(0)},\ \rho_{2k}^{(0)},\ \rho_{3k}^{(0)}, \ldots, \text{etc.,}
$$

and

$$
\text{the } \rho_{kj}^{(0)}\text{s are } \rho_{k1}^{(0)}, \rho_{k2}^{(0)}, \rho_{k3}^{(0)}, \ldots, \text{ etc.}
$$

One needs to solve for $\rho_{kj}^{(0)}(t)$ or $\rho_{jk}^{(0)}(t)$ from

$$
i\hbar \left(\frac{\partial}{\partial t} + i\omega_{jk} + \frac{1}{\tau_{jk}} \right) \rho_{jk}^{(0)}(t) = 0.
\tag{9.24}
$$

Since the zeroth-order $\rho_{jk}^{(0)}(t)$ is the diagonal matrix element of ρ before the perturbation is turned on, the solution is

$$
\begin{aligned}
\rho_{jk}^{(0)}(t) &= \rho_{jk}^{(0)}(0)\mathrm{e}^{-i\omega_{jk}t - t/\tau_{jk}} \\
&= \delta_{j1}\delta_{k1}\, \mathrm{e}^{-i\omega_{jk}t - t/\tau_{jk}}.
\end{aligned}
\tag{9.25}
$$

Substituting back into Eq. (9.23), we obtain

$$
i\hbar \overline{\frac{\partial \rho_{kk}^{(1)}}{\partial t}} = 0
$$

and hence

$$\overline{\frac{\partial \rho_{kk}^{(1)}}{\partial t}} = 0.$$

Proceeding to higher order,

$$\begin{aligned} i\hbar \frac{\partial \rho_{kk}^{(2)}}{\partial t} &= \left[H', \rho^{(1)}\right]_{kk} \\ &= \Sigma_j \left(H'_{kj}\rho_{jk}^{(1)} - \rho_{kj}^{(1)} H'_{jk}\right) \end{aligned} \tag{9.26}$$

and

$$\begin{aligned} i\hbar \left(\frac{\partial}{\partial t} + i\omega_{jk} + \frac{1}{\tau_{jk}}\right) \rho_{jk}^{(1)}(t) &= \Sigma_i \left(H'_{ji}\rho_{ik}^{(0)} - \rho_{ji}^{(0)} H'_{ik}\right) \\ &= H'_{j1}\delta_{k1} - \delta_{j1}H'_{1k}. \end{aligned} \tag{9.27}$$

The only nonzero term is $\rho_{ji}^{(1)}$, which satisfies

$$i\hbar \left(\frac{\partial}{\partial t} + i\omega_{ji} + \frac{1}{\tau_{ji}}\right) \rho_{ji}^{(1)} = H'_{ji}. \tag{9.28}$$

In the electric dipole approximation,

$$H' = -\vec{\mu} \cdot \vec{E}, \tag{9.29}$$

$$\vec{E} = \frac{\vec{E}_0}{2} \left(e^{i\omega t} + e^{-i\omega t}\right), \tag{9.30}$$

$$H' = -\frac{\vec{\mu} \cdot \vec{E}_0}{2} \left(e^{i\omega t} + e^{-i\omega t}\right) = \frac{\tilde{H}'}{2} \left(e^{i\omega t} + e^{-i\omega t}\right), \tag{9.31}$$

where

$$\tilde{H}' = -\vec{\mu} \cdot \vec{E}_0.$$

Solving for $\rho_{j1}^{(1)}(t)$ from Eq. (9.27),

$$i\hbar \left(\frac{\partial}{\partial t} + i\omega_{j1} + \frac{1}{\tau_{j1}}\right) \rho_{j1}^{(1)}(t) = \frac{\tilde{H}'_{j1}}{2} \left(e^{i\omega t} + e^{-i\omega t}\right), \tag{9.32}$$

the steady-state solution is of the form

$$\rho_{j1}^{(1)}(t) = A^{+}e^{i\omega t} + A^{-}e^{-i\omega t}.$$

Substituting back into the equation, we obtain

$$A^+ = \frac{\tilde{H}'_{j1}}{2}\frac{1}{i\hbar(i\omega + i\omega_{j1} + \frac{1}{\tau_{j1}})},$$

$$A^- = \frac{\tilde{H}'_{j1}}{2}\frac{1}{i\hbar(-i\omega + i\omega_{j1} + \frac{1}{\tau_{j1}})},$$

or

$$\rho^{(1)}_{j1}(t) = \frac{\tilde{H}'_{i1}\,\mathrm{e}^{i\omega t}}{2i\hbar(i\omega + i\omega_{j1} + \frac{1}{\tau_{j1}})} + \frac{\tilde{H}'_{i1}\,\mathrm{e}^{-i\omega t}}{2i\hbar(-i\omega + i\omega_{j1} + \frac{1}{\tau_{j1}})}. \tag{9.33}$$

Since $\rho^{(1)}_{1j}(t) = \left[\rho^{(1)}_{j1}(t)\right]^*$, we have

$$\rho^{(1)}_{1j}(t) = \frac{\tilde{H}'^*_{j1}\,\mathrm{e}^{-i\omega t}}{-2i\hbar(-i\omega - i\omega_{j1} + \frac{1}{\tau_{j1}})} + \frac{\tilde{H}'^*_{j1}\,\mathrm{e}^{i\omega t}}{-2i\hbar(i\omega - i\omega_{j1} + \frac{1}{\tau_{j1}})}. \tag{9.34}$$

All other $\rho^{(1)}_{ij}(t)$ terms with $i \neq 1$ or $j \neq 1$ are zero. We can then proceed to calculate

$$\begin{aligned} i\hbar\frac{\partial\rho^{(2)}_{kk}}{\partial t} &= \Sigma_j\left(H'_{kj}\rho^{(1)}_{jk} - \rho^{(1)}_{kj}H'_{jk}\right) \\ &= H'_{k1}\rho^{(1)}_{1k} - \rho^{(1)}_{k1}H'_{jk} \end{aligned}$$

$$= \frac{\tilde{H}'^*_{k1}}{2}\left(\mathrm{e}^{i\omega t} + \mathrm{e}^{-i\omega t}\right)\left[\frac{\tilde{H}'^*_{k1}\mathrm{e}^{-i\omega t}}{-2i\hbar(-i\omega - i\omega_{k1} + \frac{1}{\tau_{k1}})} + \frac{\tilde{H}'^*_{k1}\mathrm{e}^{i\omega t}}{-2i\hbar(i\omega - i\omega_{k1} + \frac{1}{\tau_{k1}})}\right]$$

$$-\left[\frac{\tilde{H}'_{k1}\mathrm{e}^{i\omega t}}{2i\hbar(i\omega + i\omega_{k1} + \frac{1}{\tau_{k1}})} + \frac{\tilde{H}'_{k1}\mathrm{e}^{-i\omega t}}{2i\hbar\left(-i\omega + i\omega_{k1} + \frac{1}{\tau_{k1}}\right)}\right]\frac{\tilde{H}'_{1k}}{2}\left(\mathrm{e}^{i\omega t} + \mathrm{e}^{-i\omega t}\right). \tag{9.35}$$

When the time average is taken, only DC terms contribute. There are four DC terms:

$$\frac{\mid H'_{k1}\mid^2}{4i\hbar(i\omega + i\omega_{k1} - \frac{1}{\tau_{k1}})} - \frac{\mid \tilde{H}'_{k1}\mid^2}{-4i\hbar(-i\omega - i\omega_{k1} - \frac{1}{\tau_{k1}})}$$

$$+\frac{\mid \tilde{H}'_{k1}\mid^2}{4i\hbar(-i\omega + i\omega_{k1} - \frac{1}{\tau_{k1}})} - \frac{\mid \tilde{H}'_{k1}\mid^2}{-4i\hbar(i\omega - i\omega_{k1} - \frac{1}{\tau_{k1}})}.$$

We notice that the second term is the complex conjugate of the first term and the fourth is the complex conjugate of the third. Finally, we have

$$\overline{\frac{\partial \rho_{kk}^{(2)}}{\partial t}} = \frac{\mid \tilde{H}'_{k1} \mid^2}{2\hbar^2 \tau_{k1}} \left[\frac{1}{(\omega - \omega_{k1})^2 + (\frac{1}{\tau_{k1}})^2} + \frac{1}{(\omega + \omega_{kl})^2 + (\frac{1}{\tau_{k1}})^2} \right]. \quad (9.36)$$

In the case of single-photon absorption as shown in Fig. 9.1, $\omega \approx \omega_{k1} \approx 10^{14}$–$10^{15}$ Hz, and the quantity $1/\tau_{k1} \sim 10^8$–10^{12} Hz. It is clear at resonance for optical frequencies, i.e., $\omega = \omega_{k1}$, that the second term can be neglected,

$$\begin{aligned} W_{(1)} = \overline{\frac{\partial \rho_{kk}^{(2)}}{\partial t}} &= \frac{\pi \mid \tilde{H}'_{k1} \mid^2}{2\hbar^2} \frac{1}{\pi} \frac{\frac{1}{\tau_{k1}}}{(\omega - \omega_{k1})^2 + (\frac{1}{\tau_{k1}})^2} \\ &= \frac{\pi \mid \tilde{H}'_{k1} \mid^2}{2\hbar^2} g_L(\omega - \omega_{k1}), \end{aligned} \quad (9.37)$$

where $W_{(1)}$ is the single-photon transition probability rate, and

$$g_L(\omega - \omega_{k1}) = \frac{1}{\pi} \frac{\frac{1}{\tau_{k1}}}{(\omega - \omega_{k1})^2 + (\frac{1}{\tau_{k1}})^2} \quad (9.38)$$

is a Lorentzian line shape function. In the transition probability just derived, we are assuming that there is a single photon interacting with a single quantum system which may be an ion or a molecule. In a real system there are a large number of identical photons interacting with a large number of identical quantum systems, e.g., molecules. Since all particles in this ensemble are identical (homogeneous system), the transition probability for the whole ensemble will have exactly the same line shape described by the Lorentzian line $g_L(w - w_{k1})$. To such a system, the Lorentzian line is called a homogeneously broadened line with line width (FWHM) equal to $2/\tau_{k1}$. $\tau_{k1} = T_2$ is also referred to as the transverse relaxation time. If the molecules are in a gaseous phase, they will be moving. Because of the Doppler effect the molecules with different velocities with respect to the propagation vector of the photon will cause a shift in the resonant frequency. Such a system is called an inhomogeneously broadened system. In such a system, the homogeneous line associated with a specific velocity should be integrated over its velocity distribution. The resultant line is a Gaussian line shape function. The line width is further broadened. This type of broadening is called inhomogeneous broadening. The Gaussian line function is given by

$$g_G(\omega - \omega_{kl}) = \frac{2}{\Delta W_G} \left(\frac{\ln 2}{\pi} \right)^{\frac{1}{2}} \exp\left[-4(\ln 2) \frac{(\omega - \omega_{k1})^2}{\Delta W_G} \right], \quad (9.39)$$

where Δw_G is the line width of the Gaussian line.

9.5
Two-Photon Absorption Transition Probability

In the single-photon transition case if $\omega = \frac{1}{2}\omega_{j1}$ the transition probability rate W_1 will be zero. However, if the energy-resonant denominator has the following expression:

$$\frac{1}{(2\omega - \omega_{j1})^2 + (\frac{1}{\tau_{j1}})^2},$$

one will have two-photon absorption as shown in Fig. 9.1b.

We follow a similar procedure as outlined in the previous section for the single-photon case but carry it to higher order ($p > 2$) until it contains a DC term containing the factor shown above. Since

$$i\hbar \frac{\partial \rho_{jj}^{(p)}}{\partial t} = [H', \rho^{(p-1)}]_{jj} \text{ for } p > 2,$$

we look for a term that has this factor in

$$\rho_{ij}^{(p-1)}, \rho_{ij}^{(p-2)}, \ldots \text{ or } \rho_{ij}^{(p-n)}.$$

From the equation

$$i\hbar \left(\frac{\partial}{\partial t} + i\omega_{ij} + \frac{1}{\tau_{ij}} \right) \rho_{ij}^{(p-n)} = \left[H', \rho^{(p-n-1)} \right]_{ij}, \tag{9.40}$$

we want $\rho_{ij}^{(p-n)}$ to vary as $e^{i2\omega t}$. Recall that

$$\rho_{11}^{(0)} = 1, \quad \rho_{ij}^{(0)} = \delta_{i1}\, \delta_{j1}, \quad \rho_{jj}^{(1)} = 0, \quad \rho_{ij}^{(1)} = 0 \text{ except for } \rho_{i1}^{(1)},$$
$$\rho_{i1}^{(1)} = \frac{\tilde{H}'_{i1}\, e^{i\omega t}}{2i\hbar(i\omega + i\omega_{i1} + \frac{1}{\tau_{i1}})} + \frac{\tilde{H}'_{i1}\, e^{-i\omega t}}{2i\hbar(-i\omega + i\omega_{i1} + \frac{1}{\tau_{ij}})}. \tag{9.41}$$

Substituting into Eq. (9.40) for $p = 2, n = 0$ and realizing that $H'_{11} = 0$ and $\omega_{1j} = -\omega_{j1}$,

$$i\hbar \left(\frac{\partial}{\partial t} - i\omega_{j1} + \frac{1}{\tau_{j1}} \right) \rho_{1j}^{(2)} = -\Sigma_\ell\, \rho_{1\ell}^{(1)}\, H'_{\ell j}$$

$$= -\Sigma_\ell \left[\frac{\tilde{H}'^{*}_{\ell 1}\, e^{-i\omega t}}{-2i\hbar(-i\omega - i\omega_{\ell 1} + \frac{1}{\tau_{\ell i}})} + \frac{\tilde{H}'^{*}_{\ell 1} e^{i\omega t}}{-2i\hbar(i\omega - i\omega_{\ell 1} + \frac{1}{\tau_{\ell 1}})} \right]$$

$$\times \frac{\tilde{H}'_{\ell j}}{2} \left(e^{i\omega t} + e^{-i\omega t} \right) \quad + c.c.$$

We only need to keep the terms that go with $e^{i2\omega t}$:

$$= -\Sigma_\ell \frac{\tilde{H}'^*_{\ell 1}\tilde{H}'_{\ell j}}{2} \frac{e^{i2\omega t}}{-2i\hbar(i\omega - i\omega_{\ell 1} - \frac{1}{\tau_{\ell 1}})} + c.c. + \text{ other terms}$$

$$= -\Sigma_\ell \frac{\tilde{H}'^*_{\ell 1}\tilde{H}'_{\ell j}}{4i\hbar} \frac{e^{i2\omega t}}{i\omega - i\omega_{\ell 1} + \frac{1}{\tau_{\ell 1}}} + \text{ other terms.} \tag{9.42}$$

We are looking for the complex amplitude of $\rho^{(2)}_{1j}(t)$, defined as

$$\rho^{(2)}_{1j} = \underline{\rho^{(2)}_{1j}}\, e^{i2\omega t},$$

where $\underline{\rho^{(2)}_{1j}}$ is the complex amplitude of $\rho^{(2)}_{1j}(t)$.

Substituting it into (Eq. 9.41), we find that

$$\underline{\rho^{(2)}_{1j}} = \frac{1}{4\hbar^2\left(2\omega - \omega_{j1} - \frac{i}{\tau_{j1}}\right)} \left[\Sigma_\ell \frac{\tilde{H}'_{1\ell}\tilde{H}'_{j\ell}}{\omega - \omega_{\ell j}}\right]. \tag{9.43}$$

In the case of two-photon resonance with $2\omega = \omega_{j1}$, $\rho^{(2)}_{1j}(t)$ will be the only dominant term. We may now proceed to evaluate

$$i\hbar\frac{\partial\rho^{(3)}_{jj}(t)}{\partial t} = \left[H', \rho^{(2)}\right]_{jj} = H'_{i1}\,\rho^{(2)}_{1j} - \rho^{(2)}_{j1}H'_{1j}. \tag{9.44}$$

We notice that the terms on the right-hand side of Eq. (9.44) do not contain a DC term; therefore, the time average of

$$\overline{\frac{\partial\rho^{(3)}_{jj}(t)}{\partial t}} = 0. \tag{9.45}$$

We have to go to $\partial\rho^{(4)}_{jj}/\partial t$ and $\rho^{(3)}_{1j}(t)$, etc.

In

$$i\hbar\frac{\partial\rho^{(4)}_{jj}(t)}{\partial t} = \Sigma_\ell\left(H'_{j\ell}\rho^{(3)}_{\ell j} - \rho^{(3)}_{j\ell}H'_{\ell j}\right)$$

we look for DC terms. This means that we are only interested in finding $\rho^{(3)}$ that varies as $e^{(\pm i\omega t)}$. This indeed is possible by solving

$$i\hbar\left(\frac{\partial}{\partial t} + i\omega_{ij} + \frac{1}{\tau_{ij}}\right)\rho^{(3)}_{ij}(t) = \left[H', \rho^{(2)}\right]_{ij} = H'_{i1}\rho^{(2)}_{1j}, \tag{9.46}$$

since only the $|j>$ state is resonant with 2ω. Again defining a complex amplitude of $\underline{\rho}_{ij}^{(3)}$ as $\rho_{ij}^{(3)}(t) = \underline{\rho}_{ij}^{(3)} e^{i\omega t}$, we find that

$$\underline{\rho}_{ij}^{(3)} = \frac{\tilde{H}'_{i1}}{2\hbar(\omega - \omega_{i1})} \underline{\rho}_{ij}^{(2)}, \tag{9.47}$$

where we have dropped the term $1/\tau_{ij}$ since it is much smaller than the $\omega + \omega_{ij}$ term. We also use the following relation:

$$\begin{aligned}
\frac{1}{\omega + \omega_{ij}} &= \frac{1}{\omega + \omega_i - \omega_j - \omega_1 + \omega_1} = \frac{1}{\omega + (\omega_i - \omega_1) - (\omega_j - \omega_1)} \\
&= \frac{1}{\omega + \omega_{i1} - \omega_{j1}} = \frac{1}{\omega + \omega_{i1} - 2\omega} = \frac{1}{\omega - \omega_{i1}}.
\end{aligned}$$

The expression of $\rho_{ji}^{(3)}$ can be obtained from

$$\begin{aligned}
\rho_{ji}^{(3)}(t) &= \left[\rho_{ij}^{(3)}(t)\right]^* \\
&= \left[\underline{\rho}_{ij}^{(3)}\right]^* e^{-i\omega t} \\
&= \frac{\tilde{H}'^{*}_{i1}}{2\hbar(\omega - \omega_{i1})} \left(\underline{\rho}_{ij}^{(2)}\right)^* e^{-i\omega t}.
\end{aligned}$$

Substituting $\rho^{(3)}$ back into

$$\begin{aligned}
i\hbar \frac{\partial \rho_{jj}^{(4)}}{\partial t} &= \Sigma_\ell \left(H'_{j\ell}\rho_{\ell j}^{(3)}\right) - \Sigma_k \left(\rho_{jk}^{(3)} H'_{kj}\right) \\
&= \Sigma_\ell \frac{\tilde{H}_{j\ell}}{2}\left(e^{i\omega t} + e^{-i\omega t}\right) \underline{\rho}_{\ell j}^{(3)} e^{i\omega t} \\
&\quad - \Sigma_k \underline{\rho}_{jk}^{(3)} e^{-i\omega t} \frac{\tilde{H}_{kj}}{2}\left(e^{i\omega t} + e^{-i\omega t}\right),
\end{aligned} \tag{9.48}$$

we need to keep the DC terms

$$= \Sigma_\ell \frac{\tilde{H}'_{j\ell}}{2} \frac{\tilde{H}'_{\ell 1}}{2\hbar(\omega - \omega_{\ell 1})} \underline{\rho_{ij}^{(2)}} - \Sigma_k \frac{\tilde{H}'^{*}_{k1}}{2\hbar(\omega - \omega_{k1})} \underline{\rho_{j1}^{(2)}} \frac{\tilde{H}'_{kj}}{2}.$$

Taking the time average, we have the two-photon transition probability rate,

$$
\begin{aligned}
W_{(2)} &= \overline{\frac{\partial \rho_{jj}^{(4)}(t)}{\partial t}} \\
&= \frac{1}{16 i \hbar^4} \left\{ \Sigma_\ell \frac{\tilde{H}'_{j\ell} \tilde{H}'_{\ell 1}}{\omega - \omega_{\ell 1}} \frac{1}{\left(2\omega - \omega_{j1} - \frac{i}{\tau_{j1}}\right)} \Sigma_k \frac{\tilde{H}'_{1k} \tilde{H}_{kj}}{\omega - \omega_{k1}} \right. \\
&\quad \left. - \Sigma_k \frac{1}{\left(2\omega - \omega_{j1} + \frac{i}{\tau_{j1}}\right)} \Sigma_\ell \frac{\tilde{H}'^{*}_{1\ell} \tilde{H}'^{*}_{\ell j}}{\omega - \omega_{\ell 1}} \frac{\tilde{H}'_{1k} \tilde{H}'_{kj}}{\omega - \omega_{k1}} \right\} \\
&= \frac{1}{16 i \hbar^4} \left[\frac{1}{\left(2\omega - \omega_{j1} - \frac{i}{\tau_{j1}}\right)} - \frac{1}{\left(2\omega - \omega_{j1} + \frac{i}{\tau_{j1}}\right)} \right] \left| \Sigma_\ell \frac{\tilde{H}'_{1\ell} \tilde{H}'_{\ell j}}{\omega - \omega_{\ell 1}} \right|^2 \\
&= \frac{\pi}{8 \hbar^4} \left| \Sigma_\ell \frac{\tilde{H}'_{1\ell} \tilde{H}'_{\ell j}}{\omega - \omega_{\ell 1}} \right|^2 \frac{1}{\pi} \frac{1/\tau_{j1}}{(2\omega - \omega_{j1})^2 + (\frac{1}{\tau_{j1}})^2}.
\end{aligned}
\tag{9.49}
$$

One can define the Lorentzian line shape function for two-photon absorption as

$$
g_L(2w) = \frac{1}{\pi} \frac{1/\tau_{j1}}{(2\omega - \omega_{j1})^2 + (\frac{1}{\tau_{j1}})^2}. \tag{9.50}
$$

Since $\tilde{H}' \propto E$, we note that

$$
W_{(2)} \propto | E |^4 = I^2
$$

and

$$
W_{(1)} \propto | E |^2 = I.
$$

Therefore, the single-photon transition rate is proportional to the intensity of the beam while the two-photon transition rate is proportional to the intensity squared. One can define a scattering cross section from these transition probabilities, which we do in the next section.

9.6
Scattering Cross Section

The scattering cross section is defined as the effective cross section of an absorbing atom, i.e.,

$$
\sigma_c = \frac{\text{area}}{\text{atom}} = \frac{\text{power absorbed per atom}}{\text{power per unit area}}
$$

$$= \frac{P/N}{I} = \frac{N\hbar\omega_{21} \times W/N}{I} = \frac{2\hbar\omega_{21}}{nc\epsilon_0 \mid E \mid^2} W. \tag{9.51}$$

For the single-photon cross section,

$$W_{(1)} = \frac{\pi \mid \tilde{H}'_{j1} \mid^2}{2\hbar^2} g_L(w - w_{21}),$$

$$\text{where } \mid \tilde{H}' \mid^2 = \frac{\mid \mu_{12} \mid^2}{3} \mid E \mid^2,$$

we have

$$\sigma_{c(1)} = \frac{\omega_{21}\pi}{3\hbar nc\epsilon_0} \mid \mu_{12} \mid^2 g_L(\omega - \omega_{21}), \tag{9.52}$$

$$\begin{aligned} W_{(2)} &= \frac{\pi}{8\hbar^4} \mid \Sigma_\ell \frac{\tilde{H}'_{1\ell} H'_{\ell 1}}{\omega - \omega_{\ell 1}} \mid^2 g_L(2\omega - \omega_{21}), \\ \sigma_{c(2)} &= \frac{\omega_{21}\pi}{36nc\epsilon_0\hbar^3} \mid \mu_{12} \mid^4 \mid E \mid^2 g_L(2\omega - \omega_{21}). \end{aligned} \tag{9.53}$$

Note that $\sigma_{c(1)}$ is independent of I while $\sigma_{c(2)}$ is linearly dependent on I. A typical $\sigma_{c(1)}$ is about 3×10^{-16} cm^2 while $\sigma_{c(2)} \sim 10^{-35} \mid E \mid^2$ with E in V/cm. For a laser beam of intensity 10 MW/cm^2, $E \sim 10^5$ V/cm, $\sigma_{c(2)} \sim 10^{-25}$ cm^2, much smaller than $\sigma_{c(1)}$. This shows that the two-photon process only becomes significant at high intensity. In absorption spectroscopy, single photons only connect states of opposite parity. On the other hand, the two-photon transition connects states of the same parity. Therefore, two-photon spectroscopy can yield complementary information on atomic states.

9.7 Three-Photon Absorption

The transition probability for three-photon absorption can be evaluated in a similar manner by going to $\partial\rho^{(6)}/\partial t$. We look for DC terms in the expression of $\partial\rho^{(6)}(t)/\partial t$, which has an energy denominator that shows three-photon resonance:

$$g_L(3\omega) = \frac{1}{\pi} \frac{\frac{1}{\tau_{j1}}}{(3\omega - \omega_{j1})^2 + (\frac{1}{\tau_{j1}})^2}. \tag{9.54}$$

It can be shown that [86]

$$W_{(3)} = \frac{\pi g_L(3\omega)}{32\hbar^6} \left| \Sigma_{g.k} \frac{\tilde{H}'_{jg}\tilde{H}'_{gk}\tilde{H}'_{k1}}{(2\omega - \omega_{q1})(\omega - \omega_{k1})} \right|^2. \tag{9.55}$$

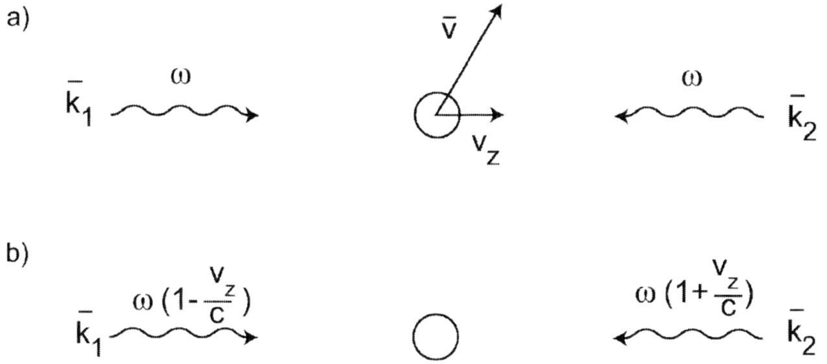

Fig. 9.2 Interaction of moving atom with velocity $\vec{v}$ with counterpropagating laser beams. (a) In the laboratory frame, (b) in the atomic frame.

Following a similar procedure, one can even calculate the m-photon transition probability, which is given in Ref. [86].

9.8 Doppler-Free Two-Photon Absorption

In a gaseous medium the single-photon transition line shape function will be inhomogeneously broadened by the Doppler effect. The line shape will be a Gaussian, instead of a Lorentzian. For two-photon absorption, even in an inhomogeneous Doppler broadened medium one can obtain Doppler-free two-photon absorption. This effect, which was first observed by Levenson and Bloembergen [87] and by Biraben et al. [88], may be understood by considering the two-photon absorption with counterpropagating laser beams as shown in Fig. 9.2. If the atomic system interacts with two photons, one each from the counterpropagating laser beams, it will sense one beam at frequency $w\,(1 - v_z/c)$ and the other at $w\,(1 + v_z/c)$; the transition line shape function will be

$$g_L = \frac{1}{\pi} \frac{\frac{1}{\tau}}{\left[\omega\left(1 - \frac{v_z}{c}\right) + \omega\left(1 + \frac{v_z}{c}\right) - \omega_{j1}\right]^2 + \left(\frac{1}{\tau}\right)^2}$$

$$= \frac{1}{\pi} \frac{\frac{1}{\tau}}{\left(2\omega - \omega_{j1}\right)^2 + \left(\frac{1}{\tau}\right)^2} = g_L(2\omega).$$

The Doppler effect is balanced out in the two-photon absorption process. Every atom in the ensemble participates, regardless of its velocity. The two-photon absorption transition coming from the atoms absorbing two photons from the same beam, on the other hand, will experience a Doppler shift, giving rise to Doppler broadening. In this case, only atoms with the same velocity contribute to the absorption at this specific frequency. Since the Doppler-free two-photon absorption involves all atoms in the medium, it will be the dominant absorption, giving rise to the Doppler-free absorption line shape with the transition probability given as

$$\begin{aligned} W_{(2)} = \frac{1}{16\hbar^4} & \left| \Sigma_n \frac{(\vec{\mu}_{jn} \cdot \vec{E}_+)(\vec{\mu}_{n1} \cdot \vec{E}_-) + (\vec{\mu}_{jn} \cdot \vec{E}_-)(\vec{\mu}_{n1} \cdot \vec{E}_+)}{\omega_{n1} - \omega} \right|^2 \\ \times & \frac{1}{\pi} \frac{\frac{1}{\tau}}{(2\omega - \omega_{j1})^2 + \left(\frac{1}{\tau}\right)^2} \\ + & \left[\left| \Sigma_n \frac{(\vec{\mu}_{jn} \cdot \vec{E}_+)(\vec{\mu}_{n1} \cdot \vec{E}_+)}{\omega_{n1} - \omega} \right|^2 + \left| \Sigma_n \frac{(\vec{\mu}_{jn} \cdot \vec{E}_-)(\vec{\mu}_{n1} \cdot \vec{E}_-)}{\omega_{n1} - \omega} \right| \right]^2 \\ \times & \exp \left[\frac{-(2\omega - \omega_{j1})^2 / \Omega_0^2}{2\sqrt{\pi}\Omega_0} \right] . \end{aligned} \tag{9.56}$$

The composite line shape is sketched in Fig. 9.3. Using this technique, the Zeeman effect for the two-photon 3S–5S transition in a Na^{23} atom has been observed [87].

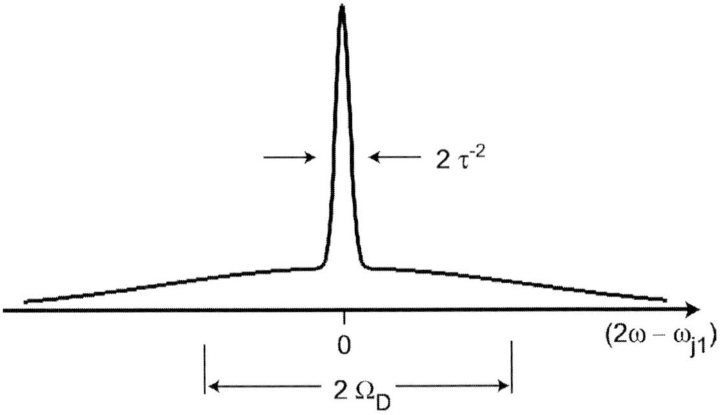

Fig. 9.3 The line-shape function of the Doppler-free two-photon absorption of an ensemble of moving atoms. The Doppler-free line is sitting on top of a Doppler pedestal.

9.9
Calculation of Linear and Nonlinear Susceptibility

Linear susceptibility is the linear response function to the external applied field. An induced polarization results,

$$\vec{P}(\omega) = \chi^{(1)}(\omega)\vec{E}(\omega) = N < \mu >= N\ \mathrm{Tr}(\rho\mu). \tag{9.57}$$

In an anisotropic medium, $\chi^{(1)}$ becomes a tensor; the relation between P, χ and E takes the tensorial form

$$P_i(\omega) = \chi_{ij}E_j(\omega). \tag{9.58}$$

For simplicity, without the loss of other physical principles, we will limit ourselves to the nontensorial relation. The relation between $\vec{P}$ and $\vec{E}$ can be extended to the nonlinear region, where one can express

$$P = P^{(0)} + P^{(1)} + P^{(2)} + \cdots,$$
$$P^{(n)} \propto \mathrm{E}^n.$$

We will extend the calculation to the optical second-harmonic nonlinear polarization,

$$\begin{aligned} P_x^{(2)} &= N < \mu_x(2\omega) >= \chi^{(2)}_{xxx}\ \mathrm{E}^2\ \mathrm{e}^{-i2\omega t} \\ &= -Ne\ \Sigma_{n,n'} x_{n'n}\rho^{(2)}_{nn'}(2\omega). \end{aligned} \tag{9.59}$$

We proceed to calculate $\chi^{(1)}$ first. Using

$$i\hbar\left(\frac{\partial}{\partial t} + i\omega_{n'n} + \frac{1}{\tau_{n'n}}\right)\rho^{(1)}_{n'n} = \Sigma_k\ \left[H'_{n'k}\rho^{(0)}_{kn} - \rho^{(0)}_{n'k}H'_{kn}\right],$$

$$H'_{nk} = -\mu E = -\mu E_0\ \mathrm{e}^{-i\omega t},$$

The $\mathrm{e}^{-i\omega t}$ component of $\rho^{(1)}_{n'n}(\omega)$ satisfies

$$\begin{aligned} i\hbar\left(-i\omega + i\omega_{n'n} + \frac{1}{\tau_{n'n}}\right)\rho^{(1)}_{n'n}(\omega) &= -E_0\ \Sigma_k\left[\mu_{n'k}\rho^{(0)}_{kn} - \rho^{(0)}_{n'k}\mu_{kn}\right] \\ &= -E_0\left(\mu_{n'n}\rho^{(0)}_{nn} - \mu_{n'n}\rho^{(0)}_{n'n}\right); \quad \text{with } \mu_{n'n} = -ex_{n'n}, \end{aligned} \tag{9.60}$$

$$\rho^{(1)}_{n'n}(\omega) = \frac{eE_0}{\hbar}\ \frac{x_{n'n}\mathrm{e}^{-i\omega t}}{\omega - \omega_{n'n} + i\tau_{n'n}}\ \left(\rho^{(0)}_{nn} - \rho^{(0)}_{n'n'}\right). \tag{9.61}$$

We have

$$\begin{aligned} P^{(1)} &= \chi^{(1)}(\omega)E_0\ \mathrm{e}^{-i\omega t} = N\ \mathrm{Tr}(\rho\mu) \\ &= -N\Sigma_{nn'}\ \rho^{(1)}_{nn'}\ ex_{n'n} \\ &= -N\frac{eE_0}{\hbar}\ \Sigma_{nn'}\ \frac{ex_{n'n}\ x_{n'n}}{\omega - \omega_{n'n} + i\tau_{n'n}}\ \left(\rho^{(0)}_{nn} - \rho^{(0)}_{n'n'}\right)\mathrm{e}^{-i\omega t}, \end{aligned} \tag{9.62}$$

$$\chi^{(1)}(\omega) = \frac{Ne^2}{\hbar} \Sigma_{nn'} \frac{x_{n'n}\, x_{n'n}}{\omega - \omega_{n'n} + i\tau_{n'n}} \left(\rho^{(0)}_{n'n'} - \rho^{(0)}_{nn}\right). \tag{9.63}$$

To calculate the nonlinear susceptibility for second-harmonic generation, one follows a similar procedure, looking for $\rho^{(2)}$ in $P^{NL}(2\omega)$ which goes with the 2ω frequency,

$$P^{NL}_x(2\omega) = \chi^{(2)}_{xx} E_x E_x\, \mathrm{e}^{-i2\omega t} = -Ne\, \Sigma_{n,n'}\, x_{n'n}\, \rho^{(2)}_{nn'}(2\omega). \tag{9.64}$$

From

$$i\hbar \frac{\partial}{\partial t} \rho^{(p)}_{jj} = \left[H', \rho^{(p-1)}\right]_{jj} + i\hbar\, \Sigma_k \left(\rho^{(p)}_{kk} W_{kj} - \rho^{(p)}_{jj} W_{jk}\right),$$

and the off-diagonal element,

$$i\hbar \left(\frac{\partial}{\partial t} + i\omega_{ij} + \frac{1}{\tau_{ij}}\right) \rho^{(p)}_{ij} = \left[H', \rho^{(p-1)}\right]_{ij}, \quad i \neq j,$$

we have $\rho^{(0)}_{nj} = 0$ for $n \neq j$, $\rho^{(1)}_{jj}(\omega) = 0$ and $\rho^{(1)}_{ij}(\omega) = \underline{\rho^{(1)}_{ij}}\, \mathrm{e}^{-i\omega t}$, where the amplitude function $\underline{\rho^{(1)}_{ij}(\omega)}$ satisfies

$$i\hbar \left(-i\omega + i\omega_{ij} + \frac{1}{\tau_{ij}}\right) \underline{\rho^{(1)}_{ij}}(\omega) = \Sigma_n \left(\tilde{H}'_{in}(\omega)\rho^{(0)}_{nj} - \rho^{(0)}_{in} \tilde{H}'_{nj}\right)$$
$$= \tilde{H}'_{ij}(\omega)\, \rho^{(0)}_{jj} - \rho^{(0)}_{ii}\, \tilde{H}'_{ij} = \tilde{H}'_{ij} \left(\rho^{(0)}_{jj} - \rho^{(0)}_{ii}\right),$$

$$\underline{\rho^{(1)}_{ij}}(\omega) = \frac{\tilde{H}'_{ij}}{i\hbar \left(-i\omega + i\omega_{ij} + \frac{1}{\tau_{ij}}\right)} \left(\rho^{(0)}_{jj} - \rho^{(0)}_{ii}\right). \tag{9.65}$$

Where $H'_{ij} = \tilde{H}'_{ij}\, \mathrm{e}^{-i\omega t} = -ex_{ij}\, E_0\, \mathrm{e}^{-i\omega t}$,

$$\rho^{(1)}_{ij}(\omega) = \frac{-ex_{ij}E_0 \mathrm{e}^{-i\omega t}}{\hbar(\omega - \omega_{ij} + i/\tau_{ij})} \left(\rho^{(0)}_{jj} - \rho^{(0)}_{ii}\right).$$

Substituting this into the equation for $\rho^{(2)}$, we have

$$i\hbar \left(-i2\omega + i\omega_{ij} + \frac{1}{\tau_{ij}}\right)\rho^{(2)}_{ij}(2\omega) = \Sigma_n \left[H'_{in}(\omega)\rho^{(1)}_{nj}(\omega) - \rho^{(1)}_{in}(\omega) H'_{nj}(\omega)\right]$$
$$= \Sigma_n \frac{\mathrm{e}^2 x_{in}\, x_{nj}\, E_0^2\, \mathrm{e}^{-i2\omega t}}{\hbar(\omega - \omega_{nj} + i/\tau_{nj})} \left(\rho^{(0)}_{jj} - (\rho^{(0)}_{nn}\right)$$
$$+ \Sigma_n \frac{\mathrm{e}^2 x_{in}\, x_{nj}\, E_0^2\, \mathrm{e}^{-i2\omega t}}{\hbar(\omega - \omega_{in} + i/\tau_{in})} \left(\rho^{(0)}_{ii} - (\rho^{(0)}_{nn}\right). \tag{9.66}$$

Solving for $\rho_{ij}^{(2)}(2\omega)$ and substituting into

$$\begin{aligned}
P_x^{NL}(2\omega) &= \chi_{xxx}^{(2)}(2\omega)E_x E_x \, \mathrm{e}^{-i2\omega t} \\
&= -Ne\Sigma_{n,n'}\, x_{n'n}\rho_{nn'}^{(2)}(2\omega) = -Ne\Sigma_{i,j}x_{ji}\rho_{ij}^{(2)} \\
&= -Ne\Sigma_{i,j}x_{ji}\Sigma_n \frac{\mathrm{e}^2 x_{in}\, x_{nj}\, E_0^2\, \mathrm{e}^{-i2\omega t}\left(\rho_{jj}^{(0)} - \rho_{nn}^{(0)}\right)}{\hbar^2\left(2\omega - \omega_{ij} + \frac{i}{\tau_{ij}}\right)\left(\omega - \omega_{nj} + \frac{i}{\tau_{nj}}\right)} \\
&\quad -Ne\Sigma_{i,j}x_{ji}\Sigma_n \frac{\mathrm{e}^2 x_{in}x_{nj}E_0^2\mathrm{e}^{-i2\omega t}}{\hbar^2\left(2\omega - \omega_{ij} + \frac{i}{\tau_{ij}}\right)\left(\omega - \omega_{in} + \frac{i}{\tau_{in}}\right)}\left(\rho_{ii}^{(0)} - \rho_{nn}^{(0)}\right), \quad (9.67)
\end{aligned}$$

we finally have

$$\begin{aligned}
\chi_{xxx}^{(2)}(2\omega) = N\Sigma_{i,j,n} &\left[\frac{\mathrm{e}^3 x_{in}\, x_{nj}\, x_{ji}\left(\rho_{nn}^{(0)} - \rho_{jj}^{(0)}\right)}{\hbar^2\left(2\omega - \omega_{ij} + \frac{i}{\tau_{ij}}\right)\left(\omega - \omega_{nj} + \frac{i}{\tau_{ij}}\right)}\right. \\
&\left. + \frac{\mathrm{e}^3 x_{in}\, x_{nj}\, x_{ji}\left(\rho_{nn}^{(0)} - \rho_{ii}^{(0)}\right)}{\hbar^2\left(2\omega - \omega_{ij} + \frac{i}{\tau_{ij}}\right)\left(\omega - \omega_{in} + \frac{i}{\tau_{in}}\right)}\right]. \qquad (9.68)
\end{aligned}$$

If there is no resonance, one can drop the i/τ_{ij} term, which is much smaller than the $(2\omega - \omega_{ij})$ or $(2\omega - \omega_{nj})$ terms:

$$\begin{aligned}
\chi_{xxx}^{(2)}(2\omega) = N\Sigma_i\Sigma_j\Sigma_n \frac{\mathrm{e}^3 x_{in}\, x_{nj}\, x_{ji}}{\hbar^2} &\left(\frac{\rho_{nn}^{(0)}}{2\omega - \omega_{ij}}\left[\frac{1}{\omega - \omega_{nj}} + \frac{1}{\omega - \omega_{in}}\right]\right. \\
&\left. - \frac{\rho_{jj}^{(0)}}{(2\omega - \omega_{ij})(\omega - \omega_{nj})} - \frac{\rho_{ii}^{(0)}}{(2\omega - \omega_{ij})(\omega - \omega_{in})}\right).
\end{aligned}$$

Since

$$\begin{aligned}
\frac{1}{\omega - \omega_{nj}} + \frac{1}{\omega - \omega_{in}} &= \frac{\omega - \omega_i + \omega_n + \omega - \omega_n + \omega_j}{(\omega - \omega_{nj})(\omega - \omega_{in})} \\
&= \frac{2\omega - \omega_{ij}}{(\omega - \omega_{nj})(\omega - \omega_{in})},
\end{aligned}$$

we have

$$\begin{aligned}
\chi_{xxx}^{(2)}(2\omega) = N\Sigma_i\Sigma_j\Sigma_n \frac{\mathrm{e}^3 x_{in}\, x_{nj}\, x_{ji}}{\hbar^2} &\left[\frac{\rho_{nn}^{(0)}}{(2\omega - \omega_{nj})(\omega - \omega_{in})}\right. \\
&\left. - \frac{\rho_{jj}^{(0)}}{(2\omega - \omega_{ji})(\omega - \omega_{nj})} - \frac{\rho_{ii}^{(0)}}{(2\omega + \omega_{ji})(\omega + \omega_{ni})}\right].
\end{aligned}$$

Exchanging the n, j indices in the second term and the i, n indices in the third summation, we finally have

$$\chi_{xxx}^{(2)}(2\omega) = N\Sigma_i\Sigma_j\Sigma_n \frac{e^3 x_{in}\, x_{nj}\, x_{ji}}{\hbar^2} \rho_{nn}^{(0)} \left[\frac{1}{(\omega+\omega_{jn})(\omega-\omega_{in})} - \frac{1}{(2\omega-\omega_{in})(\omega-\omega_{jn})} - \frac{1}{(2\omega+\omega_{jn})(\omega+\omega_{in})} \right]. \tag{9.69}$$

In most cases, only the ground state is populated, $\rho_{nn}^{(0)} = 0$ for $n \neq g, \rho_{gg}^{(0)} = 1$. Then

$$\chi_{xxx}^{(2)}(2\omega) = N\Sigma_i\Sigma_j \frac{e^3 x_{gj}\, x_{ji}\, x_{ig}\, \rho_{gg}^{(0)}}{\hbar^2} \left[\frac{1}{(\omega+\omega_{jg})(\omega-\omega_{ig})} - \frac{1}{(2\omega-\omega_{ig})(\omega-\omega_{jg})} - \frac{1}{(2\omega+\omega_{jg})(\omega+\omega_{ig})} \right], \tag{9.70}$$

where the i and j states are usually not resonant with either ω or 2ω. The second-harmonic generation process due to the interaction of material with photons is presented in Fig. 9.4.

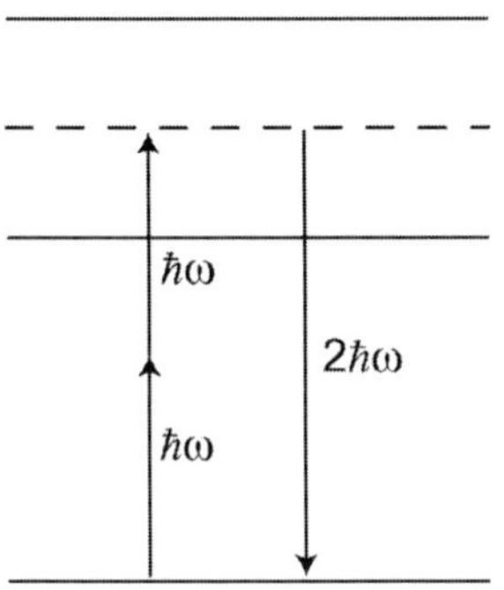

Fig. 9.4 Quantum mechanical representation of optical second-harmonic generation process.

9.10 Third-Order Nonlinear Susceptibility $\chi^{(3)}$

Following the procedure outlined in this chapter, one can calculate the third-order nonlinear susceptibility $\rho^{(3)}$. Bloembergen et al. [89] have worked out the detailed form of all 48 terms which occur in the general case. Here we will outline the procedure of the calculation using the density operator to the third-order term with the recursion formula. We start by first integrating the equation for $\rho_{ij}^{(p)}(t)$,

$$\rho_{ij}^{(p)}(t) = \int_{-\infty}^{t} \frac{-i}{\hbar} \left[H', \rho^{(p-1)}\right]_{ij} \mathrm{e}^{(i\omega_{ij}+\gamma_{ij})(t'-t)} dt', \tag{9.71}$$

where $\gamma_{ij} = 1/\tau_{ij}$ is the damping constant:

$$\rho_{nm}^{(3)}(t) = \int_{-\infty}^{t} \frac{-i}{\hbar} \left[H', (t'), \rho^{(2)}\right]_{nm} \mathrm{e}^{(i\omega_{nm}+\gamma_{nm})(t'-t)} dt'. \tag{9.72}$$

We need to first calculate $\rho^{(2)}(t)$ and $\rho^{(1)}(t)$:

$$\rho_{nm}^{(1)}(t) = \frac{-i}{\hbar} \mathrm{e}^{-(i\omega_{nm}+\gamma_{nm})t} \int_{-\infty}^{t} [H'(t'), \rho^{(0)}] \, \mathrm{e}^{(i\omega_{nm}+\gamma_{nm}(t'-t)} dt', \tag{9.73}$$

$$\rho_{nm}^{(0)} = 0 \quad \text{for } n \neq m. \tag{9.74}$$

Therefore,

$$\begin{aligned} \left[H'(t'), \rho^{(0)}\right]_{nm} &= \Sigma_\nu \, H'(t')_{n\nu} \, \rho_{\nu m}^{(0)} - \rho_{n\nu}^{(0)} \, H'(t')_{\nu m} \\ &= \Sigma_\nu \, [\mu_{n\nu}\rho_{\nu m}^{(0)} - \rho_{n\nu}^{(0)} \mu_{\nu m}] \cdot E(t') \\ &= -\left(\rho_{mm}^{(0)} - \rho_{nn}^{(0)}\right) \mu_{mm} \cdot E(t'), \end{aligned} \tag{9.75}$$

so we can rewrite this by using $E(t') = \Sigma_j E(\omega_j) \, \mathrm{e}^{-i\omega_j t'}$,

$$\begin{aligned} \rho_{nn}^{(1)} &= \frac{-i}{\hbar} \mu_{nm} \cdot \Sigma_j \, E(\omega_j) \left(\rho_{mm}^{(0)} - \rho_{nn}^{(0)}\right) \mathrm{e}^{-(i\omega_{nm}+\gamma_{nm})t} \\ &\quad \times \int_{-\infty}^{t} \mathrm{e}^{[i(\omega_{nm}+\omega_j)+\gamma_{nm}]t'} dt' \\ &= \frac{\rho_{mm}^{(0)} - \rho_{nn}^{0}}{\hbar} \sum_j \frac{\mu_{nm} \cdot E(\omega_j) \mathrm{e}^{-i\omega_j t}}{(\omega_{nm} - \omega_j) - i\gamma_{nm}}. \end{aligned}$$

This expression for $\rho_{nm}^{(1)}$ is used to calculate

$$\left[H'(t'), \rho^{(1)}\right]_{nm} = -\Sigma_\nu \left(\mu_{n\nu}\rho_{\nu m}^{(1)} - \rho_{n\nu}^{(1)} \, \mu_{\nu m}\right) \cdot E(t') \tag{9.76}$$

and

$$\rho_{nm}^{(2)} = \frac{-i}{\hbar} \mathrm{e}^{-(i\omega_{nm}+\gamma_{nm})t} \int_{-\infty}^{t} [H'(t'), \rho^{(1)}]_{nm} \, \mathrm{e}^{(i\omega_{nm}+\gamma_{nm})t'} dt'. \tag{9.77}$$

A similar procedure is used to calculate the higher-order terms. We have

$$\begin{aligned}\rho_{nm}^{(2)}(t) &= \frac{-i}{\hbar^2}\,\Sigma_{vjq}\\ &\times\left\{\left(\rho_{mm}^{(0)}-\rho_{vv}^{(0)}\right)\frac{[\mu_{nv}\cdot E(\omega_q)][\mu_{vm}\cdot E(\omega_j)]}{[\omega_{nm}-\omega_j-\omega_q-i\gamma_{nm}][(\omega_{vm}-\omega_j)-i\gamma_{vm}]}\right.\\ &\left.-\left(\rho_{vv}^{(0)}-\rho_{nn}^{(0)}\right)\frac{[\mu_{nv}\cdot E(\omega_j)][\mu_{vm}\cdot E(\omega_q)]}{[\omega_{nm}-\omega_j-\omega_q-i\gamma_{nm}][(\omega_{vm}-\omega_j)-i\gamma_{nV}]}\right\}\\ &\times \mathrm{e}^{-i(\omega_j+\omega_q)t}\end{aligned}\tag{9.78}$$

and

$$\begin{aligned}(\rho_{nm}^{(3)}(t) &= \frac{-i}{\hbar^3}\,\Sigma_{v\ell}\,\Sigma_{jg\gamma}\,\frac{(\mu_{nv}\cdot E(\omega_\gamma))\,\Omega_{nv\ell}-(\mu_{nv}\cdot E(\omega_\gamma))\,\Omega_{vm\ell}}{\omega_{nm}-\omega_j-\omega_q-\omega_\gamma-i\gamma_{nr}}\\ &\times \mathrm{e}^{-i(\omega_j+\omega_g+\omega_r)t},\end{aligned}\tag{9.79}$$

where

$$\begin{aligned}\Omega_{vm\ell} = \left(\rho_{mm}^{(0)}-\rho_{\ell\ell}^{(0)}\right)&\frac{[\mu_{v\ell}\cdot E(\omega_q)][\mu_{\ell m}\cdot E(\omega_j)]}{[(\omega_{vm}-\omega_j-\omega_q)-i\gamma_{vm}][\omega_{\ell m}-\omega_j-i\gamma_{\ell m}]}\\ -\left(\rho_{\ell\ell}^{(0)}-\rho_{vv}^{(0)}\right)&\frac{[\mu_{v\ell}\cdot E(\omega_j)][\mu_{\ell m}\cdot E(\omega_q)]}{[(\omega_{nv}-\omega_j-\omega_q)-i\gamma_{nv}][\omega_{n\ell}-\omega_j-i\gamma_{n\ell}]}.\end{aligned}\tag{9.80}$$

We can now use this result for $\rho^{(3)}$ to calculate the third-order nonlinear susceptibility by combining the following relationships:

$$\begin{aligned}P(\omega_j+\omega_q+\omega_r) &= N < \mu(\omega_j+\omega_q+\omega_r) >\\ &= N\,\mathrm{Tr}(\rho\mu)\\ &= N\,\Sigma_{nm}\,\rho_{nm}^{(3)}\,\mu_{mn}\end{aligned}$$

and

$$\begin{aligned}P_k(\omega_j+\omega_q+\omega_r) &= \Sigma_{xyz}\,\Sigma_{jgr}\,\chi_{kzyx}^{(3)}(\omega_j+\omega_q+\omega_r,\omega_r,\omega_q,\omega_j)\\ &\times E_y(\omega_r)E_x(\omega_q)E_z(\omega_j).\end{aligned}\tag{9.81}$$

Combining Eqs. (9.79), (9.80) and (9.81) and performing the summation over the input frequencies, together with the Cartesian coordinates, we finally obtain

$$\chi^{(3)}_{kijh}(\omega_j+\omega_q+\omega_r,\omega_j,\omega_q,\omega_r) = \Sigma_{\nu\ell nm}\frac{N}{\hbar^{(3)}}\rho^{(0)}_{\ell\ell}$$
$$\times\left\{\frac{\mu^k_{\ell\nu}\mu^j_{\nu n}\mu^i_{nm}\mu^h_{m\ell}}{[(\omega_{\nu\ell}-\omega_j-\omega_q-\omega_r)-i\gamma_{\nu\ell}][(\omega_{n\ell}-\omega_j-\omega_q)-i\gamma_{n\ell}][(\omega_{m\ell}-\omega_j)-\gamma_{m\ell}]}\right.$$
$$+\frac{\mu^k_{\ell v}\mu^i_{\nu n}\mu^j_{nm}\mu^h_{m\ell}}{[(\omega_{\nu\ell}-\omega_j-\omega_r-\omega_q)-i\gamma_{\nu\ell}][(\omega_{n\ell}-\omega_j-\omega_r)-i\gamma_{n\ell}][\omega_{m\ell}-\omega_j-i\gamma_{m\ell}]}$$
$$+\frac{\mu^k_{\ell\nu}\mu^h_{\nu n}\mu^j_{nm}\mu^i_{m\ell}}{[(\omega_{\nu\ell}-\omega_q-\omega_r-\omega_j)-i\gamma_{\nu\ell}][(\omega_{n\ell}-\omega_q-\omega_r)-i\gamma_{n\ell}][(\omega_{n\ell}-\omega_q)-i\gamma_{n\ell}]}$$
$$+\frac{\mu^k_{\ell\nu}\mu^j_{\nu n}\mu^h_{nm}\mu^i_{m\ell}}{[(\omega_{\nu\ell}-\omega_q-\omega_j-\omega_r)-i\gamma_{\nu\ell}][(\omega_{n\ell}-\omega_q-\omega_r)-i\gamma_{n\ell}][(\omega_{n\ell}-\omega_q)-i\gamma_{n\ell}]}$$
$$+\frac{\mu^k_{\ell\nu}\mu^i_{\nu n}\mu^h_{nm}\mu^j_{m\ell}}{[(\omega_{\nu\ell}-\omega_r-\omega_j-\omega_q)-i\gamma_{\nu\ell}][(\omega_{n\ell}-\omega_r-\omega_j)-i\gamma_{n\ell}][(\omega_{n\ell}-\omega_r)-i\gamma_{n\ell}]}$$
$$\left.+\frac{\mu^k_{\ell\nu}\mu^h_{\nu n}\mu^i_{nm}\mu^j_{m\ell}}{[(\omega_{\nu\ell}-\omega_j-\omega_r-\omega_q)-i\gamma_{\nu\ell}][(\omega_{n\ell}-\omega_j-\omega_r)-i\gamma_{n\ell}][(\omega_{n\ell}-\omega_j)-i\gamma_{n\ell}]}+\cdots\right\}, \tag{9.82}$$

plus another 42 terms of similar type with different permutations of the angular frequencies ω_j, ω_q and ω_r, along with the permutations of the Cartesian components of the dipole matrix elements. Notice that each frequency can take on a "+" or a "−" sign. The complete 48 terms have been derived by Bloembergen et al. [89]. This general expression of $\chi^{(3)}$ contains all possible third-order effects, including four-wave mixing. $\chi^{(3)}$ effects take place in all media while the $\chi^{(2)}$ effect vanishes in media with central symmetry. If there is a resonance with a particular energy level, the term representing this resonance effect will dominate. For example, in coherent anti-Stokes Raman scattering (CARS) one vibrational state $| \ t >$ is in resonance with the different frequencies of the laser photon, ω_a, and the Stokes photon, ω_e, such that $\omega_{tg} = \omega_a - \omega_e$. The anti-Stokes photon is scattered from the coherently excited phonon $\omega_A = \omega_a - \omega_c + \omega_a = 2\omega_a - \omega_e$. This situation is shown in Fig. 9.5.

In this situation, $\chi^{(3)}$ can be grouped into two terms, one resonantly enhanced and the remaining one a "nonresonant" background term,

$$\chi^{(3)} = \chi^{(3)}_{res} + \chi^{(3)}_{NR}.$$

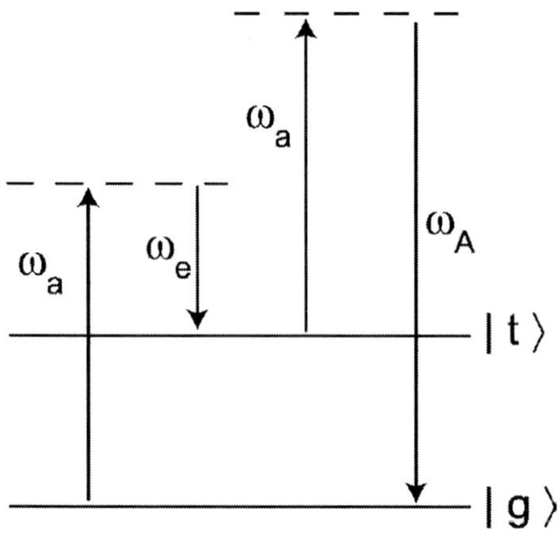

Fig. 9.5 Quantum mechanical representation of coherent anti-Stokes Raman scattering (CARS).

Problems

9.1 In the text we see examples of calculating the single-photon transition probability $W_{(1)}$ and the linear susceptibility $\chi^{(1)}$. Show that the expressions for $W_{(1)}$ and $\chi^{(1)}$ are consistent with each other by considering the power absorbed per unit volume.

9.2 Derive the expression of the three-photon transition probability $W_{(3)}$.

9.3 Derive Eq. (9.82). You may consult the paper by N. Bloembergen, H. Lotem and R. T. Lynch, Jr, Lineshapes in coherent resonant Raman scattering, *Indian J. Pure Appl. Phys.*, 16:151, 1978 for the complete 48 terms.

10
Applications of Nonlinear Optical Effects

In the last three chapters a survey of nonlinear phenomena and a simple theory of nonlinear optical effects were given. Nonlinear wave equations were illustrated mostly using only harmonic generation as examples. There are many nonlinear effects such as intensity-dependent refractive-index change, stimulated Raman and Brillouin scattering, soliton formation and propagation in optical fibers that are not addressed. These phenomena have been described extensively in the standard text books of nonlinear optics [67, 84–86, 90, 91]. Readers are referred to these references for further studies. In this chapter we shall present several nonlinear phenomena that are not commonly found in standard nonlinear optics books.

10.1
Optical Harmonic Generation at Oblique Incidence

Optical harmonic generation is commonly used for generating coherent light at twice the fundamental wave frequency. In this application, one needs to convert as much of the fundamental light waves into harmonics as possible. The conventional practice is to purchase a nonlinear optical crystal already oriented and cut for maximum frequency-conversion efficiency in the phase-matching configuration. The second-harmonic generation (SHG) is done in transmission with a normally incident laser beam. From an academic point of view there are more interesting phenomena associated with the optical harmonic generation with obliquely incident laser light. One can extend the law of optics to the nonlinear region. In the transmission experiment at normal incidence, the boundary conditions at an interface are straightforward. However, at oblique incidence they become interesting.

Extension of the fundamental optical phenomena from the linear to the nonlinear regime will not be complete, however, if the basic nonlinear interaction at the boundary between a linear and a nonlinear medium is not fully investigated [92]. This is the area that has not been fully exploited. Only a limited amount of work has been reported, in particular, when the fundamental beam is at total internal reflection [93–96]. In order to cover nonlinear counterparts

Light-Matter Interaction: Atoms and Molecules in External Fields and Nonlinear Optics.
W. T. Hill and C. H. Lee

ISBN: 978-3-527-40661-6

of total reflection, we shall assume that a light wave is incident onto a nonlinear medium from an optically denser linear medium. This situation was investigated in Refs. [93–96], where potassium dihydrogen phosphate (KDP) was immersed in 1-bromonaphthalene. In addition to SHG near total reflection, there have been a number of device applications employing nonlinear optical interactions or the electrooptic effect with total internal reflection [95,97]. The electrooptic effect with total internal reflection can be regarded as a nonlinear interaction between the low-frequency microwave and the high-frequency optical wave at the boundary of the nonlinear medium. This theoretical optical model explains critical angles well. The theoretical formalism for the electrooptic effect is essentially the same as SHG described in this book. Thus, it is important to have a good understanding of the nonlinear phenomena taking place at the boundary of the nonlinear medium, in particular, when the fundamental beam is obliquely incident. Recently, obliquely incident beams producing nonlinear optical phenomena have also gained attention because they can be used as a tool to probe solid–solid interfaces [85,98]. This is due to the fact that traditional optical spectroscopies lack interface specificity and most surface diagnostic techniques are inconvenient for studying buried interfaces. Although the work described here is for a liquid–solid interface, it can easily be extended to study buried solid–solid interfaces.

The reflection of laser light at the interface between linear and nonlinear media at an oblique angle of incidence, θ_i, near the critical angle, θ_{cr}, can provide interesting nonlinear optical effects as predicted by the Bloembergen–Pershan (BP) theory [92]. In the experimental work of Ref. [93] the investigation of second-harmonic generation in the vicinity of the critical angle, $\theta_{cr(\omega)}$, has been performed at total reflection. This leads to the verification of nonlinear optical effects [93, 96, 99]. Furthermore, using the same crystallographic cut of KDP [96], the work was extended theoretically to a nonlinear Brewster angle in ammonium dihydrogen phosphate (ADP) [100].

10.1.1
Theory of Light Wave at Boundary of Nonlinear Media

The general oblique incidence geometrical situation is shown in Fig. 10.1. For the situation just before total reflection occurs, the fundamental beam is transmitted almost parallel to the surface in the nonlinear KDP crystal. According to the BP theory [92], there are reflected harmonic beams and two transmitted harmonic beams. The driven polarization wave propagates in the same direction as the transmitted laser beam. It has a wave vector $\vec{k}_S = 2\vec{k}_L(\omega)$ and represents the particular solution of the inhomogeneous solution. In addition, there is the homogeneous solution with wave vector $k_T(2\omega)$. In the nonlinear KDP crystal, the two transmitted harmonic beams are spatially distinct and readily observed separately. The relative magnitude of θ_S and θ_T in Fig. 10.1 depends on the magnitude of the ordinary ray index of refraction $n_0(\omega)$ and the extraordinary ray index of refraction $n_e(2\omega)$, respectively. If at a particular crystallographic orientation of the KDP crystal $n_0^{\omega} > n_e^{2\omega}$, then $\theta_T > \theta_S$, and vice versa. As the angle of incidence, θ_i, in Fig. 10.1 is increased, it is intuitively clear that the beam with wave vector $k_S(2\omega)$ will disappear at the same time as the transmitted fundamental. The transmitted harmonic beam with wave vector $k_T(2\omega)$ will persist.

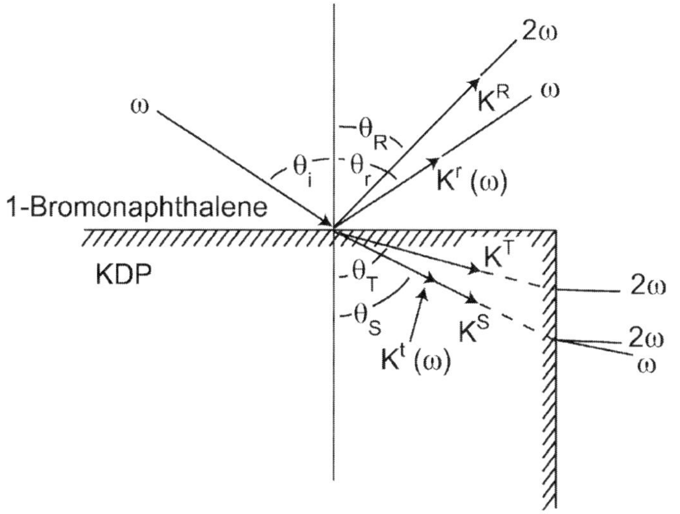

Fig. 10.1 *K* vectors of fundamental and second-harmonic waves in the vicinity of total reflection.

As θ_i becomes increasingly larger than the critical angle, this second-harmonic wave will eventually disappear and only the reflected harmonic wave remains. According to Fig. 10.1, the angles θ_R, θ_S and θ_T of the reflected, inhomogeneous and homogeneous transmitted waves are given, respectively, by

$$n_{liq}(\omega)\sin\theta_i = n_{liq}(2\omega)\sin\theta_R = n(\omega)\sin\theta_S = n(2\omega)\sin\theta_T. \tag{10.1}$$

The refractive indices without subscripts refer to the KDP crystal. The components of the nonlinear source polarization P^{NLS} along the cubic axes of the nonlinear KDP crystal are given in terms of the fundamental field components at each point inside the crystal by

$$P_z^{NLS}(2\omega) = \chi_{36}^{NL} E_x^T(\omega) E_y^T(\omega). \tag{10.2}$$

$P_x^{NLS}(2\omega)$, $P_y^{NLS}(2\omega)$ can be obtained by cyclic permutation of Eq. (10.2).

If the incident fundamental field is polarized perpendicular to the plane of incidence and along the [110] direction with respect to the crystallographic axes of KDP, P^{NLS} will be along the [001] direction and lies in the plane of incidence. Equation (10.2) can be expressed in terms of the amplitude E_0 of the fundamental wave by

$$P_z^{NLS}(2\omega) = \chi_{36}^{NL}\eta(F_T^L E_0)^2, \tag{10.3}$$

where η is a geometrical factor that depends on the orientation of the fundamental field vector and nonlinear polarization component with respect to the crystallographic axes of KDP. The linear Fresnel factor F_T^L describes the change of the amplitude of the fundamental wave on transmission at the crystal surface. If the linear polarization is perpendicular to the plane of incidence, it is given by

$$F_T^L = \frac{2\cos\theta_i}{\cos\theta_i + \sin\theta_{cr}(\omega)\cos\theta_S}, \tag{10.4}$$

where the critical angle for total reflection $\theta_{cr}(\omega)$ is defined by

$$\sin\theta_{cr}(\omega) = n(\omega)/n_{lig}(\omega) < 1.$$

The nonlinear polarization is the source of the three harmonic waves. The electric field amplitudes of the reflected and transmitted harmonic waves are given by

$$E_R(2\omega) = 4\pi P^{NLS} F_R^{NL}, \tag{10.5a}$$

$$E_S(2\omega) = 4\pi P^{NLS} F_S^{NL}, \tag{10.5b}$$

$$E_T(2\omega) = 4\pi P^{NLS} F_T^{NL}. \tag{10.5c}$$

The nonlinear Fresnel factors F_R^{NL}, F_S^{NL} and F_T^{NL} have been calculated by the BP theory. For the case of second-harmonic polarization with $P^{NLS}(2\omega)$ parallel to the plane of incidence, the nonlinear Fresnel factors, according to the BP theory, becomes

$$F_{R,||}^{NL} = \frac{\sin\theta_S \sin^2\theta_T \sin(a + \theta_S + \theta_T)}{\epsilon_R(2\omega)\sin\theta_R \sin(\theta_T + \theta_R)\cos(\theta_T - \theta_R)\sin(\theta_S + \theta_T)}, \tag{10.6}$$

$$F_{S,||}^{NL} = \frac{\sin\,\alpha}{\epsilon_S - \epsilon_T}, \tag{10.7}$$

$$F_{T,||}^{NL} = \frac{-\epsilon_S^{1/2}\sin\,\alpha}{\epsilon_T^{1/2}(\epsilon_S - \epsilon_T)} + \frac{\epsilon_R^{1/2}}{\epsilon_T^{1/2}} F_{R,||}^{NL}, \tag{10.8}$$

where $\epsilon_R^{1/2}(2\omega) = n_{liq}(2\omega)$, $\epsilon_S^{1/2} = n_{KDP}(\omega)$ and $\epsilon_T^{1/2} = n_{KDP}(2\omega)$. The angle α is the angle between the nonlinear polarization $P^{NLS}(2\omega)$ in the plane of incidence and the direction of the source vector K_S. It is emphasized that these expressions remain valid in the case of total reflection.

The time-averaged second-harmonic power carried by the harmonic beam is given by the real part of the Poynting vector times the cross-sectional area A of the respective beams:

$$I_{R,S,T}(2\omega) = (c/8\pi)\epsilon_{R,S,T}^{1/2}(2\omega) \mid E_{R,S,T}(2\omega) \mid^2 A_{R,S,T}. \tag{10.9}$$

$I_{R,S,T}(2\omega)$ in Eq. (10.9) is the intensity integrated over the beam cross section, or power. This is the experimentally observed quantity. In the remainder of this chapter intensity and power will be used interchangeably in a similar manner as in Refs. [93,96,99]. In Eq. (10.9), $A_{R,S,T}$ is the cross-sectional area of the beam and is given by

$$A_{R,S,T} = dd' \cos\theta_{R,S,T} / \cos\theta_i, \tag{10.10}$$

where dd' is the rectangular slit which defines the size of the incident laser beam. By substitution of the relevant expression into Eq. (10.9), we finally obtain

$$I_R(2\omega) = (c/8\pi)\epsilon_R^{1/2} \mid E_o \mid^4 dd'(4\pi\chi_{36}^{NL})^2\eta^2 \mid F_L \mid^4 \mid F_{R,||}^{NL} \mid^2 \times \cos\theta_R(\cos\theta_i)^{-1}, \tag{10.11}$$

$$I_S(2\omega) = (c/8\pi)\epsilon_S^{1/2} \mid E_o \mid^4 dd'(4\pi\chi_{36}^{NL})^2\eta^2 \mid F_L \mid^4 \mid F_{S,||}^{NL} \mid^2 \times \cos\theta_S(\cos\theta_i)^{-1}, \tag{10.12}$$

$$I_R(2\omega) = (c/8\pi)\epsilon_T^{1/2} \mid E_o \mid^4 dd'(4\pi\chi_{36}^{NL})^2\eta^2 \mid F_L \mid^4 \mid F_{T,||}^{NL} \mid^2 \times \cos\theta_T(\cos\theta_i)^{-1}, \tag{10.13}$$

for the intensities of reflected, transmitted source and transmitted homogeneous harmonic waves, respectively. The expression of Eqs. (10.12) and (10.13) will facilitate us in obtaining the theoretical curves of the intensity of the harmonic waves and their comparison to experimental data.

When the two transmitted second-harmonic beams are not spatially resolved, the sum of the homogeneous and inhomogeneous intensities $I_{total}^{(2\omega)} = I_S(2\omega) + I_T(2\omega)$ is observed. The total transmitted second-harmonic intensity $I_{total}^{(2\omega)}$ is equal to the average intensity of the interference pattern of the two transmitted beams, which is observed in the more common geometry that the nonlinear crystal is a plane-parallel platelet and that the light beams are nearly at normal incidence. However, for a wedge-shaped sample, as in the case discussed here, a spatial average is taken over the interference pattern in the direction normal to the surface of entry, where the two transmitted beams overlap. Therefore, the total transmitted harmonic intensity is equal to the sum of the intensities of the separated harmonic beams.

The uniaxial KDP crystal is employed for transmitted SHG at phase-matching conditions by birefringence. The phase-matching angle, θ_m, is the result of making the birefringence, $(n_0^{2\omega} - n_e^{2\omega})$, equal to the dispersion, $(n_0^{2\omega} - n_0^{\omega})$, at the phase-matching angle, θ_m, which as a consequence will give $n_e^{2\omega} = n_0^{\omega}$ for this condition. Furthermore, the determination of $n_e^{2\omega}(\theta)$ for a specific value of θ can be obtained from the equation of the index ellipsoid, given by

$$\frac{1}{[n_e^{2\omega}(\theta)]^2} = \frac{\cos^2\theta}{[n_0^{2\omega}]^2} + \frac{\sin^2\theta}{[n_e^{2\omega}(\pi/2)]^2}. \tag{10.14}$$

10.1.2
Criteria of Null Transmitted SHG

According to the BP theory for the case of $P^{NLS}(2\omega)$ lying in the plane of incidence, the nonlinear Fresnel factors in reflection and transmission are given by Eqs. (10.6), (10.7) and (10.8), respectively. As a consequence, the intensities of the reflected, transmitted inhomogeneous and homogeneous harmonic beams are given in Eqs. (10.11), (10.12) and (10.13), respectively. Analysis of Eqs. (10.6), (10.7) and (10.8) derived from the BP theory leads us to the conditions of null transmitted second-harmonic intensities. These conditions can be summarized as follows.

1. For the case of the angle of incidence, θ_i, in the neighborhood of the critical angles $\theta_{cr}^{(2\omega)}$ and $\theta_{cr}^{(\omega)}$, one can expect that the transmitted homogeneous harmonic intensity, $I_T(2\omega)$, becomes zero first when $\theta_i > \theta_{cr}^{(2\omega)}$ (64.76°). After further increasing the angle of incidence to equal the value of $\theta_{cr}^{(\omega)}$ (66.78°), the transmitted inhomogeneous harmonic intensity becomes zero as well. This situation has been confirmed by experimental results as indicated in Fig. 10.2 [101]. A more detailed description of the experimental arrangement for the observation of SHG in oblique incidence will be presented in the next section.

2. Furthermore, the condition of $I_S(2\omega) = 0$ and $I_T(2\omega) = 0$ can occur aside from the total reflection case. The situation can be achieved when $\alpha = 0°, \theta_S = 0°$ and $\theta_T = 0°$. Under this condition the fundamental beam has the angle of normal incidence ($\theta_i = 0°$) and $P^{NLS}(2w)$ must be along the face normal of the crystal, which is the optical axis in the Z direction of [001] as indicated in the inset of Fig. 10.3.

When $a = 0°$, $\theta_S = 0°$ and $\theta_T = 0°$, the nonlinear Fresnel factors $F_{R,||}^{NL}, F_{S,||}^{NL}$ and $F_{T,||}^{NL}$ in Eqs. (10.6)–(10.8) become zero and as a consequence $I_R(2\omega)$, $I_S(2\omega)$ and $I_T(2\omega)$ also become zero. It is interesting to notice that under

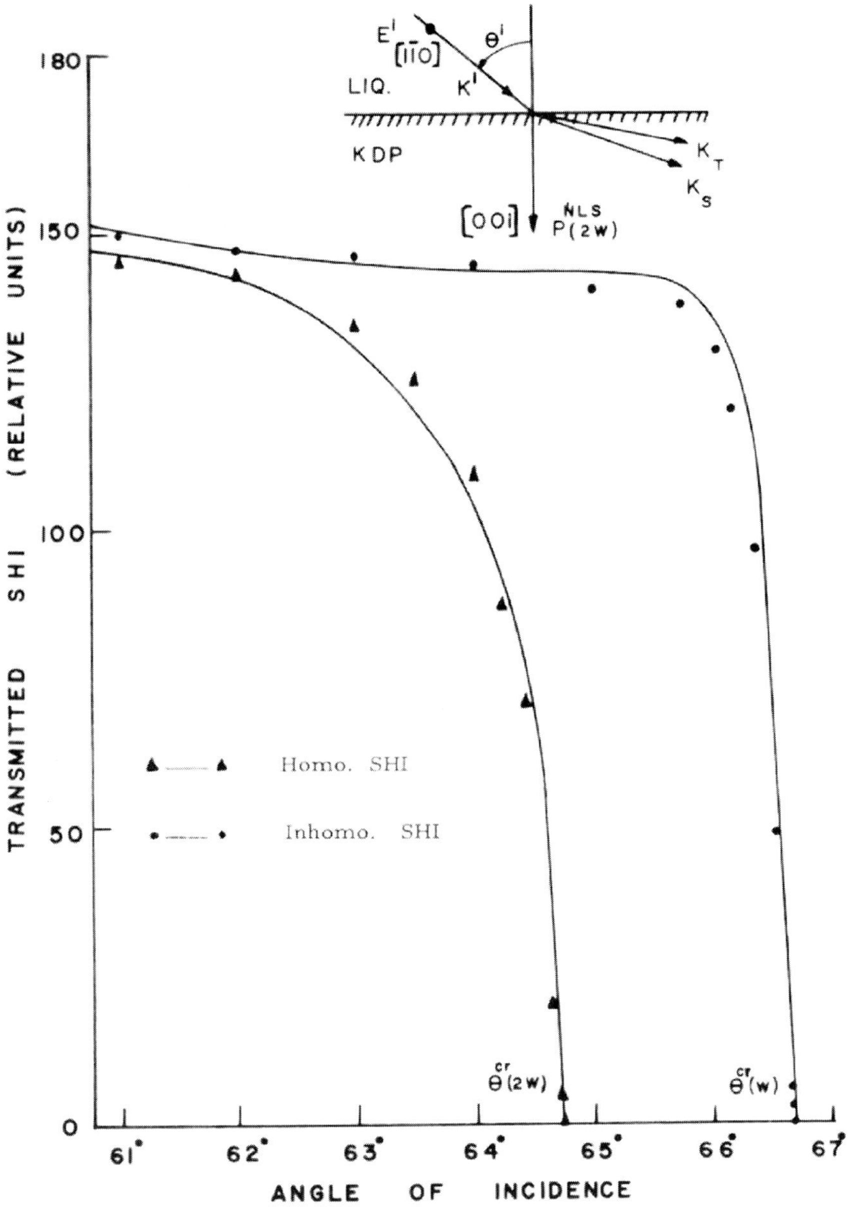

Fig. 10.2 The intensities of inhomogeneous and homogeneous transmitted second-harmonic waves in the neighborhood of the critical angles $\theta_{cr}(2\omega)$ and $\theta_{cr}(\omega)$.

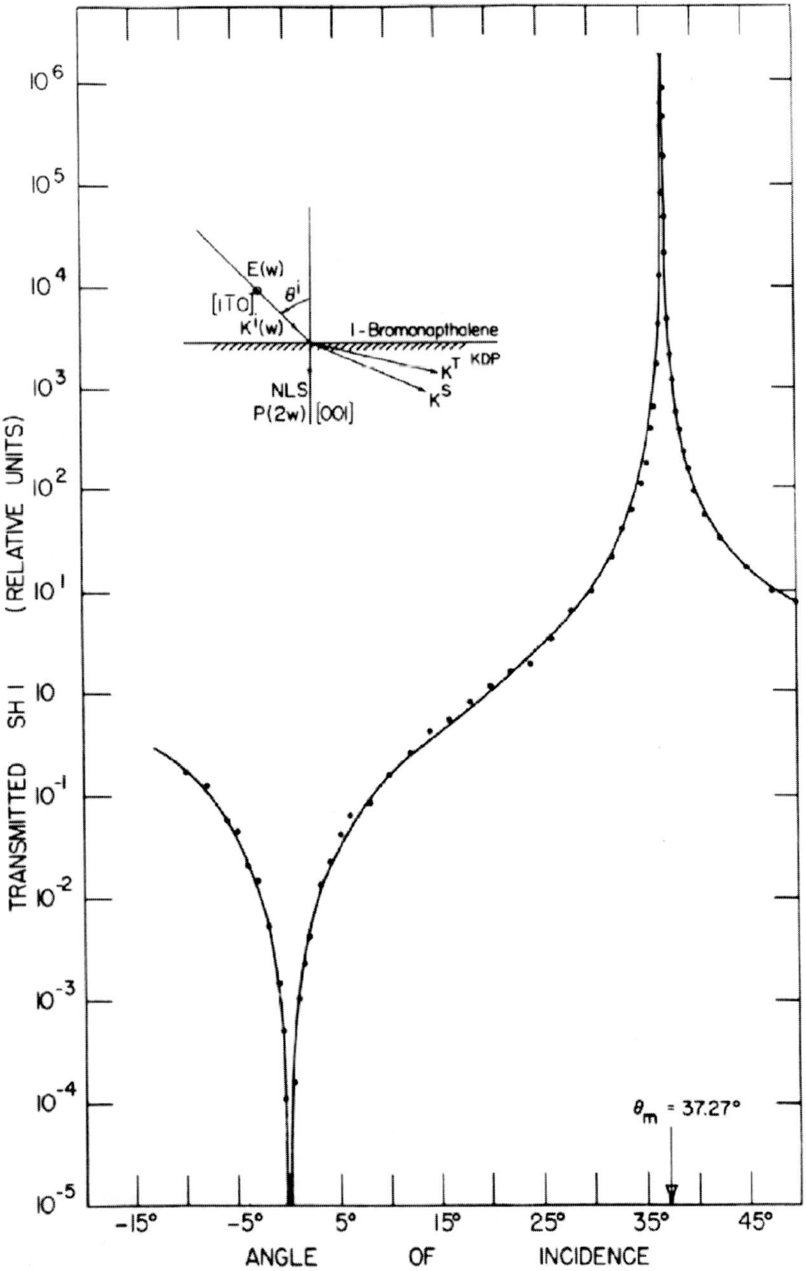

Fig. 10.3 The total transmitted second-harmonic intensity $I_{total}(2\omega) = I_s(2\omega) + I_T(2\omega)$ as a function of the angle of incidence θ_i. $I_s(2\omega) = I_T(2\omega) = 0$ at $\theta_i = 0$ and the phase-matching angle $\theta_i^m = 37.27°$.

this condition not only do the inhomogeneous and homogeneous harmonic intensities become zero, but also the reflected harmonic intensities. Therefore, no second-harmonic generation occurs at normal incidence for the crystal that has a particular crystallographic orientation as indicated in Fig. 10.3. The physical interpretation of the null transmitted second-harmonic inhomogeneous, homogeneous and reflected intensities is that the nonlinear polarization, $P^{NLS}(2\omega)$, cannot radiate inside the medium in the direction which otherwise would yield transmitted inhomogeneous homogeneous, and reflected harmonic rays. This situation reflects an uncommon phenomenon in which one cannot obtain transmitted second-harmonic light with transmission from a nonlinear optical medium as compared to the simple normal incidence case. It has been demonstrated for the null of $I_S(2\omega)$ and $I_T(2\omega)$ and they are in good agreement with the BP theory.

10.1.3
Experimental Observations of the Second-Harmonic Generation at Oblique Incidence

The experimental arrangement to verify the points mentioned in the previous section is shown in Fig. 10.4. The KDP crystal with various cuts for different experimental purposes was immersed in an optically dense fluid, 1-bromonapthalene, which has larger indices than the KDP crystal at both frequencies ω and 2ω. The wavelength for the fundamental wave is 1.05 μm. The indices for KDP are $n_0^{\omega} = 1.4943$, $n_0^{2\omega} = 1.5130$ and $n_e^{2\omega} = 1.4707$; for 1-bromonapthalene, $n_{liq}(\omega) = 1.6262$ and $n_{liq}^{2\omega} = 1.6701$. The fluid is transparent in the wavelength range of 0.4–1.6 μm. The critical angles are $\theta_{cr}^{\omega}(\omega) = 66.78°$ and $\theta_{cr}^{2\omega}(2\omega) = 64.76°$, respectively.

A. Transmitted Homogeneous and Inhomogeneous SHG in the Neighborhood of Critical Angles

According to the BP theory, there are two transmitted harmonic beams. The driven polarization wave propagates in the same direction as the transmitted laser beam. It has a wave vector $K_S = 2\ K_{laser}(\omega)$. Furthermore, there is a homogeneous harmonic wave with a wave vector $K_T(2\omega)$. In the experiments, the KDP crystal has right-angled corners and the two transmitted harmonic beams are spatially distinct and readily observed separately. According to the KDP crystallographic orientation as indicated in Fig. 10.2 and also the investigation of transmitted SHG in the neighborhood of critical angles, the KDP crystal employed in the experiment under this situation has the value $n_0^{\omega} > n_e^{2\omega}(\theta)$. Therefore, in the neighborhood of the critical angle, for a given angle of incidence, θ_i, there exists θ_S and θ_T which can be determined by Eq. (10.1), such that $\theta_T > \theta_S$. Therefore, from the experiment it is found that

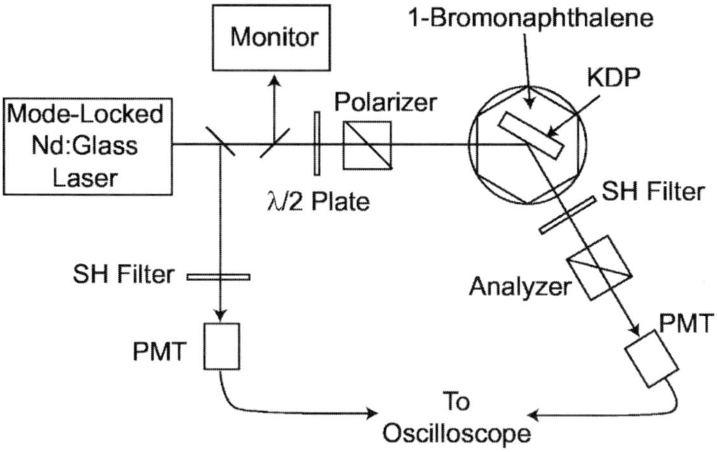

Fig. 10.4 Experimental arrangement for the oblique incidence SHG experiment. To detect SHG in transmission, the detector arm needs to swing in the direction of transmission.

$\theta_{cr}(\omega) > \theta_{cr}(2\omega)$. It is clear that the beam with wave vector $K_T(2\omega)$ will disappear at $\theta_i = \theta_{cr}(2\omega)$ and the ray with wave vector K_S will disappear at the same time as the fundamental beam. As the angle of incidence, θ_i, becomes larger than $\theta_{cr}(\omega)$, there will be no transmitted second-harmonic beam.

The inhomogeneous and homogeneous harmonic intensities are given by Eqs. (10.11) and (10.12), respectively. The solid curves drawn in Fig. 10.2 are theoretical curves predicted by the BP theory and are calculated from the last four factors, respectively,

$$| F^L |^4 | F^{NL}_{S,T} |^2 \cos\theta_{S,T}(\cos\theta_i)^{-1}.$$

The experimental dotted points are in excellent agreement with the theoretical prediction that the homogeneous and inhomogeneous harmonic intensities will be terminated at $\theta_{cr}(2\omega)$ and $\theta_{cr}(\omega)$, respectively. The reason that the homogeneous and inhomogeneous harmonic intensities vanish at $\theta_{cr}(2\omega)$ and $\theta_{cr}(\omega)$ is that when θ_i is greater than $\theta_{cr}(2\omega)$ and $\theta_{cr}(\omega)$ the values of $\cos\theta_T$ and $\cos\theta_S$ become purely imaginary. As a consequence, the two harmonic intensities will become imaginary, which is not physically allowed. It is worthwhile to notice that, from Fig. 10.2, there is an interval $\theta_{cr}(2\omega) < \theta_i < \theta_{cr}(\omega)$ that is about 2.02°. In this region, there exists only an inhomogeneous harmonic intensity that has a direct association with the nonlinear polarization, $P^{NL}(2\omega)$, and with the nonlinear susceptibility, χ^{NLS}. The knowledge of inhomogeneous second-harmonic generation in this particular region will directly facilitate the study of $P^{NLS}(2\omega)$ and χ^{NL} of a nonlinear medium.

B. Second-Harmonic Generation in Reflection Near the Critical Angle.

The orientation of KDP is the same as the one shown for the transmission experiment. The polarization of the fundamental beam is along the [110] direction with respect to the crystallographic axes of the crystal. The reflected second-harmonic intensity generated from the KDP crystal immersed in 1-bromonapthalene was observed as a function of the angle of incidence, θ_i. The theoretical curve was calculated from the expression

$$| F_T^L |^4 F_{R,11}^{NL} |^2 \cos\theta_R (\cos\theta_i)^{-1}$$

of Eq. (10.11). The result of the experiment is shown in Fig. 10.5. The solid curve represents the result of the theoretical calculation. The figure shows excellent agreement between the experimental data and the result predicted by the BP theory. There are two cusps at $\theta_{cr}(\omega)$ and $\theta_{cr}(2\omega)$. The theoretical curve reflects the influence of the linear and nonlinear Fresnel factors given by Eqs. (10.4) and (10.6) when θ_i approaches $\theta_{cr}(\omega)$ and $\theta_{cr}(2\omega)$. Nonanalytical singularities occur at $\theta_i = \theta_{cr}(\omega)$ and $\theta_i = \theta_{cr}(2\omega)$, as predicted by the BP theory. The reason for the appearance of two cusps at critical angles $\theta_{cr}(\omega)$ and $\theta_{cr}(2\omega)$ is that at these critical angles the values of $\cos\theta_S$ and $\cos\theta_T$ change from real to imaginary. This experimental feature was not clearly observed in earlier experiments [93] of this sort. We attribute the success of the work in Ref. [96] to the fact that in the experiment, a mode-locked laser beam was used and, therefore, multimode effects in the nonlinear process were minimized.

The enhancement of the reflected second-harmonic intensity in the vicinity of the critical angles $\theta_{cr}(\omega)$ and $\theta_{cr}(2\omega)$ arises mainly from two sources. First, the linear Fresnel factor F^L near the critical angle $\theta_{cr}(\omega)$ is larger than it is away from this angle, by a factor of approximately two. This gives an increase by a factor of 16 in the reflected second-harmonic intensity $I_R(2\omega)$ in Eq. (10.9). Second, the nonlinear Fresnel factor $F_{R,11}^{NL}$, defined in Eq. (10.6), is dominated in the vicinity of $\theta_{cr}(\omega)$ and $\theta_{cr}(2\omega)$ by the term $[\sin\ (\theta_S + \theta_T)]^{-1}$, and it is larger than that away from these points by a factor of three. Therefore, after all factors are accounted for, the additional enhancement of $I_R(2\omega)$ in the vicinity of the critical angle will be about two orders of magnitude, as indicated in Fig. 10.5.

It is interesting to consider the limiting case when $\theta_{cr}(\omega)$ and $\theta_{cr}(2\omega)$ coalesce into a single value [96]. Under this condition, the wave vectors K_S and K_T will propagate along the same direction, i.e., along the crystal surface, because of total internal reflection. Since the crystal was cut with the phase-matching direction along the crystal surface, the two critical angles $\theta_{cr}(\omega)$ and $\theta_{cr}(2\omega)$ merged into one and the additional enhancement of $I_R(2\omega)$ occurs as a result of phase matching, as indicated in Fig. 10.6. One can consider the curve shown in Fig. 10.6 as a limiting case of that of Fig. 10.5. When the phase-matching

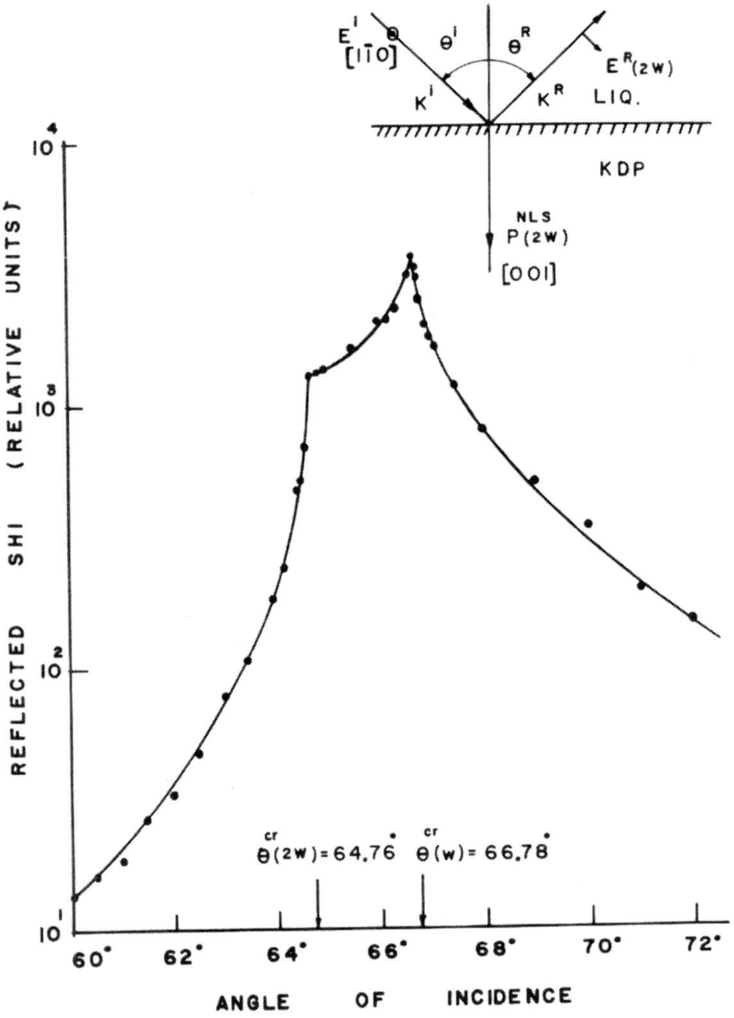

Fig. 10.5 Reflected second-harmonic intensity (SHI) from KDP crystal of the non-phase-matchable orientation in the neighborhood of the critical angle for total reflection.

condition of total reflection prevails, the two cusps in Fig. 10.5 collapsed into a single maximum peak of $I_R(2\omega)$ at $\theta_{cr}(\omega)$.

C. Nonlinear Brewster Angle

According to the BP theory, a nonlinear Brewster angle is predicted for second-harmonic generation from a nonlinear medium. The nature of a nonlinear Brewster angle is considered as a counterpart to the Brewster angle in the

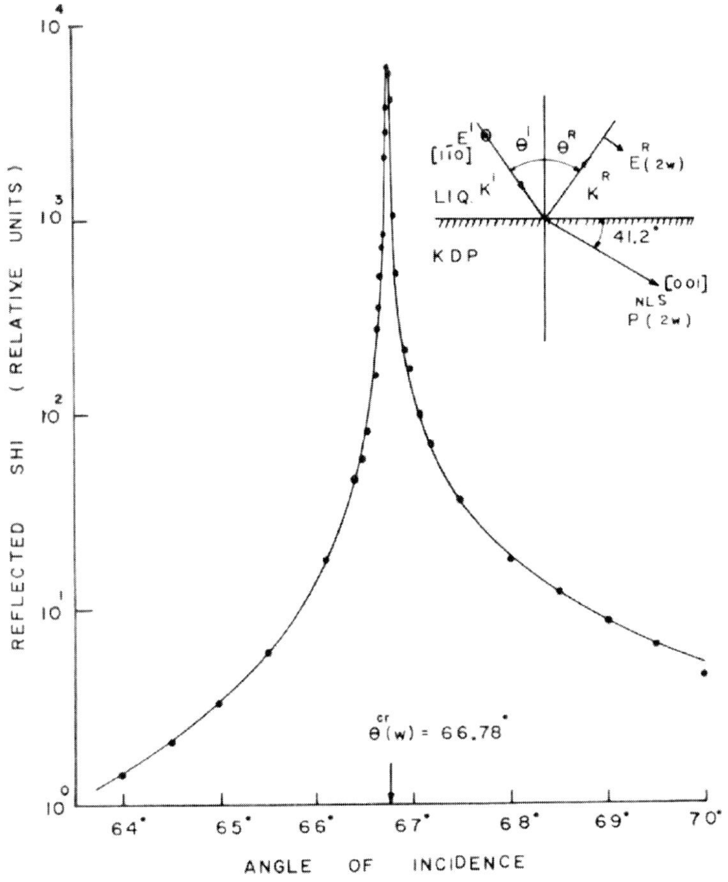

Fig. 10.6 Reflected second-harmonic intensity (SHI) from a KDP crystal with phase matching at total reflection.

linear optical case. When the fundamental beam is incident upon a nonlinear medium of a specific crystallographic orientation, the reflected second-harmonic intensity $I_R(2\omega)$ vanishes at a particular angle of incidence called the nonlinear Brewster angle. The origin and physical interpretation of the nonlinear Brewster angle [92,93] can be understood in terms of classical dipole radiation. At the nonlinear Brewster angle, the fundamental beam will create a nonlinear polarization $P^{NLS}(2\omega)$ inside the medium in the direction of propagation of the reflected second-harmonic wave. According to classical dipole radiation theory, there is no radiation in this direction. This nonradiative wave upon refraction back into the linear medium would otherwise give rise to the reflected second-harmonic wave in the direction of K^R. The reflected second-harmonic intensity $I_R(2\omega)$ vanishes at the nonlinear Brewster angle when the nonlinear polarization source $P^{NLS}(2\omega)$ lies in the plane of incidence. Since the reflected second-harmonic intensity $I_R(2\omega)$ depends on the components of $P^{NLS}(2\omega)$ in the plane of incidence, measurement of the nonlinear Brewster angle, at which $I_R(2\omega) = 0$, will directly yield the ratio of the components of $P^{NLS}(2\omega)$ in the plane of incidence, which could be helpful in determining the relative signs of various components of the $\chi^{(2)}$ tensor [102]. Hence, measurement of the nonlinear Brewster angle could be applicable as a "null" method, which is similar to the work of Heinz et al. [103] for the precise measurement of the nonlinear optical susceptibility tensor. It is interesting to point out that if the orientation of the KDP crystal as indicated in Fig. 10.5 is employed for studying the reflected second-harmonic intensity, $I_R(2\omega)$, for the range $20° \leq \theta_i < 50°$, one can obtain a dip of $I_R(2\omega)$ at $\theta_i^{NB} = 42.83°$, which corresponds to the nonlinear Brewster angle of KDP that was observed experimentally [95] as shown in Fig. 10.7.

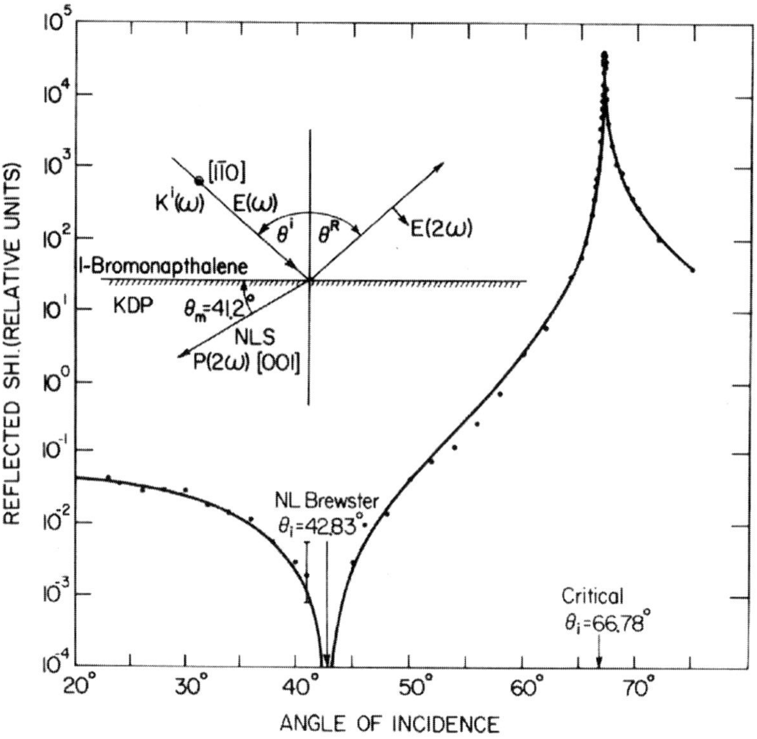

Fig. 10.7 Reflected second-harmonic intensity from KDP, showing nonlinear Brewster angle of KDP at 42.83°.

10.2 Optical Second-Harmonic Generation due to Reflection from Media with Inversion Symmetry

The third-rank $\chi^{(2)}$ tensor vanishes in media with inversion symmetry, such as Ag, Au, Si and Ge. However, second-harmonic generation in reflection has been observed in these media. In early experiments, interests were in identifying the physical mechanism responsible for the observed effects. It was found that both conduction and bound electrons contribute to the observed effect. From the beginning, it was pointed out that such phenomena may be utilized to probe the surface property of materials. Surface optical harmonic generation indeed becomes a powerful tool for probing interfaces between media with inversion symmetry. Second-harmonic generation processes are forbidden by symmetry in media with an inversion center but are allowed on the surface because of a lack of inversion symmetry of the surface layers. They are therefore highly surface specific and can be used as surface probes. In comparison with other surface techniques, such as coherent anti-Stokes Raman scattering (CARS) and stimulated Raman gain, surface SHG has the advantages of being much simpler in the experimental arrangement and yields much stronger signals.

The theory of SHG at an interface between linear and nonlinear media with inversion symmetry follows from the work of Bloembergen and Peshan, which was outlined in Section 10.1.1 with the following nonlinear polarization:

$$\begin{aligned} P_S^{NL}(2\omega) &= (\delta - \beta)(E(\omega) \cdot \nabla)E(\omega) + \beta E(\omega)(\nabla \cdot E(\omega)) \\ &\quad + \alpha E(\omega) \times H(\omega) + \chi_S^{(2)}\delta(z) : E(\omega)E(\omega) \\ &\quad + \chi_b^{(3)} : E(\omega)E(\omega)E_{dc}(z). \end{aligned}$$

The first two terms on the right are of quadrupole character. The third term is a magnetic dipole. Following the work of Shen, a surface dipole layer has been added with surface nonlinear susceptibility $\chi_S^{(2)}$, which can model the submonolayer of adsorbate at the interface between two dense media. Finally, a DC-field-induced SHG (EFISHG) was added. The last term is of technological significance since it can be utilized to probe the solid interface in Si microelectronic materials and devices. For a given sample, it may only have a contribution from a few terms in $P_S^{NL}(2\omega)$. In rare cases there are contributions from all terms. Using the appropriate term for P_S^{NL} as the driving source for the nonlinear wave equation, one can solve for the solution of the reflected SH field. In the case of quadrupolar source terms only, the solution has been worked out by Bloembergen et al. [104].

In general, the reflected harmonic wave has components polarized in the plane of incidence $(E_{2\|}^R)$ as well as perpendicular to the plane of incidence

($E^R_{2\perp}$). In terms of the incident field E_0, one obtains

$$\begin{aligned} | E^R_{2_\perp} | &= (32\pi\omega/c)E_0^2 \sin\theta\cos^2\theta \sin\phi\cos\phi, \\ &\times \left| \frac{[\epsilon(\omega)-1]\bar{\beta}}{[\epsilon(\omega)]^{1/2} g_0(\omega) g_1(\omega) g_1(2\omega)} \right| \end{aligned} \tag{10.15}$$

and

$$\begin{aligned} | E^R_{2\|} | &= (32\pi\omega/c)E_0^2 \sin\theta\cos^2\theta \\ &\times | \{[\epsilon(2\omega)]^{1/2}\epsilon(\omega) g_0(2\omega) g_0^2(\omega)\}^{-1} | \\ &\times \Big| [-\gamma\epsilon(\omega)[\cos^2\phi + \sin^2\phi g_0^2(\omega)/g_1^2(\omega)] \\ &- \frac{1}{2}\bar{\delta}[\epsilon^2(\omega)-1]\epsilon(2\omega)\cos^2\phi \\ &+ \bar{\beta}[\epsilon(\omega)-1][\epsilon(\omega)-\sin^2\theta]^{1/2} \\ &\times [\epsilon(2\omega)-\sin^2\theta]^{1/2}\cos^2\phi] \Big|, \end{aligned} \tag{10.16}$$

where

$$\begin{aligned} g_0(\omega) &= [\epsilon(\omega)]^{1/2}\cos\theta + [1-\epsilon^{-1}(\omega)\sin^2\theta]^{1/2} \\ g_0(2\omega) &= [\epsilon(2\omega)]^{1/2}\cos\theta + [1-\epsilon^{-1}(2\omega)\sin^2\theta]^{1/2}, \\ g_1(\omega) &= \cos\theta + [\epsilon(\omega)-\sin^2\theta]^{1/2}, \\ g_1(2\omega) &= \cos\theta + [\epsilon(2\omega)-\sin^2\theta]^{1/2}. \end{aligned}$$

These equations give the SH amplitude and polarization in terms of the incident fundamental amplitude E_0, as a function of the angle of incidence θ and of the direction of the incident polarization ϕ. For semiconductors and insulators one expects that $\beta = -2\gamma = 2\bar{\beta}$. For materials with a high dielectric constant, the dominant term in the expression of $E^R_{2\|}$ is the term involving $\bar{\beta}$. The drawn curve in Fig. 10.8 describes the theoretical dependence of the square of this term on the angle of incidence, θ. The behavior is largely determined by the function $\sin^2\theta\cos^4\theta$. The behavior is common to silicon, where the contribution due to valence electrons dominates. However, in metals both conduction and bound electrons contribute, leading to a more complicated angular dependence. The dominant nonlinear surface terms are also responsible for the dependence of the harmonic intensity on ϕ. The last two terms in the expression of $E^R_{2\|}$ predict a $\cos^4\phi$ dependence for both Ag and Si as shown in Fig. 10.9. For metals, the conduction electron (plasma) contribution to the nonlinearity is usually larger than that arising from interband transitions of

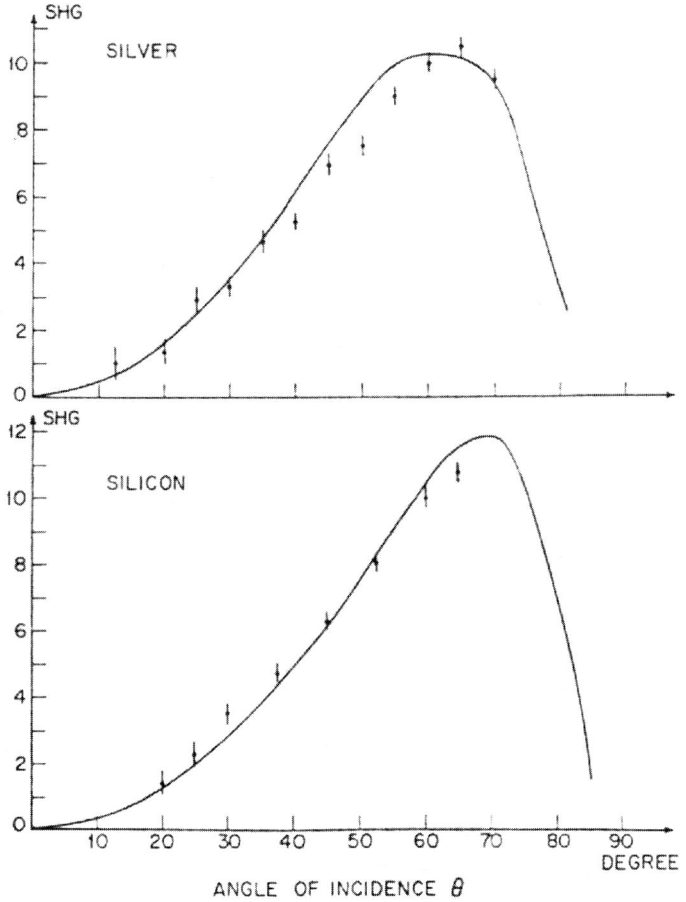

Fig. 10.8 The variation of the SH intensity in Si and Ag as a function of the incidence angle. The fundamental electric field vector lies in the plane of incidence, $\phi = 0°$. The drawn curves are derived from theory.

the valence electrons. In this case $\alpha, \beta, \delta, \gamma$ should be replaced by

$$\beta_{pl} = \frac{e}{\delta\pi m^* \omega^2}, \quad \gamma_{pl} = \frac{c}{2i\omega} \; \alpha_{pl} = \beta_{pl} \left(\frac{\omega_p^2}{4\omega^2} \right)$$

$$\delta_{pl} = \beta_{pl} + 2\gamma_{pl},$$

with $\omega_p^2 = 4\pi n_0 e^2 / m^*$ being the plasma frequency.

From earlier studies, it has been shown that SHG due to reflection from media with inversion symmetry is described rather well by the quadrupole-type nonlinear properties calculated for the homogeneous bulk material with an abrupt discontinuity at the boundary. The order of magnitude can be correctly

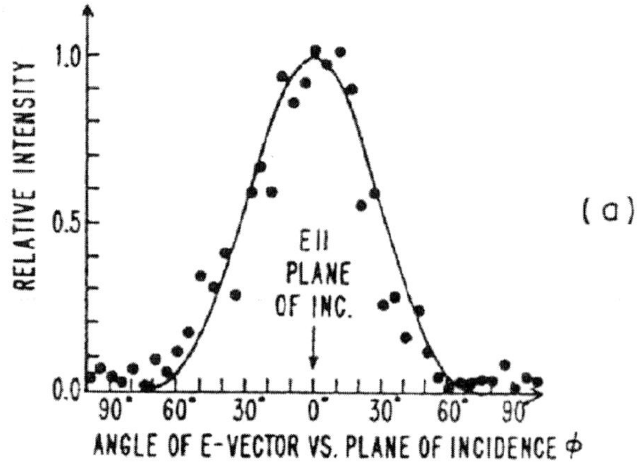

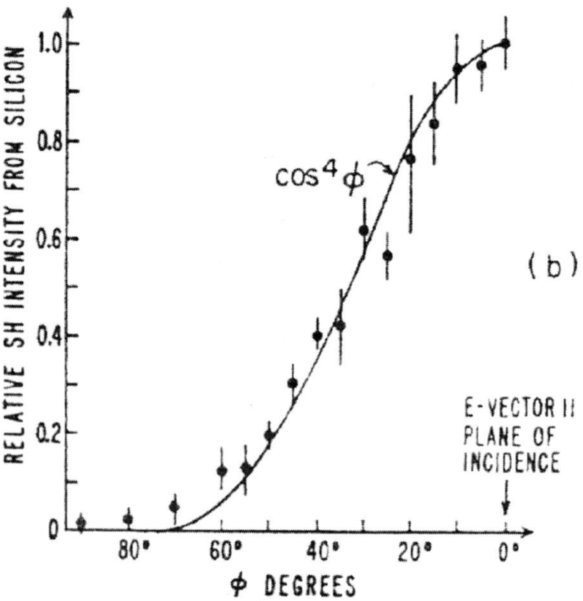

Fig. 10.9 Variation of the SH intensity as a function of angle ϕ between the fundamental electric vector and the plane of incidence. The angle of incidence is 45°. (a) Data for Ag; (b) data for Ge. The solid line representing a $\cos^4 \phi$ dependence is drawn for comparison.

related to the linear dielectric constant in insulators. There is a marked trend of increasing nonlinearity with increasing dielectric constant. The bound electrons also contribute significantly to the nonlinearity in metals. The directional and polarization properties are well described by a combination of linear and nonlinear Fresnel factors. In materials with a high dielectric constant, the dominant contribution comes from the first layer of atoms in the bulk material.

The second-harmonic generation from a submonolayer of adsorbates can be described by the surface susceptibility term,

$$\chi_S^{(2)} = \chi_{SA}^{(2)} + \chi_{SS}^{(2)}, \tag{10.17}$$

where $\chi_{SA}^{(2)}$ denotes the contribution from adsorbed atoms or molecules, and $\chi_{SS}^{(2)}$ that from the surface layer of the adjoining medium. If $| \chi_{SA}^{(2)} | \gg | \chi_{SS}^{(2)}$, the surface SHG can be used to probe the adsorbates. This effect has been utilized by Shen [85] to study the nonlinear spectroscopy of dye molecules on a fused-quartz substrate. Using a tunable laser, resonant-enhanced SHG was observed from rhodamine 6G and rhodamine 110 dyes, which show distinct peaks of the SH signal when the second-harmonic frequency 2ω is in resonance with the $S_0 \rightarrow S_2$ transition. The resonant-enhanced SH signal was very strong, several orders of magnitude stronger than the SH signal from the fused-quartz substrate. With input and output beam polarizations and geometries properly chosen, it is possible to measure a particular tensor element of $\chi_{SA}^{(2)}$ selectively. The average orientation of p-nitrobenzoic acid (PNBA) molecules on fused quartz has been determined in this fashion at solid–air and solid–ethanol interfaces to be 70° and 40°, respectively [85].

The SHG technique offers a number of advantages as a surface-specific probe: simple experimental setup, *in situ* probing of the interface and time-resolved dynamic properties of interfaces obtainable with a pump–probe method using a femtosecond laser. There is a major advantage in using a high repetition rate femtosecond laser for the surface SHG experiment. In these experiments, a photon-counting technique is used to detect the SH photons. The data-acquisition rate and signal-to-noise ratio depend on the photon-counting rate, i.e., photons/s. It has been shown [105] that the photon-counting rate, C, is given by

$$C = S(2\omega)\, f_{rep} - 10^{-9}\, F^2\, \frac{A}{T_p}\, f_{rep},$$

where $S(2\omega)$ is SH photons/pulse, f_{rep} the pulse-repetition rate, F the fluence in mJ/cm^2, A the area of the laser spot in cm^2 and T_p the pulse duration in seconds. For a Ti:sapphire laser with $A \sim 10^{-4}$ cm^2, $t_p \sim 100$ fs, $F \sim 0.1$ mJ/cm^2 and $f_{rep} \sim 10^6$/s the counting rate is about 10^6 SH photons/s. With this counting rate, near "real-time" monitoring of subsecond surface kinetic processes, e.g., CVD growth, can be obtained using a SHG surface probe [105].

With the improved sensitivity due to the use of a femtosecond high repetition rate laser, it is now possible to study the nonlinear electroreflectance or electric-field-induced second-harmonic generation (EFISHG) at silicon–solid dielectric interfaces [106].

10.3
Nonlinear Electroreflectance

Nonlinear electroreflectance (NER) was first observed in 1967 [107]. It has been shown that the reflected optical second-harmonic generation depends quadratically on the externally applied bias voltage perpendicular to the interface. Its precise physical mechanism remains uncertain due to the lack of sensitivity in detecting SHG signals from these early experiments. Thanks to the advent of high repetition rate femtosecond lasers, sensitive gated photon counters and data-acquisition instruments, systematic and detailed studies of NER or electric-field-induced SHG (EFISHG) are now possible. This technique has been applied to silicon microelectronic devices operating at multiple gigahertz frequencies. The internal electric field distribution at the solid–silicon interface can be probed and mapped out with this new technique. We shall review some of these works in this section.

Optical SHG was widely observed from interface-specific probes of centrosymmetric materials, because $\chi^{(2)}$ vanishes in the bulk of these materials. Si(001) interfaces are among the most important and the simplest for such nonlinear optical analysis. Their importance derives from their ubiquitous presence in metal–oxide–semiconductor (MOS) devices. Their simplicity derives from the small number of tensor elements required to describe the interfacial nonlinear response. Downer and coworkers [108] have systematically studied the SH spectroscopy in which they have varied and independently measured various parameters affecting SHG: subsurface DC electric fields, carrier-induced screening of DC EFISHG, surface hydrogen coverage and doping concentration.

By varying the DC E fields, one can distinguish the surface, bulk and DC-field-induced contributions due to SHG and validate the theoretical framework for NER in semiconductors with unprecedented completeness. Phenomenological description of NER for Si(001)/SiO_2 interfaces is much simpler than that for the Si(111)/SiO_2 interfaces. For a p-polarized fundamental light wave and a p-polarized SH field (Pin–Pout) configuration, the second-harmonic intensity from a Si(001) interface can be written as [108]

$$I_{pp}(2\omega) \sim | P^S_{isotropic} + P^B_{iso} + P^B_{aniso} \cos(4\Psi) + P^{BE}(\phi - \phi_{fb}) |^2,$$

where Ψ is the azimuthal angle between the incident plane and the $< 110 >$ direction of Si. P^S_{is0} and P^B_{is0} are due to the isotropic surface and the bulk nonlinear polarizations. P^B_{anis0} and P^{BE} are the anisotropic and field-induced bulk polarizations. Also, ϕ_{fb} is the flat-band voltage and ϕ is the applied voltage. If we substitute $2P^{BE} \times P^{B}_{aniso} \sim a_4$, $\mid P^{BE} \mid^2 \sim a_0^{FD}$, $\mid P^{B}_{aniso} \mid^2 /2 \sim a_8$, $I_{pp}(2\omega, \Psi)$ can be expressed as [108]

$$I_{pp}(2\omega, \Psi) = a_0^{FI} + a_0^{FD}(\phi - \phi_0)^2 + a_4(\phi - \phi_0)\ \cos(4\Psi) + a_8\ \cos(8\Psi), \tag{10.18}$$

where a_0^{FI} is a field-independent parameter, a_0^{FD} is the coefficient for the field-dependent term, $\cos(4\Psi)$ reflects the four-fold symmetry of the Si(001) surface and ϕ_0 is the applied voltage that yields a minimum SH intensity when $\phi_0 \sim \phi_{fb}$. By measuring I_{pp} as a function of the applied bias voltage, azimuthal angle Ψ and second-harmonic photon energy, all a parameters in Eq. (10.18) can be extracted from experimental data. Since EFISHG is strongly related to the intrinsic potential at the silicon surface, carrier-induced screening of the DC field at the Si(001)–SiO_2 interface plays an important role in determining the SH intensity. Downer's group [108] studied this screening effect using a pump beam to illuminate the area which generates the EFISHG. It was observed that the SH signal amplitude decreases as the pump laser fluence increases. This can be understood by rewriting

$$P(2\omega) \sim \vec{\chi}_B^{(3)} : E(\omega)E(\omega)E_{DC}(z) + \vec{\chi}_S^{(2)} : E(\omega)E(\omega) + \vec{\chi}_B^{(2)Q} : E(\omega)\nabla E(\omega),$$

where $\vec{\chi}_B^{(2)Q}$ is the four-fold anisotropic bulk quadrupole susceptibility, E_{DC} is the space-charge field normal to the interface which is sensitive to screening and $\vec{\chi}_S^{(2)}$ is the isotropic surface dipole susceptibility. It should be noted that $\chi_S^{(2)}$ and $\chi_B^{(3)}$ are frequency dependent. $\vec{\chi}_B^{(2)}$ dominates near the $2\hbar\omega \sim 3.26$ eV resonance, and is responsible for the field-independent SHG; while $\chi_B^{(3)}$ dominates near the $2\hbar\omega \sim 3.4$ eV resonance associated with the E_1 critical point. Fluence-independent SHG was observed at $2\hbar\omega \sim 3.26$ eV, while the SH intensity at 3.4 eV decreases quadratically with increasing fluence, illustrating the carrier-induced screening of the DC electric field. Since doping density affects the band bending near the surface, SH signals show a strong influence from doping concentration and hydrogen coverage. This effect has been utilized for *in situ* "real-time" monitoring of the silicon epitaxial growth rate and growth condition on Si(001).

Since Si is the dominant semiconductor for advanced microelectronic devices that are fabricated on a single chip, there is a strong demand for new noninvasive measurement techniques with high temporal and spatial resolution. The EFISHG method described above can conveniently fill this need. It can be used to map out the electrical field distribution at various internal nodes on the chip and at the same time measure the high-frequency electrical waveform or electrical impulses propagating through each node. Broadband electrical impulse measurement with 1-ps time resolution with the EFISHG technique has been reported [106].

10.4
Near-Field Second-Harmonic Microscopy

In the previous section we have shown examples of harmonic generation in reflection. These are all far-field effects. We have learned that optical second-harmonic generation is an extremely sensitive technique for characterization and investigation of a wide variety of material surfaces and absorbates. Second-harmonic generation is known to be affected by crystal structure, magnetic and ferroelectric order, mechanical strain, etc. Second-harmonic generation is especially useful for recovery of the symmetry properties of the sample under investigation. Far-field SH microscopy has already demonstrated great potential in such fields as magnetic imaging and characterization of metal–oxide–semiconductor (MOS) devices [106]. In many applications, spatial resolution below the diffraction limit of far-field optics is required. For example, better than 100-nm resolution is required for characterization of thin ferroelectric films such as $BaTiO_3$ and $PbZr_xTi_{1-x}O_3$ (PZT), which form the basis of a new thin-film technology for data storage. As a result, many attempts have been made recently to implement various near-field optical techniques for SH imaging. In this section we shall review these topics.

The first example to be discussed is the near-field SH imaging of $BaTiO_3$ and PZT. They are ferroelectric materials. The interest of these materials lies in their ability to switch from one stable polarization state to another. This property forms the basis of a new thin-film technology for data storage. Thin PZT films are used in prototype nonvolatile ferroelectric random-access memory (NVFRAM) and dynamic random-access memory (DRAM). PZT materials also have numerous applications in actuators, transducers, resonators and sensors. Crucial parameters of piezoceramic performance in different applications are hysteresis, nonlinearity, polarization retention (or the loss of polarization), etc. For proper application of the material, it is imperative to understand this nonlinear behavior on a microscopic level. Optical microscopy of SHG in piezoceramic PZT is a natural tool to study this local nonlinearity.

Far-field observations of SHG in PZT ceramic near the ferroelectric phase transition demonstrate a sharp drop in SHG in the paraelectric state. The SH intensity vs temperature curve corresponds qualitatively to the temperature dependence of the spontaneous polarization, which is an order parameter in the transition to the polar phase. $PbZr_xTi_{1-x}O_3$ (PZT) ceramic is a strongly scattering medium consisting of individual submicrometer-size crystallites. The elastic scattering lengths of fundamental and second-harmonic light in PZT are of the order of the characteristic size of a crystallite. Theory of SHG in strongly scattering media shows that generation of SH light in PZT should occur in a thin layer (of the order of the elastic scattering length) near the illuminated surface. Thus near-field second-harmonic microscopy is ideally suited for studies of PZT nonlinearity and poling properties at the microscopic level (the level of individual crystallites and crystallite boundaries). Its main

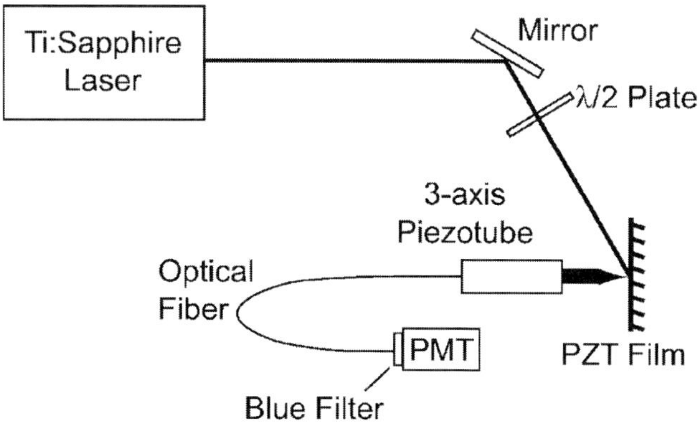

Fig. 10.10 Schematic of the near-field second-harmonic microscopy setup. p- and s-polarization directions of the fundamental light are indicated.

advantages in comparison with other scanning probe techniques such as the recently developed piezo response atomic force microscopy are the possibility of fast time-resolved measurements and substantially smaller perturbation of the sample under investigation caused by the optical probe. It is a noninvasive probe technique. The experimental setup for near-field SH imaging of the surface of PZT ceramic is described in Fig. 10.10. Typically, a Ti–sapphire laser system consisting of an oscillator and a regenerative amplifier at 810 nm is used as the light source. The laser has a repetition rate up to 250 kHz and a pulse duration of 100 fs with a pulse energy of 10 μJ. The SH from the surface of a PZT sample is excited at an angle of incidence around 60° by a weakly focused beam of the Ti–sapphire laser. The excitation power should be kept well below the ablation threshold. The local SHG is collected using either a coated or an uncoated adiabatically tapered fiber tip. The fiber tip is drawn at the end of a single-mode fiber by a standard heating and pulling procedure. The fiber tip is scanned over the sample surface with a constant tip–surface distance of a few nanometers using a shear force feedback control system [109]. Therefore, surface topography images can be obtained with a resolution on the nanometer scale, while simultaneously recording the SH near-field image. The SH signal is measured with a gated photon-counting photodetector. The characteristic SH photon counting rate is about one photon count per 30 laser pulses. With a laser repetition rate of 250 kHz, a typical SH signal at every point of the image requires averaging over 30–100 ms, or 7500 to 25 000 pulses. The image acquisition time for an image size of $1 \times 3\ \mu m^2$ is about 20–30 min. This shows a clear advantage for using high repetition rate, high peak power Ti–sapphire regenerative amplifier lasers. In comparison, a similar image would require 200–300 min to acquire if a Q-switched kilohertz repetition rate Nd:YAG laser were used.

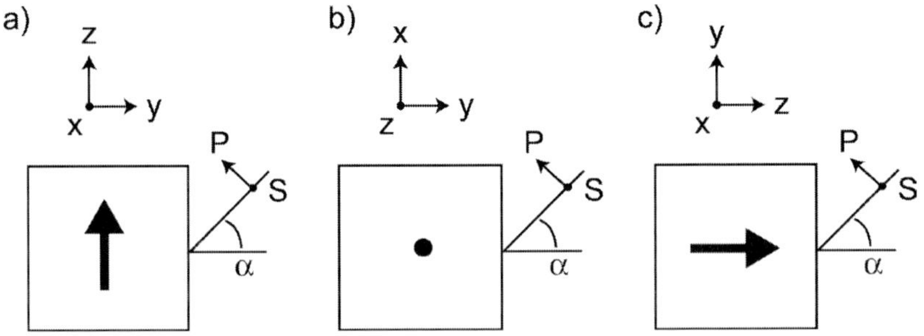

Fig. 10.11 Orientation of the $BaTiO_3$ crystal with respect to the illuminating fundamental light for different configurations.

Smolyaninov et al. [110] have recently demonstrated the near-field SH imaging of the c/c/a/a polydomain structure of epitaxial PZT thin films. In order to investigate the symmetry property of the nonlinear effect, they first performed an experiment on a poled single crystal of $BaTiO_3$. Since $BaTiO_3$ and PZT have a similar perovskite structure and belong to the same tetragonal symmetry class, symmetry characteristics of $BaTiO_3$ can be applied directly to the crystallites of the PZT thin film. Furthermore, phase-matched SHG is prohibited in the bulk of $BaTiO_3$ because of its strong dispersion. As a result, the measured SH signal originates at the surface, which makes the experimental situation look very similar to the case of a thin PZT film. Near-field SH signal dependences on the polarization of the fundamental light for three different poling (optical axis) directions of the $BaTiO_3$ crystal can be analyzed according to the configuration shown in Fig. 10.11. Let us consider fundamental light illuminating the tip of the microscope located near the $BaTiO_3$ crystal surface. Three different cases of the poling vector orientation are shown in Fig. 10.11, where α is the incident angle, and the z-axis is chosen to coincide with the poling direction. All three cases may be considered in a similar way, so we will concentrate on case (b) and only present the final results for the cases (a) and (c). Let us consider the fundamental optical field distribution in the tip–sample region at distances much smaller than the wavelength λ of the fundamental light. In this region the quasielectrostatic approximation may be used. The tip shape may be approximated as an ellipsoid with a very high aspect ratio. For p-polarized excitation light along the y component (the component perpendicular to the sample surface) of the crystal the exact analytic solution of the Laplace equation [111] gives $E_{tip}(\omega) = \varepsilon E_{y0}(\omega)$, where ε is the dielectric constant of the tip, E_{y0} is the incoming field and E_{tip} is the y component of the field just below the tip apex. As a result, for the incidence angle α around $\pi/2$ the local SHG under the tip is enhanced by a factor of n^8 (since $I(2\omega)_{tip} = E^2_{tip}(2\omega) = E^4_{tip}(\omega) = n^8 E^4_{y0}(\omega)$), where n is the tip refractive index. Assuming that $n = 1.6$, the enhancement factor equals about 40. It may be even higher (of the order of $(2\varepsilon)^4$) in the vicinity of a sample surface with a high dielectric constant, such as a metal or a ferroelectric, due to the image potential. In our discussion to follow, we will not consider any particular value of the field enhancement factor. Instead, we denote the fundamental field enhancement factor by γ. The fundamental optical field components under the tip apex at a polarization angle ϕ can be written as

$$E_x = E\cos\phi\cos\alpha, \quad E_y = (\gamma/\varepsilon_b)E\cos\phi\sin\alpha, \quad E_z = E\sin\phi, \tag{10.19}$$

where ε_b is the dielectric constant of $BaTiO_3$ and $\phi = 0$ and $\phi = 90°$ correspond to the p-polarized and s-polarized excitation light, respectively. Taking into account the nonzero components of the second-harmonic susceptibility tensor for $BaTiO_3$: $\chi^{(2)}_{yyz} = \chi^{(2)}_{xxz} = \chi^{(2)}_{yzy} = \chi^{(2)}_{xzx} = 17.7 \times 10^{-12}$ m/V, $\chi^{(2)}_{zyy} = \chi^{(2)}_{zxx} = -18.8 \times 10^{-12}$ m/V and $\chi^{(2)}_{zzz} = -7.1 \times 10^{-12}$ m/V (where z is the poling direction [76]), the second-harmonic field components at the tip apex may be written as

$$\begin{aligned} D^{(2)}_x &= \chi^{(2)}_{xzx}E^2\sin^2\phi\cos\alpha, \qquad D^{(2)}_y = (\gamma/\varepsilon_b)\chi^{(2)}_{xzx}E^2\sin^2\phi\cos\alpha, \\ D^{(2)}_z &= (\gamma^2/\varepsilon_b^2)\chi^{(2)}_{zxx}E^2\cos^2\phi\sin^2\alpha \\ &\quad + \chi^{(2)}_{zxx}E^2\cos^2\phi\cos^2\alpha + \chi^{(2)}_{zzz}E^2\sin^2\phi. \end{aligned} \tag{10.20}$$

If we assume that the microscope tip collects only dipole radiation, the second-harmonic optical signal will be proportional to

$$I^{(2)} = D^{(2)2}_z + D^{(2)2}_x. \tag{10.21}$$

(Second-harmonic dipoles that oscillate along the y direction do not radiate toward the tip.) Similar calculations of the field at the tip apex are easy to perform for cases (a) and (c). Let us point out that within this model the angular dependences of the field E for the points in space around the tip apex should be approximately the same as in Eq. (10.19). Only the enhancement factor γ is different for these points (its value changes from a maximum just under the tip apex to $\gamma = 1$ far from the apex). So, in the final result one must replace the factors γ^4 and γ^2 by the average values $< \gamma^4 >$ and $< \gamma^2 >$ taken over all the area around the tip apex where the SH signal is collected.

The final results of the calculations are presented in Fig. 10.12. They should be compared to the experimental result shown in Fig. 10.13. The SH signal exhibits four-fold symmetry with respect to a polarization angle rotation in the case (a) of the poling (optical axis) direction located in the incidence plane of the fundamental light parallel to the crystal surface. The symmetry becomes two-fold in case (b) when the optical axis is perpendicular to the incidence plane. In case (c) when the poling direction is perpendicular to the crystal surface, the SH signal is much weaker than in cases (a) and (b). Big differences

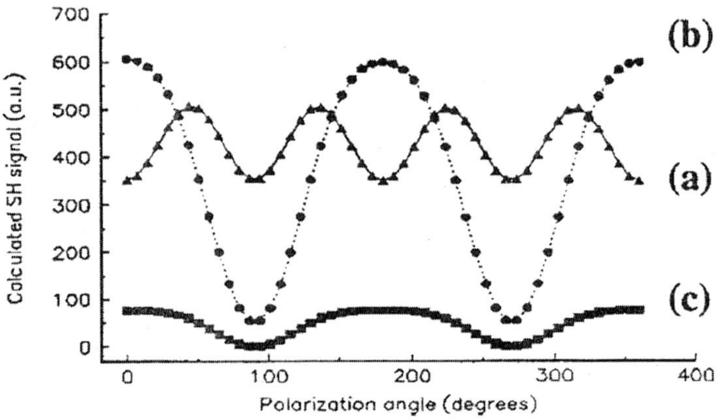

Fig. 10.12 Theoretical near-field SH signal dependences on the polarization angle of fundamental light corresponding to the geometry shown in Fig. 10.11.

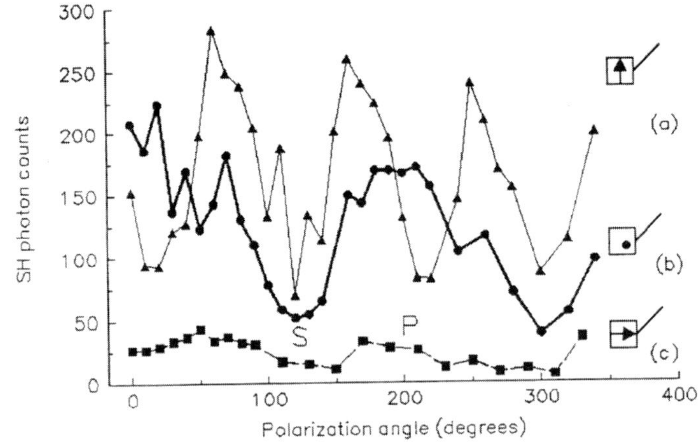

Fig. 10.13 Near-field SH signal dependences on the polarization of the fundamental light for the three different poling directions of the $BATiO_3$ crystal (shown by the arrows) with respect to the incoming fundamental light (shown by the straight lines) for the geometry in Fig. 10.11.

between these dependences indicate that the near-field SHG may be used to recover the local poling direction in the thin ferroelectric films. The best agreement with experimentally measured symmetry properties of the near-field SH polarization curves was achieved in the $\gamma = 5.7$ range for the field enhancement factor. This is reasonably close to the expected value of $\gamma = 2\varepsilon_{tip}$. A much smaller value of the SH signal in case (c) is a consequence of the fact that in the cases (a) and (b), $I^{(2)}$ contains terms proportional to $< \gamma^4 >$, but in case (c) the SH signal is proportional to $< \gamma^2 >$:

$$\begin{aligned} I^{(2)} &= D_x^{(2)2} + D_y^{(2)2} \\ &= (< \gamma^2 > / \varepsilon_b^2) \chi_{xzx}^{(2)2} E^4 (\sin^2 2\phi \sin^2 \alpha + \cos^4 \phi \sin^2 2\alpha). \end{aligned} \tag{10.22}$$

The transition from the near-field to the far-field mode of the microscope operation (which corresponds to an increase in the tip–sample separation from a few nanometers to a distance of a few wavelengths of light) can be described mathematically as a transition from $\gamma = 5$–7 to $\gamma = 1$ (no field enhancement by the tip). Results of numerical calculations for such a transition performed using Eq. (10.21) are shown in Fig. 10.14. Experimental data shown in the same figure are in good qualitative agreement with the calculations: both theoretical and experimental data have the same symmetry of the SH polarization dependences in the near-field and far-field zones. This indicates that this simple model provides an adequate description of the essential physics involved in near-field observation of SHG from ferroelectric samples.

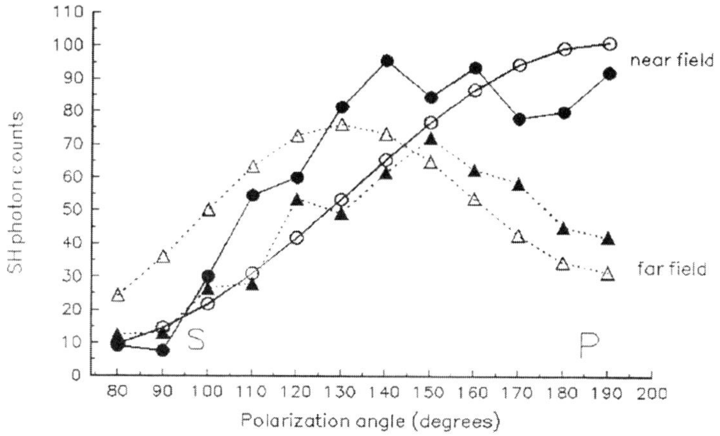

Fig. 10.14 Theoretical (open circles and triangles) and experimentally measured (filled circles and triangles) data for the transition from the near-field to the far-field behavior of the SH signal polarization dependency for geometry (b) in Fig. 10.11.

Based on the developed understanding of the SHG contrast due to differences in ferroelectric poling directions, one can calculate the spatial distribution of the local SHG from a typical $PbZr_xTi_{1-x}O_3$ film showing c/a/c/a polydomain structure assuming 100-nm spatial resolution of the microscope. Such calculations have been carried out by Smolyaninov et al. [110]. Their experimental observation agreed qualitatively with the theoretical analysis.

10.5
Terahertz Pulse Generation by Optical Rectification

Most examples given so far has been on second-harmonic generation, which is the degenerate case for the sum-frequency generation. To generate low-frequency or even DC waves one can use difference-frequency generation with $\chi^{(2)}(\omega_1 - \omega_2)$. When $\omega_1 = \omega_2$ this leads to an optical rectification effect which was first observed in 1962. Interest in difference-frequency generation was primarily centered on the generation of coherent far-infrared waves. With the advent of the femtosecond optical pulse, optical rectification provides a convenient means of generating ultrashort electrical impulses by the optical rectification process [85]

$$P^{NL}(0;\omega,-\omega) = \chi^{(2)}(0;\omega,-\omega)E(\omega)E^*(\omega).$$

The beat wave has zero frequency. The nonlinear process essentially strips off the high-frequency component, leaving only the envelope function of the pulse. In principle, the electric impulse thus generated should follow the envelope function of the original optical pulse. For an optical pulse of a few hundred femtoseconds in duration, the electric impulse will last about 1 ps. When such pulses are radiated into free space, they are referred to as terahertz radiation since their central frequency lies in the terahertz spectral range. A pure DC electric signal cannot radiate; however, carrierless short electrical waves can radiate, generating ultrawideband signals with a small number of RF cycles. Previous work with this type of terahertz radiation has reported that the radiation typically consists of the one and one-half cycles – Mexican hat – pulse shape [112].

Pulsed terahertz radiation emits ultrawideband signals covering the spectral region from 300 GHz to 12 THz (or 10–400 cm^{-1}). This spectral region fills the so-called "gap" (or void) between microwave and mid-infrared. The difficulties of conventional far-IR spectroscopy lie in the lack of efficient radiation sources and detectors. Pulsed terahertz radiation is a clean coherent THz source enabling scientists to perform THz spectroscopy by using a time-domain spectroscopy technique.

By sampling the precise waveform emitted from the radiator and by measuring the transmitted or scattered signal from the sample, one can extract all necessary information about the sample. This is possible because the change in the waveform is the result of interaction between the pulsed terahertz radiation and the sample. The time-domain measurement provides complete information on the amplitude and phase (time delay or variation) change of the pulsed waveform as it interacts with the sample. Thus, by performing a Fourier transform of the time-domain result one obtains the frequency-domain information just like that obtained in standard spectroscopy.

Time-domain spectroscopy (TDS) techniques can yield both the real and imaginary parts of the dielectric constant of the material as a function of frequency without resorting to the Kramers–Kronig relations. THz emission from materials can also be used to study the physical properties of the materials, such as semiconductor interfaces, lattice and carrier dynamics via coherent phonon oscillation, velocity overshoot and the coherent charge oscillator. Optical pump-THz probe spectroscopy techniques have also been applied to study carrier transport in semiconductors. Strong THz pulses have been used to ionize Rydberg atoms for atomic study.

THz impulse radiation can also be generated by exciting a photoconductive switch with femtosecond optical pulses [113]. In this case the photoconductive semiconducting material is transformed instantaneously from a semi-insulating state (assume that a high dark resistivity semiconductor is used) to a quasimetallic state when the femtosecond pulses illuminate the semiconductor with a photon energy greater than the semiconductor band-gap energy. If a voltage is applied between the two electrodes of the dipole antenna, a transient current will suddenly appear when the semiconducting gap is illuminated, causing radiation of an ultrawideband electrical impulse signal. Among the two methods of generating THz pulses, the photoconductive dipole antenna is in general more efficient. However, optical rectification enables one to study nonlinear optical characteristics of the sample. It serves as the only technique to generate THz pulses from a dielectric. In this section we will limit our discussion to the optical rectification method.

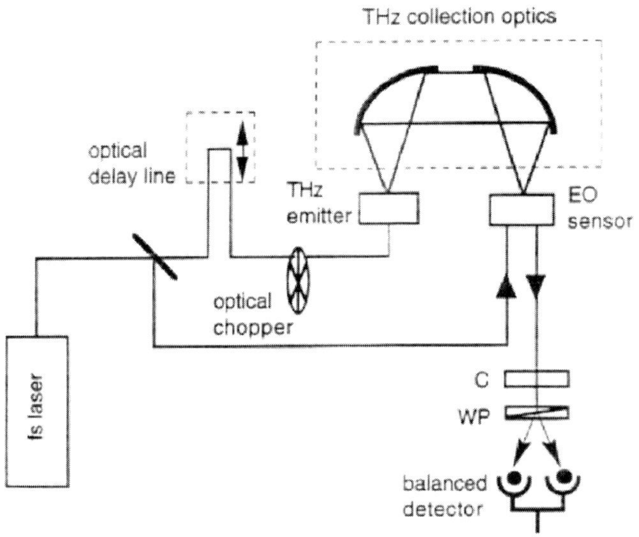

Fig. 10.15 Experimental setup for THz pulse generation and detection.

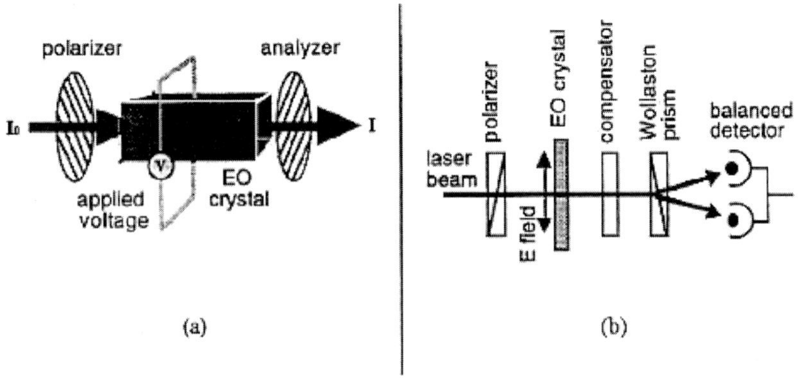

Fig. 10.16 Electrooptic THz field sensor in a balanced detection scheme.

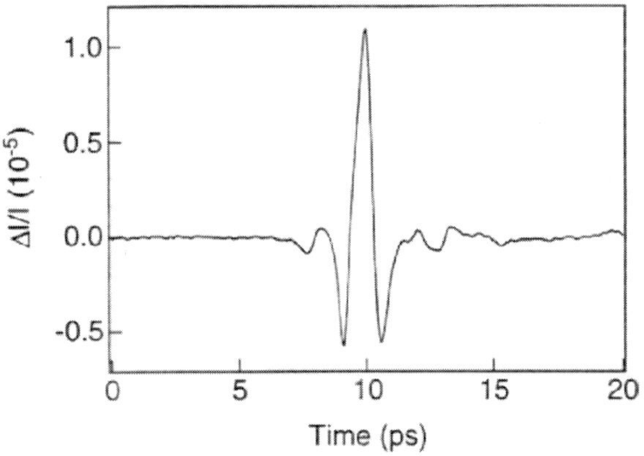

Fig. 10.17 THz signal detected by $TiTaO_3$ EO sensor.

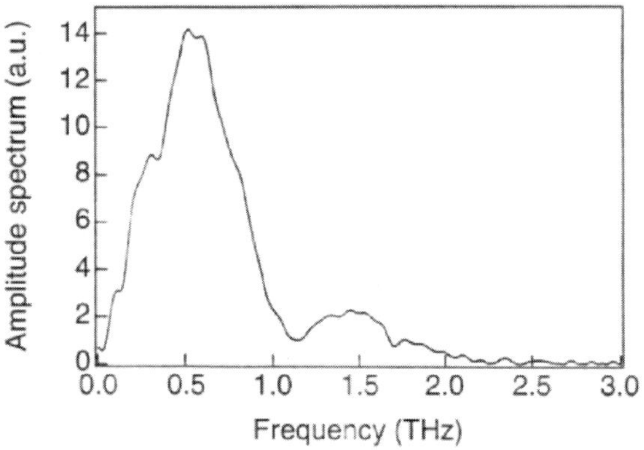

Fig. 10.18 Fourier transform of the waveform shown in Fig. 10.17.

A typical THz experimental setup is shown in Fig. 10.15 [114]. It consists of a laser-driven THz emitter and a laser-actuated, gated THz detector (sampler or sensor). THz detectors can either be a PDA (photoconductive dipole antenna) type or an electrooptic sensor type. Impulse THz is generated in the nonlinear optical crystal by the optical rectification effect. It radiates in the forward direction. THz radiation is then collected by an off-axis paraboloidal mirror or THz lens onto a THz detector or sensor (EO sensor, for example). The temporal waveforms of the THz radiation from the emitter without the sample and the waveforms after transmission through the sampler are obtained by a pump and probe technique. The EO sensor to detect the free-space THz radiation is shown in detail in Fig. 10.16 [115]. The THz beam and optical probe beam interact in the EO sensor, which should be properly designed and cut with velocity matching between THz pulses and probe optical pulses, yielding the maximum EO modulation effect. Without THz radiation illuminating the EO sensor, the probe laser beam polarization, EO crystal orientation, compensator and Wollaston prism are adjusted so that the balanced detector gives zero signal. When the THz pulse overlaps with the probe laser pulse in the EO crystal, the transient electric field associated with the THz beam induces an EO effect and the two beams emerging from the Wollaston prism become unequal in amplitude; consequently, the detector becomes unbalanced and registers a signal. The system is set up and calibrated such that the signal from the balanced detector is proportional to the amplitude of the electric field of the THz beam. Since the EO crystal senses the electric field, the signal changes polarity when the electric field polarity is changed. To map out the THz waveform the probe pulse is temporally delayed by an optical delay line driven by a stepping motor. The THz field induced intensity modulation of the probe beam is picked up by the balanced detector and is mapped out as a function of the delay between the pump and probe pulses. To detect a small change in modulation a phase sensitive detection technique as typically employed in a lock-in amplifier is usually required.

A typical THz signal generated by a $LiTaO_3$ crystal is shown in Fig. 10.17; a Mexican hat shape temporal waveform is clearly displayed. The Fourier transform of the waveform in Fig. 10.18 renders frequency-domain information.

The coherent THz detection technique can be used to study the mid-infrared property of the EO crystal. Using a 10-fs laser pulse, this technique is capable of reaching 44 THz. Recently, this has been demonstrated by Wu [114] and Wu and Zhang [115]. They studied the mid-IR optical property using well-characterized EO sensors of 10 and 30 μm in thickness. Each sensor has been tested to operate between 10 and 50 THz. For example, the mid-IR frequency responses of 10- and 30-μm ZnTe sensors are shown in Fig. 10.19. These EO sensors have been used to study the mid-IR radiation from a GaAs emitter

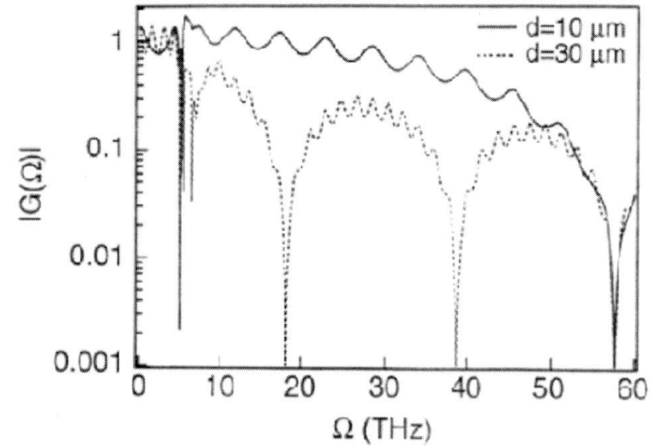

Fig. 10.19 Mid-IR frequency response functions of ZnTe sensors with thicknesses of 10 μm and 30 μm, respectively.

by optical rectification. Figure 10.20 shows a typical waveform from a GaAs emitter. The EO sensor is a 30-μm-thick ZnTe crystal. A Fourier transform of the waveform is shown in Fig. 10.21. The measured frequency characteristics of the THz pulses emitted from GaAs are the convolution of the frequency responses of the emitter and detector (sensor). A spectral modulation period of about 2 THz is due to the multiple reflections from the ZnTe sensor (30 μm) and the dips at 17 THz and 37 THz are also due to the response of the 30-μm ZnTe sensor. It would be more desirable to use a thinner sensor (10 μm) to avoid these big dips.

The coherent THz detection technique can also be applied to determine the ratios between various nonlinear optical coefficients. The THz field due to optical rectification is

$$\begin{aligned} E^{rd}(t) &\sim \frac{\partial^2}{\partial t^2} P(t) \\ &\sim d_{ijk}^{dc} \int_0^\infty \exp(i\Omega t)\Omega^2 \, d\Omega \int_{\omega_0-\Delta\omega/2}^{\omega_0+\Delta\omega/2} E_j(\omega+\Omega)E_k^*(\omega)\, d\omega. \end{aligned}$$

When the detection of this field is done with a dipolar photoconductive antenna (DPA), the antenna will measure the component of the radiated electric field along its axis. The signal measured depends on the crystalline symmetry, the orientation of the sample crystal DPA and the polarization of the laser and THz fields. Ma and Zhang have figured out a way to sort out the various configurations [117]. They have derived theoretically the THz field as a function of angle, θ, between the axis of the detector and the x'-axis of the crystal sample for various crystal symmetries and geometrical orientations.

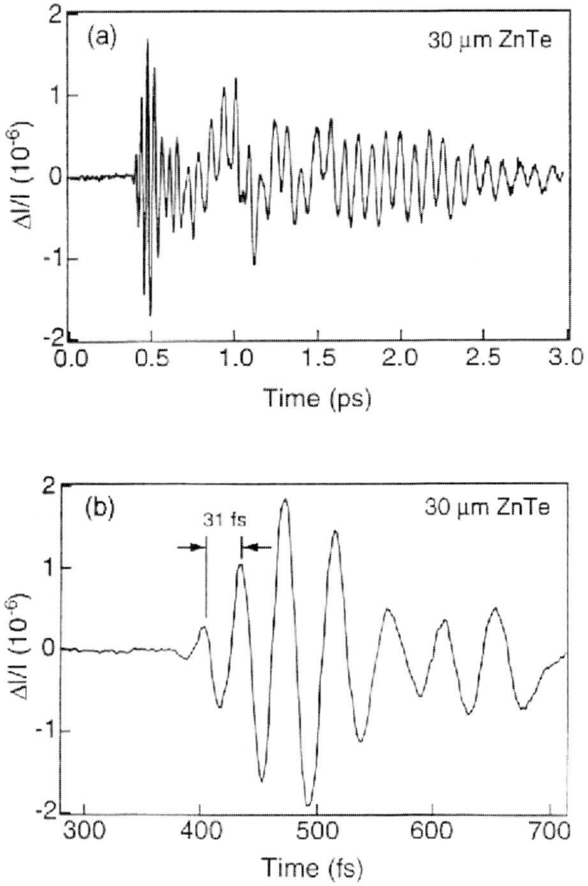

Fig. 10.20 (a) Temporal waveform of the THz radiation from a GaAs emitter measured by a 30-μm ZnTe sensor. (b) Expanded view of the first wave packet.

For the cubic structure GaAs with point group 43m, there is only one independent nonvanishing second-order nonlinear optical coefficient, namely $d_{14} = d_{25} = d_{36}$. It can be shown and experimentally confirmed that there is no optical rectification fields when a GaAs(100) sample is incident normally. However, there are rectification fields with three-fold rotational symmetry from a GaAs(111) sample at normal incidence. The experimental result is shown in Fig. 10.22. The experimental curve can be explained by

$$E_p = -E_s = \frac{A}{\sqrt{6}} \cos(3\theta),$$

where E_p is the component of the rectified field parallel to the polarization of the incident laser light and E_s is the perpendicular component. For crystals

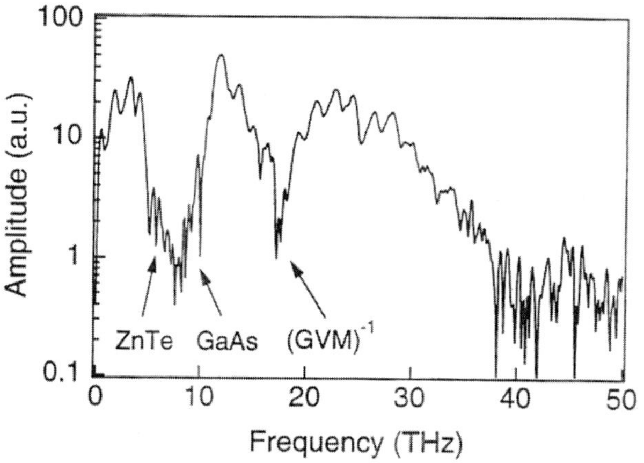

Fig. 10.21 Spectrum of the waveform in Fig. 10.20.

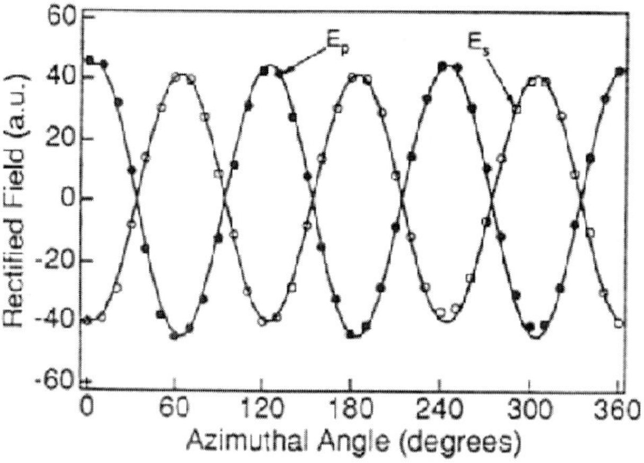

Fig. 10.22 Rectification field vs GaAs(111) crystal rotation angle, showing three-fold rotation symmetry of both E_p and E_s.

with lower symmetry, such as $LiTaO_3$, (trigonal structure), there will be more independent d coefficients; one has

$$\begin{aligned} E_p(\theta) &= A\big[(3d_{33} + d_{31} + 2d_{15})\ \cos\theta \\ &\qquad + (d_{33} - d_{31} - 2d_{15}) \cos 3\theta\big], \\ E_s(\theta) &= A'\big[(3d_{31} + d_{33} - 2d_{15})\ \cos\theta \\ &\qquad - (d_{33} - d_{31} - 2d_{15})\ \cos 3\theta\big]. \end{aligned}$$

Similar expressions can be derived for monoclinic structures such as the organic crystal dimethyl amino 4-N-methylstilbazolium tosylate or DAST. By measuring $E_p(\theta)$ and $E_s(\theta)$ of the rectified fields as a function of the azimuthal angle, θ, the ratio of the d coefficients can be determined. For example, $d_{31} = 0.21d_{33}$ for $LiTaO_3$ and $d_{12} = -0.4d_{11} = 295d_{26}$ for DAST.

Problems

10.1 Derive Eqs. (10.11)–(10.13).

10.2 Terahertz (THz) pulses can be generated by optical rectification in a poled polymer. If the optical pulses used for THz generation have a pulse width of 1 ps, calculate the bandwidth of the THz pulses.

Appendix A
Atomic Physics Definitions

In this appendix we present the most recent accepted values for the fundamental constants used in this volume. All of these can be obtained from a web site maintained by the National Institute for Standards and Technology (NIST). At the writing of this text, the web address is

http://physics.nist.gov/cuu/Constants/.

It must be understood that no physical quantity can be measured with absolute certainty; there will always be uncertainties! The uncertainties in the last digits are given in parentheses and are the *standard uncertainties*.

Uncertainties come in two principal flavors, statistical and systematic. Systematic uncertainties can be difficult to ascertain, requiring intimate knowledge of the instrumentation and the technique used in the measurement as well as a reliable way to determine the deviation *a priori*. Often one can correct for systematic uncertainties. Statistical uncertainties are not correctable *per se*, and are thus handled differently. Consider a measurement that yields a value y. The uncertainty in the measurement is statistical when the distribution that characterizes the probability for measuring a specific value y is approximately *normal* (Gaussian). We can define a standard deviation, σ, such that if we were to measure y many many times, approximately 68% of the distribution would lie in a range

$$\overline{y} - \sigma \text{ to } \overline{y} + \sigma,$$

where $\overline{y}$ is the mean value of the measurements. By definition, the *standard uncertainty* is the estimated standard deviation. Typically, this uncertainty is expressed as

$$\overline{y} = 1234.56789 \pm 0.00011 \text{ [units]}.$$

A more concise form used in this appendix, as well as the NIST tables, is

$$\overline{y} = 1234.56789(11) \text{ [units]}.$$

Here, (11) is understood to be the numerical value for the standard uncertainty referred to the last two digits of the quoted result.

Light-Matter Interaction: Atoms and Molecules in External Fields and Nonlinear Optics.
W. T. Hill and C. H. Lee

ISBN: 978-3-527-40661-6

A.1
Air and Vacuum Wavelengths

The index of refraction of a medium changes the wavelength of light traversing through it. The difference between the vacuum and air wavelengths near 600 nm, for example, is approximately 0.16 nm in dry air. Given the air wavelength, the vacuum wavelength is calculated by

$$\lambda_{air} = \frac{\lambda_{vac}}{n_{air}}. \tag{A.1}$$

At 288 K and 1 atm, the index of refraction of air can be approximated with the Cauchy formula given by

$$n_{air} = 1 + 2.72643 \times 10^{-4} + \left(1.2288 + \frac{3.555 \times 10^4}{\lambda^2}\right)\frac{1}{\lambda^2}, \tag{A.2}$$

where λ is in nanometers.

A.2
Wavenumber

The wavenumber, $\widetilde{\nu}$, is a convenient way to specify the energy of an electromagnetic wave as well as the transition energy between states. Recall that the modulus of the wave vector, $\overrightarrow{k}$, of an electromagnetic wave is given by

$$\left|\overrightarrow{k}\right| = \frac{2\pi}{\lambda_{air}} = \frac{2\pi n_{air}}{\lambda_{vac}} = \frac{2\pi n_{air}\nu}{c}, \tag{A.3}$$

$$\widetilde{\nu} = \frac{1}{\lambda_{vac}} = \frac{1}{n_{air}\lambda_{air}} = \frac{\left|\overrightarrow{k}\right|}{2\pi n_{air}}, \tag{A.4}$$

where c is the speed of light in vacuum and ν is its frequency. Typically, $\widetilde{\nu}$ has the units of cm^{-1}, which is sometimes referred to as a kayser. One can think of the wavenumber as an energy unit with $h = c = 1$. Note that ν is usually thought of as independent of the index of refraction. An example where this is not true is standing waves generated by a high-finesse, fixed-length optical cavity. When the index of refraction is changed in this case, the wavelength remains fixed and so the frequency changes.

A.3
Fine-Structure Constant

$$\alpha = \frac{1}{4\pi\varepsilon_0}\frac{e^2}{\hbar c} = \frac{1}{137.035\,999\,76(50)} = 7.297\,352\,533(27) \times 10^{-3}. \tag{A.5}$$

Recently, Gabrielse et al. [1] reported a new value for $\alpha^{-1} = 137.035\,999\,710(96)$.

1) G. Gabrielse, D. Hanneke, T. Kinoshita, M. Nio, and B. Odom. Phys. Rev. Lett. 97, 030802 (2006).

A.4
Atomic Energy Unit (Hartree)

The atomic energy unit (a.u.), also known as a Hartree, is e times the electric potential associated with two elementary charges (e) separated by the Bohr radius (a_0),

$$1\ \text{a.u.} = e\left(\frac{e}{4\pi\varepsilon_0 a_0}\right) = \frac{\alpha\hbar c}{a_0} = 4.35974381(34)\times 10^{-18}\ \text{J}. \tag{A.6}$$

In Eq. (A.6), α $\left(\equiv e^2/4\pi\varepsilon_0\hbar c \sim 1/137\right)$ is the fine-structure constant, $\hbar$ is Planck's constant and c is the speed of light. These units are equivalent to setting $e = \hbar = m_e = 1$, where m_e is the electron rest mass.

A.5
Rydberg Energy Unit

The Rydberg energy unit is given by

$$1\ \text{Ry} = \frac{1}{2}\ \text{Hartree} = 13.60569172(53)\ \text{eV}, \tag{A.7}$$

and is the binding energy of the hydrogen atom with an infinitely massive nucleus. This unit is equivalent to setting $\hbar = e^2/2 = 2m_e = 1$. The atomic energy unit (a.u.) and Rydberg energy unit (cm^{-1}) are also related:

$$1\ \text{a.u.} = 4\pi\hbar c\mathcal{R}_\infty, \tag{A.8}$$

where $\mathcal{R}_\infty$ is the Rydberg constant:

$$\mathcal{R}_\infty \equiv \frac{1}{2}\frac{e^2}{4\pi\varepsilon_0 a_o\hbar c} = \frac{\alpha}{4\pi a_0} = 109737.31568549(83)\ \text{cm}^{-1}. \tag{A.9}$$

A.6
eV Energy Unit

Since the quantity in parentheses in Eq. (A.6) has units of volts, we can also define a new energy unit called an electron volt, eV. An atomic unit of energy is related to the eV through

$$1\ \text{a.u.} = 1\ \text{Hartree} = 27.2113834(11)\ \text{eV}. \tag{A.10}$$

The conversion between Rydberg energy units and electron-volt units (eV) is

$$1\text{eV} = 8065.54477(32)\ \text{cm}^{-1}. \tag{A.11}$$

It is helpful to know that the NIST web site also offers a conversion factor applet that was used here to convert from a.u. to eV.

A.7
Mass

The atomic mass, as found in the periodic table, is typically expressed in atomic mass units (amu) and based on the definition of carbon 12 being exactly 12 amu. Consequently,

$$1\ \text{amu} = 1.66053873(13) \times 10^{-24}\ \text{g}. \tag{A.12}$$

Another convenient mass unit is the rest mass, mc^2, which is expressed as eV. The conversion between amu and eV is given by

$$1\ \text{amu} \cdot c^2 = 931.494013(37)\text{MeV}. \tag{A.13}$$

The mass of the electron is then

$$\begin{aligned} m_e &= 9.10938188(72) \times 10^{-28}\ \text{g} \\ &= 5.485799110(12) \times 10^{-4}\ \text{amu} \\ &= 0.510998902(21)\text{MeV}, \end{aligned} \tag{A.14}$$

while the mass of the proton is

$$\begin{aligned} m_p &= 1.67262158(13) \times 10^{-24}\ \text{g} \\ &= 1.00727646688(13)\ \text{amu} \\ &= 938.271998(38)\text{MeV}. \end{aligned} \tag{A.15}$$

A.8
Length

There are several lengths that characterize atomic systems.

- The classical electron radius,

$$r_e \equiv \frac{e^2}{4\pi\varepsilon_0 m_e c^2} = 2.817940285(31) \times 10^{13}\ \text{cm}, \tag{A.16}$$

  is the radius at which the electrostatic potential is equal to the electron rest mass.

- The Compton wavelength,

$$\lambda_c \equiv \frac{r_e}{\alpha} = \frac{\hbar}{m_e c} = 386.1592642(28) \times 10^{-11}\ \text{cm}, \tag{A.17}$$

  is the wavelength where the photon energy is equal to the electron rest mass.

- The Bohr radius,

$$a_o \equiv \frac{4\pi\varepsilon_o\hbar^2}{m_e e^2} = \frac{r_e}{\alpha^2} = 5.291772083(19) \times 10^{-9}\ \text{cm}, \qquad (A.18)$$

is the smallest orbit in the hydrogen atom. The typical atomic size for a 1s electron in a hydrogen-like ion is given by a_o/Z, where Ze is the charge of the nucleus.

A.9 Atomic Velocity and Momentum

In Section 1.1 we developed the Bohr energy spectrum by equating the Coulomb potential with the centripetal acceleration (Eq. (1.4a)). Solving for v in Eq. (1.4a) with $r = a_o$, the Bohr radius, and $Z = 1$ defines the atomic velocity; the velocity with which the electron moves in the first Bohr orbit. Thus,

$$\begin{aligned} v_o &= \sqrt{\left(\frac{e^2}{4\pi\varepsilon_o}\right)\frac{1}{m_e a_o}} \\ &= \alpha c = 2.1876912529(80) \times 10^8\ \text{cm/s}, \end{aligned} \qquad (A.19)$$

where α is the fine-structure constant and c is the speed of light. The last step was made with the aid of Eq. (A.18). Multiplying this expression by the mass of the electron defines the atomic momentum

$$p_o = m_e \alpha c = 1.99285151(16) \times 10^{-19}\ \text{g cm/s}, \qquad (A.20)$$

which can also be written as

$$p_o = \frac{\hbar}{a_o} = 1.2438 \times 10^{-7}\ e\text{V s/cm}. \qquad (A.21)$$

A.10 Atomic Time Scale

The scale is set by the electron's period in the Bohr orbit,

$$\tau_{a_o} \equiv \frac{a_o}{\alpha c} = 2.418884326500(18) \times 10^{-17}\ \text{s}, \qquad (A.22)$$

where αc is the atomic velocity (Eq. (A.19)).

A.11 Atomic Field Strength

The characteristic electric field strength is the atomic field associated with the field produced by the proton at the Bohr distance,

$$F_{a_0} \equiv \frac{e}{4\pi\varepsilon_0 a_0^2} = 5.142\,206\,24(20) \times 10^9 \text{ V/cm}.$$

A.12 Atomic Unit of Dipole Moment

$$ea_0 = 8.47835267(33) \times 10^{-28} \text{ C cm}.$$

A.13 Magnetic Moments

Magnetic moments play a central role in the details of atomic structure. We will summarize the key features for the electron and the nucleons in this appendix.

A.13.1 Electron Magnetic Moment

The magnetic moment of the electron has two contributions – orbital angular momentum and spin angular momentum. The contribution due to its orbital motion about the nucleus is given by

$$\vec{\mu}_l = -\mu_B g_{e,l} \vec{l}, \tag{A.23}$$

with an expectation value in a state of the form $|\gamma l m_l\rangle$ of

$$\langle \vec{\mu}_l \rangle = -\mu_B g_{e,l} \sqrt{l\,(l+1)}. \tag{A.24}$$

The z component is given by

$$\vec{\mu}_{l,z} = -\mu_B g_{e,l} \vec{l}_z, \tag{A.25}$$

$$\langle \vec{\mu}_{l,z} \rangle = -\mu_B g_{e,l} m_l. \tag{A.26}$$

In these expressions μ_B is the Bohr magneton,

$$\mu_B = \frac{e\hbar}{2m_e} = 5.788\,381\,804(39) \times 10^{-5}\, eV/T, \tag{A.27}$$

and $g_{e,l}$ is known as the *orbital g-factor* for the electron. In the flavor of the oscillator strength, the g-factor is a measure of the degree to which the quantum mechanical orbital dipole moment is determined by the mechanical (*classical*) orbital angular momentum – $g_{e,l} = 1$.[2]

The spin contribution to the magnetic moment is expressed similarly as

$$\vec{\mu}_s = -\mu_B g_{e,s} \vec{s}, \tag{A.28}$$

$$\langle \vec{\mu}_s \rangle = -\mu_B g_{e,s} \sqrt{s\,(s+1)}. \tag{A.29}$$

Its z component is given by

$$\vec{\mu}_{s,z} = -\mu_B g_{e,s} \vec{s}_z, \tag{A.30}$$

$$\langle \vec{\mu}_{s,z} \rangle = -\mu_B g_{e,s} m_s, \tag{A.31}$$

with $g_{e,s} \approx -2$, given by the Dirac equation. The total magnetic moment is given by

$$\begin{aligned} \vec{\mu} &= \vec{\mu}_l + \vec{\mu}_s \\ &= -\mu_B [g_{e,l} \vec{l} + g_{e,s} \vec{s}] \\ &= -\mu_B [g_{e,l} \vec{j} + (g_{e,s} - g_{e,l}) \vec{s}]. \end{aligned} \tag{A.32}$$

Since $g_{e,s}/g_{e,l} \approx 2$, $\vec{\mu}$ will not point in the same direction as $\vec{j}$ in general and, thus, will not always be a constant of the motion. A more useful quantity that is a constant of the motion is the projection of $\vec{\mu}$ onto $\vec{j}$,

$$\mu_j \equiv \vec{\mu} \cdot \vec{u}_j = -\mu_B [g_{e,l} \vec{j} \cdot \vec{u}_j + (g_{e,s} - g_{e,l}) \vec{s} \cdot \vec{u}_j] = -\mu_B g_j \vec{j} \cdot \vec{u}_j, \tag{A.33}$$

where $\vec{u}_j = \vec{j}/|\vec{j}|$ is a unit vector in the $\vec{j}$ direction. With the aid of Eq. (E.23), it is possible to show that the matrix element of μ_j in the vector $|\gamma l s j m_j\rangle$ is

$$\begin{aligned} \langle \vec{\mu}_j \rangle = -\mu_B \Bigg[& g_{e,l} \sqrt{l\,(l+1)} \frac{j\,(j+1) + l\,(l+1) - s\,(s+1)}{2\sqrt{j\,(j+1)}\sqrt{l\,(l+1)}} \\ & + g_{e,s} \sqrt{s\,(s+1)} \frac{j\,(j+1) + s\,(s+1) - l\,(l+1)}{2\sqrt{j\,(j+1)}\sqrt{s\,(s+1)}} \Bigg]. \end{aligned} \tag{A.34}$$

Exploiting Eqs. (E.26) and (E.27) allows us to write this as

$$\langle \vec{\mu}_j \rangle = -\mu_B \left[g_{e,l} \langle \cos(\vec{l}, \vec{j}) \rangle \sqrt{l(l+1)} + g_{e,s} \langle \cos(\vec{s}, \vec{j}) \rangle \sqrt{s(s+1)} \right]. \tag{A.35}$$

This can be written as

$$\langle \vec{\mu}_j \rangle = -\mu_B g_j \sqrt{j(j+1)}, \tag{A.36}$$

2) We point out that caution is required when comparing equations involving the g-factor of the electron, as some authors define this quantity as being negative. See also, footnote 5 on p. 60.

from which we define $\vec{\mu}_j$ as

$$\vec{\mu}_j = -\mu_B g_j \vec{\mathbf{j}}. \tag{A.37}$$

It is straightforward to show that g_j is given by

$$g_j = \frac{3}{2} + \frac{1}{2}\left[\frac{s(s+1) - l(l+1)}{j(j+1)}\right]. \tag{A.38}$$

Before we leave this section, a few words about the anomalous value of $g_{e,s}$ are in order. Both experiment and quantum electrodynamics require the spin g factor to be anomalous, $g = 2(1+a)$, where $a > 0$ is called the g-factor anomaly. The best measured values of the anomaly for free electrons and positrons are

$$a_{e^-} = 1.159621884(43) \times 10^{-3}, \tag{A.39}$$

$$a_{e^+} = 1.159621879(43) \times 10^{-3}. \tag{A.40}$$

These results were obtained by comparing the spin-precession and cyclotron frequencies of isolated charges in a Penning trap, which contains both electric and magnetic fields [118]. Theoretically, Schwinger showed that the first correction term contributing to the anomaly is $0.5\alpha/\pi$ [119]. Additional terms of higher powers of α/π contribute to a as well. Hughes and Kinoshita, for example, gave [120]

$$a_e = \frac{1}{2}\left(\frac{\alpha}{\pi}\right) - 0.328478965\left(\frac{\alpha}{\pi}\right)^2 + 1.181241456\left(\frac{\alpha}{\pi}\right)^3 - 1.5098\left(\frac{\alpha}{\pi}\right)^4 + 4.393 \times 10^{-12}. \tag{A.41}$$

The current best theoretical value is $1.159652201(27) \times 10^{-3}$. It is important to note that the coefficient of the second term in Eq. (A.41) is given erroneously as 2.973 in Bethe and Salpeter.[3] The best accepted value for the anomalous spin g factor is

$$g_{e,s} = 2.002\,319\,304\,3718(75). \tag{A.42}$$

Recently, Odom et al.[4] reported a new value $g_{e,s}/2 = 1.001\,159\,652\,180\,85(76)$.

A.13.2
Proton Magnetic Moment

Since the proton is a spin-1/2 particle with positive charge we would expect gyromagnetic ratios of $g_{p,l} = 1$ and $g_{p,s} = 2$. Although the orbital portion is

3) See p. 93, Eq. (18.5) of Ref. [3].
4) B. Odom, D. Hanneke, B. D'Urso and G. Gabrielse Phys. Rev. Lett. 97, 030801 (2006).

correct, experimentally $g_{p,s} = 5.585\,694\,701(56)$. One must remember that the proton is not a simple point charge. The magnetic moment of the proton then becomes

$$\vec{\mu}_p = \mu_N \left(\vec{l} + g_{p,s}\vec{s}\right), \tag{A.43}$$

where μ_N is the nuclear magneton and has the value

$$\mu_N \equiv \frac{e\hbar}{2m_p} = \mu_B \frac{m_e}{m_p} = 3.152\,451\,259(21) \times 10^{-8}\ \mathrm{eV/T}. \tag{A.44}$$

A.13.3
Neutron Magnetic Moment

Even though the neutron has no charge, its magnetic moment is not zero. Its contribution from orbital motion is zero but its spin contribution leads to $g_{n,s} = 3.826\,085, 46(90)$ and $\vec{\mu}_n = -g_{n,s}\mu_N\vec{s}$.

A.13.4
Magnetic Moment of the Nucleus

It is convenient to write a g factor for the entire nucleus parallel to the total nuclear spin $\vec{I}$ as

$$\vec{\mu}_I = -g_I\mu_N\vec{I}.$$

The value of g_I depends on how the spins are coupled.

A.14
Quadrupole Moment of the Nucleus

In electrodynamics the quadrupole moment tensor is given by

$$Q_{ij} = \int \rho \left(3r_ir_j - \delta_{ij}r^2\right) d^3r, \tag{A.45}$$

with units of C m^2. It is convenient to classify the strength of this moment quantum mechanically by the expectation value of the Q_{zz} component in a state $|\gamma IM\rangle$,

$$Q \equiv \langle \gamma II | Q_{zz} | \gamma II \rangle. \tag{A.46}$$

Converting to spherical tensors, where $Q_{zz} = 2Q_{2,0}$, we have

$$Q = 2\langle \gamma I || Q_2 || \gamma I \rangle \begin{pmatrix} I & 2 & I \\ -I & 0 & I \end{pmatrix}. \tag{A.47}$$

The reduced matrix element is evaluated with the aid of the Wigner–Eckart theorem, Eq. (E.14). Values for Q/e (i.e., in units of cm^2) for a few nuclei can be found in Ref. [27].

A.15
Frequently Used AMO Quantities

Many of the quantities given in this appendix are used frequently in calculations in AMO[5] science. For quick and dirty estimations, the reader will find it convenient to commit several of the following to memory:

Quantity		Value
Atomic momentum, $p_o = \hbar/a_o$	$\simeq$	1.2438×10^{-7} eV s/cm
1 Bohr, a_o	$\simeq$	0.0529 nm
Distance light travels in 1 ns	$\simeq$	1 foot
Electron mass, m_e	$\simeq$	0.511 MeV
Fine-structure constant, α	$\simeq$	1/137
Refractive index of glass (visible light)	$\simeq$	1.5
Proton mass, m_p	$\simeq$	937 MeV
Speed of light in vacuum, c	$\simeq$	3×10^8 m/s
	$\equiv$	$1/\sqrt{\varepsilon_o \mu_o}$
Vacuum permeability, μ_o	$\equiv$	$4\pi \times 10^{-7}$ N/A^2
$k_B T$ at room temperature	$\simeq$	1/40 eV
$\hbar c$	$\simeq$	197 eV nm

Further Reading

The National Institute of Standards and Technology maintains a comprehensive web source.

1. For the most up to date values, see http://physics.nist.gov/cuu/Constants/.

2. For a detailed discussion of units and uncertainties, see http://physics.nist.gov/cuu/.

5) Atomic, Molecular and Optical

Appendix B
Mathematics Related to AMO Calculations

B.1
Kronecker Delta, δ_{ij}

$$\delta_{ij} = \begin{cases} 1, \ i = j, \\ 0, \ i \neq j, \end{cases} \tag{B.1}$$

where

$$\sum_i f_i \delta_{ij} = f_j \tag{B.2}$$

and

$$\sum_i \delta_{ij} = 1. \tag{B.3}$$

B.2
Dirac Delta Function, $\delta(x - x_o)$

The formal definition of the delta function introduced by Dirac is

$$\int_{-\infty}^{\infty} dx \, f(x) \, \delta(x - x_o) = f(x_o), \tag{B.4}$$

where

$$\int_{-\infty}^{\infty} dx \, \delta(x - x_o) = 1 \tag{B.5}$$

and

$$\delta(x - x_o) = \begin{cases} \infty, \ x = x_o, \\ 0, \ x \neq x_o. \end{cases} \tag{B.6}$$

A very useful representation of the delta function is

$$\delta(x - x_o) = \frac{1}{2\pi} \int_{-\infty}^{\infty} dk e^{i(x - x_o)k}. \tag{B.7}$$

Light-Matter Interaction: Atoms and Molecules in External Fields and Nonlinear Optics.
W. T. Hill and C. H. Lee

ISBN: 978-3-527-40661-6

However, it must be understood that the left-hand side of Eq. (B.7) is only well defined under the integral sign. There are several approximations for delta functions. One particularly useful one is

$$\delta(x-x_o)=\frac{1}{2\pi}\lim_{n\to\infty}\int_{-n/2}^{n/2}dke^{i(x-x_o)k}. \tag{B.8}$$

The nth derivative of the delta function, $(d^n/dx^n)\delta(x-x_o)$, leads to

$$\int_{-\infty}^{\infty}dxf(x)\frac{d^n}{dx^n}\delta(x-x_o)=(-1)^n\left.\frac{d^n}{dx^n}f(x)\right|_{x=x_o}, \tag{B.9}$$

while

$$(x-x_o)\frac{d}{dx}\delta(x-x_o)=\delta(x-x_o). \tag{B.10}$$

B.3
Hypergeometric Series

The hypergeometric functions, which are sometimes called Gauss functions, are defined by[1]

$$\begin{aligned}F(\alpha,\beta,\gamma;x)&=1+\frac{\alpha\beta}{\gamma}\frac{x}{1!}+\frac{\alpha(\alpha+1)\beta(\beta+1)}{\gamma(\gamma+1)}\frac{x^2}{2!}+\cdots,\quad\gamma\neq0,-1,-2,\ldots\\&=\frac{\Gamma(\gamma)}{\Gamma(\alpha)\Gamma(\beta)}\sum_{k=0}^{\infty}\frac{\Gamma(\alpha+k)\Gamma(\beta+k)}{\Gamma(\gamma+k)}\frac{x^k}{k!}.\end{aligned} \tag{B.11}$$

The hypergeometric functions obey

$$F(\alpha,\beta,\gamma;x)=(1-x)^{-\gamma}F\left(\alpha,\gamma-\beta,\gamma;\frac{-x}{1-x}\right), \tag{B.12}$$

and are related to the Legendre and associated Legendre functions (see Section B.6):

$$P_n(x)=F\left(-n,n+1,1;\frac{1-x}{2}\right), \tag{B.13}$$

$$\begin{aligned}P_n^m(x)&=\frac{(n+m)!}{(n-m)!}\frac{(1-x^2)^{m/2}}{2^m m!}\\&\times F\left(m-n,m+n+1,m+1;\frac{1-x}{2}\right).\end{aligned} \tag{B.14}$$

1) For integer n, $\Gamma(n+1)=n!$.

B.4 Confluent Hypergeometric Series

In the limit as $\beta \rightarrow \infty$ in Eq. (B.11), the confluent hypergeometric functions, which are also known as Kummer functions, are defined as

$$
\begin{aligned}
F(\alpha, \gamma; x) &= 1 + \frac{\alpha}{\gamma}\frac{x}{1!} + \frac{\alpha(\alpha+1)}{\gamma(\gamma+1)}\frac{x^2}{2!} + \cdots, \qquad \gamma \neq 0, -1, -2, \ldots \\
&= \frac{\Gamma(\gamma)}{\Gamma(\alpha)} \sum_{k=0}^{\infty} \frac{\Gamma(\alpha+k)}{\Gamma(\gamma+k)} \frac{x^k}{k!}.
\end{aligned} \tag{B.15}
$$

The confluent hypergeometric functions obey

$$F(\alpha, \gamma; x) = \mathrm{e}^x F(\gamma - \alpha, \gamma; -x), \tag{B.16}$$

and have an asymptotic behavior for large $x > 0$ given by

$$F(\alpha, \gamma; x) \rightarrow \frac{\Gamma(\gamma)}{\Gamma(\alpha)} \mathrm{e}^x x^{\alpha-\gamma}. \tag{B.17}$$

B.5 Associated Laguerre Polynomials

The Laguerre polynomials are defined as

$$
\begin{aligned}
L_n^m(x) &= \mathrm{e}^x \frac{x^{-m}}{n!} \frac{d^n}{dx^n} \mathrm{e}^{-x} x^{n+m} \\
&= \sum_{k=0}^{n} \frac{\Gamma(n+m+1)}{\Gamma(k+m+1)} \frac{(-x)^k}{k!\,(n-k)!},
\end{aligned} \tag{B.18}
$$

and are associated with the confluent hypergeometric functions

$$L_{n-l-1}^{2l+1}(x) = \frac{(n+l)!}{(n-l-1)!(2l+1)!} F(l+1-n, 2l+2; x). \tag{B.19}$$

B.6 Legendre and Associated Legendre Functions

The Legendre functions $P_l(x)$ are the solutions to Legendre's equation,

$$\left(1 - x^2\right) \frac{d^2 P_l}{dx^2} - 2x \frac{dP_l}{dx} + l(l+1) P_l = 0, \tag{B.20}$$

which are finite at the two singular points, $x = \pm 1$.[2] The first three Legendre functions are

$$P_0(x) = 1, \quad P_1(x) = x \quad \text{and} \quad P_2(x) = \tfrac{1}{2}\left(3x^2 - 1\right), \tag{B.21}$$

and can be generated from *Rodrigues' formula*,

$$P_l = \frac{1}{2^l l!}\frac{d^l}{dx^l}\left(x^l - 1\right)^l. \tag{B.22}$$

The Legendre functions obey

$$\int_{-1}^{+1} P_l(x) P_{l'}(x) dx = \frac{2}{2n+1}\delta_{ll'}. \tag{B.23}$$

Thus, any *well-behaved* function, $f(x)$, can be expanded in terms of them,

$$f(x) = \sum_{l=0}^{\infty} a_l P_l(x). \tag{B.24}$$

Finally, the parity of the functions depends on the index, l,

$$P_l(-x) = (-1)^l P_l(x). \tag{B.25}$$

The associated Legendre functions, $P_l^m(x)$,[3] are solutions to the associated Legendre equation

$$\left(1 - x^2\right)\frac{d^2 P_l^m}{dx^2} - 2x\frac{dP_l}{dx} + \left[l(l+1) - \frac{m^2}{1-x^2}\right] P_l^m = 0. \tag{B.26}$$

It is straightforward, via direct substitution, to show that $P_l^m(x)$ and P_l are related,

$$P_l^m(x) = \left(1 - x^2\right)^{m/2}\frac{d^m}{dx^m} P_l(x). \tag{B.27}$$

The associated Legendre functions obey

$$P_l^{-m}(x) = (-1)^m \frac{(l-m)!}{(l+m)!} P_l^m(x), \tag{B.28}$$

$$P_l^m(-x) = (-1)^{l+m} P_l^m(x) \tag{B.29}$$

and

$$P_l^0(x) = P_l(x). \tag{B.30}$$

2) The $P_l(x)$ functions are solutions of the first kind. Solutions of the second kind, $Q_l(z)$, are discussed, for example, in Refs. [1] and [121].

3) See Ref. [121] for a discussion of solutions that are singular at $x = \pm 1$.

Thus, m can be positive or negative with a range $-l \leqslant m \leqslant l$. A more useful expression for $P_l^m(x)$ is then

$$P_l^m(x) = \frac{1}{2^l l!}\left(1 - x^2\right)^{m/2} \frac{d^{l+m}}{dx^{l+m}}\left(x^2 - 1\right)^l. \tag{B.31}$$

Two associated Legendre functions with the same m obey the following orthogonality property:

$$\int P_l^m(x) P_{l'}^m(x) dx = \frac{(l+m)!}{(l-m)!}\frac{2}{2l+1}\delta_{ll'}. \tag{B.32}$$

Finally, the first three associated Legendre functions for $m \neq 0$ are

$$\begin{aligned} P_1^1(x) &= \left(1 - x^2\right)^{1/2} &&= \sin\vartheta, \\ P_2^1(x) &= 3x\left(1 - x^2\right)^{1/2} &&= 3\cos\vartheta\sin\vartheta, \\ P_2^2(x) &= 3\left(1 - x^2\right) &&= 3\sin^2\vartheta, \end{aligned} \tag{B.33}$$

where in the last column it is assumed that $x = \cos\vartheta$.

B.7 Spherical Harmonics

It is customary to combine the associated Legendre functions of the previous section with the azimuthal phase, $e^{im\varphi}$, to define the spherical harmonics,

$$Y_{lm}(\vartheta, \varphi) = (-1)^{(m+|m|)/2}\sqrt{\frac{(2l+1)}{4\pi}\frac{(l-|m|)!}{(l+|m|)!}} P_l^{|m|}(\cos\vartheta)\, e^{im\varphi}, \tag{B.34}$$

for all positive and negative values of m with $P_l^m(\cos\vartheta)$ given by Eq. (B.27). The spherical harmonics can also be written as

$$Y_{lm}(\vartheta, \varphi) = (-1)^m\sqrt{\frac{(2l+1)}{4\pi}\frac{(l-m)!}{(l+m)!}} P_l^m(\cos\vartheta)\, e^{im\varphi}, \tag{B.35}$$

with $P_l^m(\cos\vartheta)$ given by Eq. (B.31). In either case,

$$Y_{l,-m}(\vartheta, \varphi) = (-1)^m Y_{lm}^*(\vartheta, \varphi) \tag{B.36}$$

and

$$\int Y_{lm}^*(\vartheta, \varphi)\, Y_{l'm'}(\vartheta, \varphi) \sin\vartheta d\vartheta d\phi = \delta_{ll'}\delta_{mm'}. \tag{B.37}$$

B.8
Mathematical Formalism of Quantum Mechanics

Extracting measurable quantities in quantum mechanics requires a mathematical formalism that allows us to track the state of a system and to perform various calculations. This formalism, introduced by Dirac in the late 1920s, and discussed in detail in his book *The Principles of Quantum Mechanics* [122] as well as elsewhere (see for example Condon and Odabaşi [123]), involves linear algebra, abstract vectors and an infinite-dimensional complex vector space.[4] In this appendix we will outline some of the key principles through worked examples that are used in the main body of the book. We end the appendix with a summary of the main issues of the algebra necessary for quantum mechanics.

The complex vector space, which we will call the *system space*, includes linear operators, vectors and inner products. To distinguish vectors in physical space, symbolized by $\vec{r}$ or $\mathbf{r}$, from the *state vectors* in the system space, we symbolize the latter by $|S\rangle$; Dirac called these *kets*. Every state vector in this abstract vector space has a so-called dual, $\langle S|$, which Dirac called a *bra* and exists in a corresponding vector space. The bras and kets are related by their *Hermitian adjoint*,

$$\langle S| \equiv |S\rangle^{\dagger}, \tag{B.38}$$

where the Hermitian adjoint of the state vector is the complex conjugate of its transpose. Just as a vector is used to describe the state of a point in physical space – its location in space – the state vectors describe the state of the system in the system space, for a one-electron atom this might include the values for n, l, m_l, etc.

To write an expression for a vector requires a *complete basis* – a set of *linearly independent* vectors. For example, in three-dimensional Cartesian space the basis set is the three unit vectors $\vec{x}$, $\vec{y}$, $\vec{z}$. These vectors are linearly independent because no linear combination of any two will give the third. The set is complete because any vector in the space can be expressed by a linear combination of just these three vectors,

$$\vec{V} = V_x\vec{x} + V_y\vec{y} + V_z\vec{z}, \tag{B.39}$$

where V_x, V_y and V_z are real numbers. It is often convenient to let $x \to x_1$, $y \to x_2$ and $z \to x_3$ and $V_x \to V_1$, $V_y \to V_2$ and $V_z \to V_3$, so that the vector is

4) This complex vector space is often referred to in quantum mechanics texts as a Hilbert space [124, 125]. A discussion of a Hilbert space is beyond the scope of this text but we do point out that technically, as Dirac mentioned, the abstract space of quantum mechanics is more general than the standard Hilbert space. Extensions, such as the *rigged* Hilbert space, have been shown to include continuous spectra, for example, which are not included in the standard Hilbert space.

expressed as[5]

$$\vec{V} = \sum_{i=1}^{3} V_i x_i. \tag{B.40}$$

In the system space, an arbitrary state vector is similarly given by a linear combination of the basis states (vectors),

$$|S\rangle = \sum\!\!\!\!\!\!\int_i s_i |\sigma_i\rangle, \tag{B.41}$$

where s_i are complex numbers and $|\sigma_i\rangle$ are the basis vectors. A similar expression exists for a state vector $\langle S|$ involving bras. The compound symbol composed of a summation sign with the superimposed integral sign means to sum over discrete parameters and integrate over continuous parameters. For example, the energy of the bound electron in the hydrogen atom is discrete and given by the value of n. However, when the atom is ionized, the energy of the electron is determined by a continuous parameter. The complete state, then, must span both the bound and continuous parts of the spectrum because these are all the possible states of the electron. In this text we will not use this compound symbol. Rather, we will apply the sum to discrete parameters and an integration to continuous parameters. For the general case like Eq. (B.41) that could include both types of parameters, we will simply write a summation sign: but it should be understood that this will mean that we sum over the discrete parameters and integrate over the continuous parameters.

The inner product in this algebra is defined between a bra and a ket. Thus, the inner product between $|A\rangle$ and $|B\rangle$ is

$$\langle A|B\rangle. \tag{B.42}$$

The inner product in this complex vector space, like in physical space, generates a number, albeit complex in this case, so that

$$\langle A|B\rangle = \langle B|A\rangle^*. \tag{B.43}$$

Below, we will see that such an inner product can also correspond to an amplitude of the wavefunction in physical space. It follows that $\langle A|A\rangle$ is a real number and is called the *norm* of $|A\rangle$. Setting $\langle A|A\rangle = 1$ is convenient because this means that the state vector is of unit length. However, the state vector is not fully specified when normalized because it can still be multiplied by a phase, $\mathrm{e}^{i\phi}$, without changing its length,

$$\left(\langle A|\mathrm{e}^{-i\phi}\right)\left(\mathrm{e}^{i\phi}|A\rangle\right) = \langle A|A\rangle. \tag{B.44}$$

5) Einstein introduced the more compact convention of summing over repeated indices, $\sum_{i=1}^{3} V_i x_i \rightarrow V_i x_i$, thus allowing the summation symbol to be dropped.

Thus, only the direction of the state vector is important, not its length! When $\langle A|B\rangle = 0\,(= \langle B|A\rangle)$, the two states are said to be orthogonal. The basis vectors in this space are orthogonal and normalized to 1, so that

$$\langle \sigma_i|\sigma_j\rangle = \delta_{ij}. \tag{B.45}$$

It is straightforward to see that the inner product is tantamount to matrix multiplication of a row and a column vector of the coefficients when $|A\rangle$ and $|B\rangle$ are expressed in terms of an orthonormal set of basis vectors, $|A\rangle = a_1|\sigma_1\rangle + a_2|\sigma_2\rangle + \cdots + a_n|\sigma_n\rangle$ and $|B\rangle = b_1|\sigma_1\rangle + b_2|\sigma_2\rangle + \cdots + b_n|\sigma_n\rangle$.[6] In this case, the inner product can be written as

$$\langle A|B\rangle = \begin{pmatrix} a_1^* & a_2^* & \cdots & a_n^* \end{pmatrix} \begin{pmatrix} b_1 \\ b_2 \\ \vdots \\ b_n \end{pmatrix} = \sum_{i=1}^{n} a_i^* b_i, \tag{B.46}$$

where Eq. (B.45) has been used. If $|\sigma_i\rangle$ represents a continuous parameter, the sum would be an integration.

The inner product is not the only possible product between state vectors; $|A\rangle|B\rangle$ or $\langle A|\langle B|$, for example, defines a separable state in the spirit of the radial and angular parts of the wavefunctions (see Eq. (1.21)) and can be written as $|AB\rangle$ or $\langle AB|$. The product $|A\rangle\langle B|$, called the outer product, is an operator because when it acts on a bra or a ket, it changes the state of the system. For example, when $|A\rangle\langle B|$ is applied to $|C\rangle$,

$$(|A\rangle\langle B|)\,|C\rangle = |A\rangle \cdot \langle B|C\rangle, \tag{B.47}$$

it changes the state to $|A\rangle$ since $\langle B|C\rangle$ is just a complex number. The special outer product

$$|A\rangle\langle A| \tag{B.48}$$

is known as the projection operator, because it will extract the contribution of $|A\rangle$ from an arbitrary state vector. When a projection operator is formed from the basis vectors, $|\sigma_i\rangle$, it obeys a property called completeness, where[7]

$$\sum_{\text{all states}} |\sigma_i\rangle\langle\sigma_i| = \mathbf{1}. \tag{B.49}$$

6) The Gram–Schmidt procedure can be used to construct an orthonormal set of basis vectors from an arbitrary initial set (see Ref. [121]).

7) When $|\sigma_i\rangle$ represents a discrete parameter then this would be an actual sum. If, however, it were continuous then the sum would really be an integral. Clearly, if $|\sigma_i\rangle$ were a mixture we would have to sum and integrate over the appropriate parts.

Now we are ready to discuss the extraction of physics from this abstract vector space. Of chief concern is the measurement of physical quantities and explicit representations of operators and wavefunctions. Measurable quantities, dynamical parameters like position, energy, quadrupole moment, etc., are called observables and exist in physical space. These quantities, which are real numbers and variables in classical mechanics, become tensor operators in quantum mechanics because they can modify the state of the system, in general. If the system were in an eigenstate of the operator, however, the operator simply extracts the eigenvalue, the observable, while leaving the system in the same state. Consequently, in quantum mechanics the extraction of observables can be reduced to eigenvalue problems. What makes quantum physics so interesting is the fact that not all eigenvalues are extractable simultaneously.

To begin, we will look at an example of a simple eigenvalue problem in the abstract space. Consider an operator $\hat{A}$ acting on a state $|\alpha\rangle$ with eigenvalue a. The relationship between the operator, eigenstate and eigenvalue can be written as[8]

$$\hat{A}|\alpha\rangle = a|\alpha\rangle. \tag{B.50}$$

Applying $\langle\alpha|$ to the left, defines the expectation value of $\hat{A}$

$$\langle\alpha|\hat{A}|\alpha\rangle = a\langle\alpha|\alpha\rangle = a, \tag{B.51}$$

where we have assumed that $|\alpha\rangle$ is normalized and we used the fact that a, a number, commutes with the state vector. The expectation value, a, can be thought of as the average value that can be measured by experiment. This average is in the statistical sense of repeated measurement. When there are several eigenvalues of the operator, the problem can be solved with matrices. In general, we can define a *matrix element*,

$$\begin{aligned}\langle\alpha_2|\hat{A}|\alpha_1\rangle &= A_{21}\\ &= a_2\langle\alpha_2|\alpha_1\rangle\\ &= a_2\delta_{21}.\end{aligned} \tag{B.52}$$

In the second step we used the facts that $\langle\alpha_2|\hat{A} = a_2$ and that operators associated with the physical observables are Hermitian – eigenvalues are real. If α were to represent a continuous observable, the Kronecker delta would

8) We will use a hat over a variable to indicate an operator. Quantities without hats will simply be complex numbers, which in some cases will be purely real. John von Neumann is given credit for inventing operator algebra in his 1932 book *Mathematische Grundlagen der Quantenmechanik* (*Mathematical Foundations of Quantum Mechanics*). His work allowed the Schrödinger and Heisenberg approaches to quantum mechanics to be shown to be mathematically equivalent.

have to be replaced by the Dirac delta function (Appendix B.7). Since $|\alpha\rangle$ is an eigenstate of $\hat{A}$, only diagonal terms exist. In general, for an arbitrary state vector off-diagonal elements will exist as well. The off-diagonal elements are associated with transitions between eigenstates. A state of the system can also be an eigenstate of two or more observables simultaneously. In this case, the operators commute.

To make this more concrete, let us apply this to the motion of a free, spinless point particle. Observables that we can extract include position ($\mathbf{x}$), momentum ($\mathbf{p}$) and energy (E). The corresponding operators are $\hat{\mathbf{x}}$, $\hat{\mathbf{p}}$ and $\hat{H} = \hat{\mathbf{p}}^2/2m$. For this problem there are two convenient sets of basis vectors, $|x_1 x_2 x_3\rangle \equiv |\mathbf{x}\rangle$ and $|p_1 p_2 p_3\rangle \equiv |\mathbf{p}\rangle$,[9] that can be used to describe the state of the particle. In general, the explicit expression for an operator depends on the basis. To see this, we will find the expectation value of the energy operator in both bases. In the momentum basis, for example, $\hat{H}$ is diagonal since it can operate on $|\mathbf{p}\rangle$ without changing the state of the system:

$$\begin{aligned}
\langle \mathbf{p}'|\hat{H}|\mathbf{p}\rangle &= \frac{1}{2m}\langle \mathbf{p}'|\hat{\mathbf{p}}^2|\mathbf{p}\rangle \\
&= \frac{1}{2m}\int d^3p''\langle \mathbf{p}'|\hat{\mathbf{p}}|\mathbf{p}''\rangle \cdot \langle \mathbf{p}''|\hat{\mathbf{p}}|\mathbf{p}\rangle \qquad \text{(B.53)} \\
&= \frac{1}{2m}\int d^3p''\mathbf{p}'' \cdot \mathbf{p}\delta(\mathbf{p}' - \mathbf{p}'')\delta(\mathbf{p}'' - \mathbf{p}) \\
&= \frac{\mathbf{p}^2}{2m}\delta(\mathbf{p}' - \mathbf{p}). \qquad \text{(B.54)}
\end{aligned}$$

Since the momentum of a free particle is a continuous variable, the Kronecker delta in Eq. (B.52) has been replaced by a delta function. The delta function means that $\hat{H}$ indeed does not alter the state of the system so that $\mathbf{p}' = \mathbf{p}$. This expectation value is precisely what we expected.

Now we will do the problem in the position basis. The problem is a bit more involved because these basis vectors are not eigenstates of $\hat{H}$ and so $|\mathbf{x}\rangle$ will be modified when $\hat{H}$ is applied. To solve this problem we can use completeness twice to surround $\hat{H}$ by its eigenstates,

$$\begin{aligned}
\langle \mathbf{x}'|\hat{H}|\mathbf{x}\rangle &= \iint d^3p'd^3p\langle \mathbf{x}'|\mathbf{p}'\rangle\langle \mathbf{p}'|\hat{H}|\mathbf{p}\rangle\langle \mathbf{p}|\mathbf{x}\rangle \\
&= \iint d^3p'd^3p\left(\frac{\mathbf{p}'^2}{2m}\delta(\mathbf{p}' - \mathbf{p})\right)\langle \mathbf{x}'|\mathbf{p}'\rangle\langle \mathbf{p}|\mathbf{x}\rangle \\
&= \int d^3p\left(\frac{\mathbf{p}^2}{2m}\right)\langle \mathbf{x}'|\mathbf{p}\rangle\langle \mathbf{p}|\mathbf{x}\rangle.
\end{aligned}$$

To go further, we must digress and discuss the meaning of $\langle \mathbf{x}'|\mathbf{p}\rangle$ and $\langle \mathbf{p}|\mathbf{x}\rangle$. From Eq. (B.43) we know that both of these are complex numbers or an am-

9) It is typical to use just one symbol to represent several states of the system that are of the same nature.

plitude. But, they are more than this. Consider the arbitrary state $|\psi\rangle$ and the basis states $|\mathbf{x}\rangle$. The set of complex numbers $\langle\mathbf{x}|\psi\rangle$, which are essentially the coefficients in Eq. (B.41), can be thought of as the projection of the state into the position representation. These coefficients provide a way to obtain an explicit representation of the state in physical space. That is,

$$\langle\mathbf{x}|\psi\rangle = \psi(\mathbf{x}), \tag{B.55}$$

where $\psi(\mathbf{x})$ is the representation of the state in physical space. We then recognize $\langle\mathbf{x}'|\mathbf{p}\rangle$ as the physical space wavefunction of the particle. For a free particle the Schrödinger solution is a plane wave,

$$\psi(\mathbf{x}) = \langle\mathbf{x}|\mathbf{p}\rangle = e^{i\mathbf{p}\cdot\mathbf{x}/\hbar}/\sqrt{2\pi}.$$

It follows then that the matrix element of $\hat{H}$ in the position basis is

$$\langle\mathbf{x}'|\hat{H}|\mathbf{x}\rangle = \frac{1}{2\pi}\int d^3p e^{i\mathbf{p}\cdot(\mathbf{x}'-\mathbf{x})/\hbar}\left(\frac{\mathbf{p}^2}{2m}\right). \tag{B.56}$$

The expectation value of the energy operator in position eigenstates yields the Fourier transform of the momentum space solution and takes the system from $|\mathbf{x}\rangle$ to $|\mathbf{x}'\rangle$.

Before we summarize, we point out that it is possible to determine the representation of $\hat{p}$ in the position representation if we go back to Eq. (B.53) and insert a complete set of position basis states in $\langle\mathbf{p}'|\hat{\mathbf{p}}|\mathbf{p}''\rangle$. This leads to

$$\frac{1}{2m}\int d^3p''\langle\mathbf{p}'|\hat{\mathbf{p}}|\mathbf{p}''\rangle\cdot\int d^3x\langle\mathbf{p}''|\hat{\mathbf{p}}|\mathbf{x}\rangle\langle\mathbf{x}|\mathbf{p}\rangle,$$
$$\frac{1}{2m}\int d^3p''\langle\mathbf{p}'|\hat{\mathbf{p}}|\mathbf{p}''\rangle\cdot\int d^3x\langle\mathbf{p}''|\hat{\mathbf{p}}\left(e^{i\mathbf{p}\cdot\mathbf{x}/\hbar}/\sqrt{2\pi}\right)|\mathbf{x}\rangle.$$

If we let $p \rightarrow -i\hbar\nabla$, perform the differentiation and write $\langle\mathbf{p}''|\mathbf{x}\rangle$ in terms of the plane wave again, we will have

$$\frac{\mathbf{p}}{2m}\cdot\langle\mathbf{p}'|\hat{\mathbf{p}}|\mathbf{p}\rangle.$$

Repeating this process with the remaining matrix element leads to Eq. (B.54). Consequently, $\hat{\mathbf{p}}$ takes a different form in different representations.

B.8.1
Summary

Here is a summary of some rules associated with the algebra of bras and kets used in quantum mechanical calculations. Most of these can be found in Dirac's text [122] and many are found elsewhere such as in Condon and Odabaşi [123], and Messiah [4] as well as standard linear algebra texts. The reader is encouraged to read these texts for a more thorough understanding.

1. Any linear operator can be represented by a matrix.
2. The unit operator is represented by the unit matrix.
3. A real linear operator is represented by a Hermitian matrix, $\hat{A} = \hat{A}^{\dagger}$.
4. Eigenvalues are represented by diagonal matrices.
5. The matrix representing the product of two linear operators is the product of the matrices representing the two factors.
6. A state can be an eigenstate of two observables (operators) simultaneously, if the two observables commute.
7. A linear operator that commutes with an observable also commutes with any function of the observable.

B.9
Schrödinger's Equation in Parabolic Coordinates

In this appendix we define parabolic coordinates and use them to express the Schrödinger equation along with its solutions for the hydrogen atom. The three parabolic coordinates, ξ, η and ϕ, are related to the more familiar spherical and rectangular coordinates through

$$
\begin{aligned}
\xi &= r(1 + \cos\theta), \\
\eta &= r(1 - \cos\theta), \\
\phi &= \phi, \\
x &= \sqrt{\xi\eta}\cos\phi, \\
y &= \sqrt{\xi\eta}\sin\phi \quad \text{and} \\
z &= (\xi - \eta)/2.
\end{aligned}
$$

The coordinate surfaces for ξ and η are paraboloids with their foci at the origin. Those for constant ξ open toward $z \to -\infty$ while those for constant η open toward $z \to \infty$; both take values from 0 to ∞. The coordinate surfaces for ϕ are the meridian planes and range, as usual, from 0 to 2π. Figure B.1 shows a cross section of the coordinate system in the yz-plane.

The length in spherical coordinates is related to that in parabolic coordinates through

$$r = (\xi + \eta)/2,$$

while the differential volume element is given by

$$dV = \frac{1}{4}(\xi + \eta)d\xi d\eta d\phi.$$

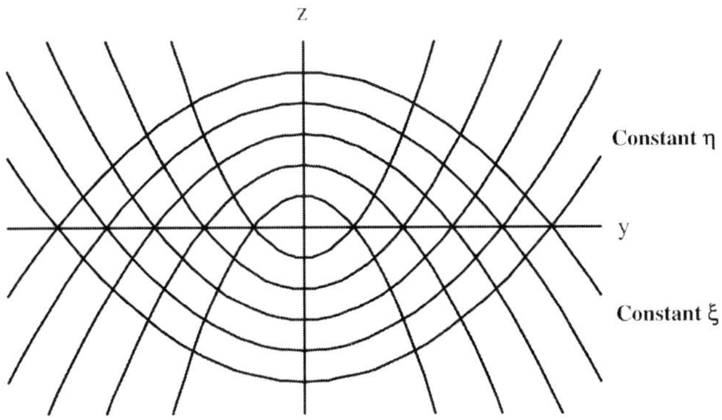

Fig. B.1 A cross section of the parabolic coordinate system. The coordinate surfaces for η and ξ are paraboloids with the z-axis as the axis of rotation. The surfaces for ϕ are the meridian planes, one of which is the plane of the page.

The wave equation can be written for $E < 0$ as

$$\frac{1}{\xi+\eta}\left[\frac{\partial}{\partial\xi}\left(\xi\frac{\partial}{\partial\xi}\psi(\xi,\eta,\phi)\right)+\frac{\partial}{\partial\eta}\left(\eta\frac{\partial}{\partial\eta}\psi(\xi,\eta,\phi)\right)\right]+\frac{1}{\xi\eta}\frac{\partial^2}{\partial\phi^2}$$
$$+\frac{2\mu}{\hbar^2}\left(E+\frac{2}{\xi+\eta}\frac{Ze^2}{4\pi\varepsilon_o}\right)\psi(\xi,\eta,\phi)=0. \quad \text{(B.57)}$$

The wavefunction is separable, so that $\psi(\xi,\eta,\phi) = f(\xi)g(\eta)\Phi(\phi)$. As with spherical coordinates, we can let

$$\Phi(\phi) = \frac{1}{\sqrt{2}}e^{im\phi}, \quad \text{where} \quad m = 0, \pm 1, \pm 2, \ldots.$$

If we let ν be the separation constant, f and g are solutions to

$$\frac{\partial}{\partial\xi}\left(\xi\frac{\partial f}{\partial\xi}\right)-\left(\frac{m^2}{4\xi}+\frac{\mu|E|\xi}{2\hbar^2}-\frac{\mu}{2\hbar^2}\frac{Ze^2}{4\pi\varepsilon_o}+\nu\right)f = 0, \quad \text{(B.58)}$$

$$\frac{\partial}{\partial\eta}\left(\eta\frac{\partial g}{\partial\eta}\right)-\left(\frac{m^2}{4\eta}+\frac{\mu|E|\eta}{2\hbar^2}-\nu\right)g = 0, \quad \text{(B.59)}$$

respectively. We can simplify these equations if we let

$$\begin{aligned} a^2 &= \frac{\mu|E|}{2\hbar^2}, \\ b_1 &= \frac{1}{a}\left(\frac{\mu}{\hbar^2}\frac{Ze^2}{4\pi\varepsilon_o}-\nu\right), \\ b_2 &= \frac{\nu}{a}. \end{aligned} \quad \text{(B.60)}$$

If we let $\rho = a\xi$ in Eq. (B.58) and $\rho = a\eta$ in Eq. (B.59) both equations take the same form:

$$\frac{1}{\rho}\frac{\partial}{\partial\rho}\left(\rho\frac{\partial u}{\partial\rho}\right) + \left(\frac{b}{\rho} - \frac{1}{4} - \frac{m^2}{4\rho^2}\right) u = 0, \tag{B.61}$$

where $b = b_1(\text{or } b_2)$ when $u = f(\text{or } g)$.

The approach we can take to solve Eq. (B.61) is the same we took for the hydrogen atom problem in spherical coordinates in Section 1.2.2. Namely, for large ρ,

$$u(\rho) \simeq \mathrm{e}^{-\rho/2}.$$

Seeking a power-series solution that must be truncated for $m > 0$ leads to

$$u(\rho) = \mathrm{e}^{-\rho/2}\rho^{m/2}U(\rho).$$

The equation to be solved becomes

$$\rho\frac{d^2U}{d\rho^2} + (m+1-\rho)\frac{dU}{d\rho} + \left[b - \frac{1}{2}(m+1)\right] u = 0. \tag{B.62}$$

The solutions can be written in terms of associated Laguerre polynomials or confluent hypergeometric functions as we had before:

$$L^m_{n+m}(\rho), \qquad \text{where} \qquad n = b - \frac{m+1}{2}.$$

Since m can be positive or negative, our solutions become

$$f(\rho) = \mathrm{e}^{-\rho/2}\rho^{|m|/2}L^{|m|}_{n_1+|m|}(\rho), \qquad n_1 = b_1 - \frac{|m|+1}{2}, \tag{B.63}$$

$$g(\rho) = \mathrm{e}^{-\rho/2}\rho^{|m|/2}L^{|m|}_{n_2+|m|}(\rho), \qquad n_2 = b_2 - \frac{|m|+1}{2}, \tag{B.64}$$

respectively, where both n_1 and n_2 must be positive integers or zero. The principal quantum number from our spherical solution is given by

$$n = b_1 + b_2 = n_1 + n_2 + |m| + 1 = \frac{\mu}{\hbar^2 a}\frac{Ze^2}{4\pi\varepsilon_0},$$

which leads to

$$|E_n| = \frac{1}{2}\left(\frac{Ze^2}{4\pi\varepsilon_0}\right)^2 \frac{4\mu}{\hbar^2 n^2},$$

as it must.

Each stationary state of the discrete spectrum ($E < 0$) is determined by the three integers in parabolic coordinates, the parabolic quantum numbers n_1, n_2

and m. For a given n, $|m|$ can range from 0 to $n-1$. For a given n and $|m|$, n_1 and n_2 can be chosen in $n-|m|$ ways such that $n_1+n_2=n-|m|-1-n_1$ and n_2 sort of play the role of l.[10] As in spherical coordinates, the energy is n^2 degenerate.

The total wavefunction is given by

$$\psi_{n_1 n_2 m}(\xi,\eta,\phi) = C e^{-(\xi+\eta)/2a_{n_1 n_2 m}}(\xi\eta)^{|m|/2} e^{im\phi} \times L^{|m|}_{n_1+|m|}(\xi/a_{n_1 n_2 m}) L^{|m|}_{n_2+|m|}(\eta/a_{n_1 n_2 m}), \quad (B.65)$$

where

$$a_{n_1 n_2 m} = \frac{4\pi\varepsilon_o}{Ze^2} \frac{\hbar^2(n_1+n_2+|m|+1)}{\mu}.$$

If we look at the ground state we will immediately see how Eq. (B.65) is related to R_{10} of Eq. (1.45). The ground state in parabolic coordinates, $n_1 = n_2 = m = 0$, takes the form

$$\psi_{000} \propto e^{-(\xi+\eta)/2a_{000}}.$$

In spherical coordinates, $n=1$ and $l=m=0$, leading to

$$\psi_{100} \propto e^{-r/a_o}.$$

Clearly, these are the same if we recognize a_{000} as the Bohr radius. In this case, the charge distribution is symmetric. For other choices, the charge distribution is asymmetric with respect to a plane through $z=0$. For example, when $n_1 > n_2$ the distribution is skewed to positive z.

B.10
Voigt Line Profile

In cases were the Gaussian and Lorentzian widths of the line are of the same order of magnitude, we must treat them on an equal footing. The appropriate line shape, $S(\omega, v)$, is given by

$$S(\omega, v) = \frac{2\Gamma_o}{(\omega_o - \omega + v\omega/c)^2 + \Gamma_o^2}. \quad (B.66)$$

To determine the line shape as a function of ω, $S(\omega)$, we must integrate over the normalized velocity distribution, $f(v)$.

$$S(\omega) = \int_{-\infty}^{\infty} dv S(\omega, v) f(v). \quad (B.67)$$

10) In spherical coordinates we had $|m| \le l < n-1$.

For atoms or molecules in thermal equilibrium with a temperature bath,

$$f(v) = (M/2\pi RT)^{1/2}\, e^{-Mv^2/2RT},$$

where R is the gas constant (Boltzmann constant per mole). Substituting $S(\omega, v)$ and $f(v)$ into Eq. (B.67) and changing variables, we can write the line shape in the well-known *Voigt profile* form as

$$S(\omega) = \frac{1}{\pi^{3/2}} \frac{b^2}{\Gamma_o} \int_{-\infty}^{\infty} \frac{dy e^{-y^2}}{(x+y)^2 + b^2}, \qquad \text{(B.68)}$$

where

$$x = \frac{\omega_0 - \omega}{\Gamma_D} \sqrt{4 \ln 2},$$

$$b = \frac{\Gamma_o}{\Gamma_D} \sqrt{4 \ln 2}$$

and Γ_D is the Doppler width. When nonthermal velocity distributions are more appropriate it will be necessary to redo the integrations to find a new expression for $S(\omega)$.

Further Reading

1 G. Arfken and H. Weber, *Mathematical Methods for Physicists*, Academic, New York, NY, 2000.

2 J. D. Jackson, *Classical Electrodynamics*, Wiley, New York, NY, 1999.

3 J. Mathews and R. L. Walker, *Mathematical Methods of Physics*, Benjamin Cummings, San Francisco, CA, 1970.

4 N. N. Lebedev, *Special Functions and Their Applications*, Dover, New York, NY, 1972.

5 P. A. M. Dirac, *The Principles of Quantum Mechanics*, Oxford University Press, Oxford, UK, 1981.

6 E. U. Condon and H. Odabaşi, *Atomic Structure*, Cambridge University Press, New York, NY, 1980.

Appendix C
Atomic and Molecular Data

C.1
NIST Online Data

The National Institute of Standards and Technology (NIST) maintains a very valuable online compilation of atomic and molecular data. The general link to this portal is

http://physics.nist.gov/products.html.

Below, we give the links to some specific key information.

C.1.1
NIST Online Atomic Data

The link to the Physical Reference Data page is

http://physics.nist.gov/PhysRefData/contents.html.

Fundamental constants can be obtained from

http://physics.nist.gov/PhysRefData/contents-constants.html.

C.1.2
NSIT Online Molecular Data

Chemical data, including constants for diatomic molecules, can be accessed through

http://webbook.nist.gov/.

C.2
Molecular Constants

Another source for constants for diatomic molecules is the compilation by Huber and Herzberg [45].

Light-Matter Interaction: Atoms and Molecules in External Fields and Nonlinear Optics.
W. T. Hill and C. H. Lee

ISBN: 978-3-527-40661-6

C.3
Filling Subshells

The table below shows the order in which subshells are filled. Energy increases down and to the left.

$1s$															
	$2s$	$2p$													
			$3s$	$3p$		$3d$									
					$4s$		$4p$		$4d$			$4f$			
								$5s$		$5p$			$5d$		
											$6s$			$6p$	
															$7s$

C.4 Electronic Configurations

The following table shows the ground-state configurations for the currently known elements in the periodic table.

H	–	He	1s	–	1s^2
Li	–	Be	1$s^2$2s	–	1$s^2$2s^2
B	–	Ne	1$s^2$2$s^2$2p	–	1$s^2$2$s^2$2p^6
Na	–	Mg	[Ne]3s	–	[Ne]3s^2
Al	–	Ar	[Ne]3$s^2$3p	–	[Ne]3$s^2$3p^6
K	–	Ca	[Ar]4s	–	[Ar]4s^2
Sc	–	Zn	[Ar]4$s^2$3d	–	[Ar]4$s^2$3d^{10}
Ga	–	Kr	[Ar]4$s^2$3d^{10}4p	–	[Ar]4$s^2$3d^{10}4p^6
Rb	–	Sr	[Kr]5s	–	[Kr]5s^2
Y	–	Cd	[Kr]5$s^2$4d	–	[Kr]5$s^2$4d^{10}
In	–	Xe	[Kr]5$s^2$4d^{10}5p	–	[Kr]5$s^2$4d^{10}5p^6
Cs	–	Ba	[Xe]6s	–	[Xe]6s^2
La			[Xe]6$s^2$5d		
Ce	–	Lu	[Xe]6$s^2$5d4f	–	[Xe]6$s^2$5d4f^{14}
Hf	–	Hg	[Xe]6$s^2$4f^{14}5d^2	–	[Xe]6$s^2$4f^{14}5d^{10}
Tl	–	Rn	[Xe]6$s^2$4f^{14}5d^{10}6p	–	[Xe]6$s^2$4f^{14}5d^{10}6p^6
Fr	–	Ra	[Rn]7s	–	[Rn]7s^2
Ac			[Rn]7$s^2$6d		
Th	–	Lr	[Rn]7$s^2$6d5f	–	[Rn]7$s^2$6d5f^{14}
Rf	–		[Rn]7$s^2$5f^{14}6d^2(?)	–	

Appendix D
Coupling Angular Momenta

D.1
Two Angular Momenta and 3-j Symbols

Consider a system described by two angular momenta j_1 and j_2. We can define all states of the system by combining two state vectors $|j_1 m_1\rangle$ and $|j_2 m_2\rangle$. If the states of the system these vectors represent not interact then we can write the total state vector as

$$|j_1 m_1 j_2 m_2\rangle = |j_1 m_1\rangle\,|j_2 m_2\rangle\,. \tag{D.1}$$

We could equally well have coupled j_1 and j_2 to give a total angular momentum J and have written a total state vector in terms of the total angular momentum as

$$|JM\rangle = |(j_1 j_2)JM\rangle\,, \tag{D.2}$$

where in $(j_1 j_2)J$ we coupled the two angular momenta within the parentheses to give the angular momentum just to the right outside the parentheses. Depending upon the problem, one representation is often more convenient than the other. Since either representation can be used to completely define the system, we can write one representation in terms of the other. Using the fact that

$$1 = \sum_{m_1,m_2} |j_1 m_1 j_2 m_2\rangle\,\langle j_1 m_1 j_2 m_2|\,, \tag{D.3}$$

often called the sum over states, we can write

$$|JM\rangle = \sum_{m_1,m_2} |j_1 m_1 j_2 m_2\rangle\,\langle j_1 m_1 j_2 m_2|JM\rangle\,, \tag{D.4}$$

which is subject to the constraint $M = m_1 + m_2$. In Eq. (D.4), $\langle j_1 m_1 j_2 m_2|JM\rangle$ is a real number, called a Clebsch–Gordon coefficient, that can be rewritten in terms of 3-j symbols:

$$\langle j_1 m_1 j_2 m_2|JM\rangle = (-1)^{-j_1+j_2-M}\sqrt{2J+1}\begin{pmatrix} j_1 & j_2 & J \\ m_1 & m_2 & -M \end{pmatrix}. \tag{D.5}$$

Light-Matter Interaction: Atoms and Molecules in External Fields and Nonlinear Optics.
W. T. Hill and C. H. Lee

ISBN: 978-3-527-40661-6

A consequence of Eqs. (D.2), (D.4) and (D.5) allows us to write $|JM\rangle$ as

$$|JM\rangle = (-1)^{j_2-j_1-M}\sqrt{2J+1}\sum_{m_1,m_2}|j_1m_1j_2m_2\rangle\begin{pmatrix} j_1 & j_2 & J \\ j_1 & m_2 & -M \end{pmatrix}. \quad \text{(D.6)}$$

D.2
Properties of 3-j Symbols

In this section we summarize a few of the symmetry properties and sum rules of 3-j symbols that are useful in manipulating equations. Additional properties and numerical values for specific 3-j symbols can be found in Edmonds [126] and Appendix C.8 of Cowan [2]. An algebraic expression for the 3-j symbol can be found in Chapter 5 of Cowan [2] and Appendix 2 of [126].

The 3-j symbol is a real number and has the property that the three elements in the upper row must add as vectors. That is, all permutations of the elements obey $j_1 + j_2 \geq J$. At the same time, the sum of the three elements in the lower row must equal zero. If these conditions are not met, the 3-j symbol is identically zero. The value of the 3-j symbol is unchanged upon any cyclic or even permutation of columns. Any odd permutation, however, or a change of all the signs of the magnetic quantum numbers, leads to

$$\begin{aligned}\begin{pmatrix} j_1 & J & j_2 \\ m_1 & -M & m_2 \end{pmatrix} = \begin{pmatrix} J & j_2 & j_1 \\ -M & m_2 & m_1 \end{pmatrix} &= (-1)^{j_1+j_2+J}\begin{pmatrix} j_1 & j_2 & J \\ m_1 & m_2 & -M \end{pmatrix} \\ &= (-1)^{j_1+j_2+J}\begin{pmatrix} j_1 & J & j_2 \\ -m_1 & M & -m_2 \end{pmatrix}.\end{aligned} \quad \text{(D.7)}$$

The 3-j symbols are orthogonal and obey the sum rules

$$\sum_{j,m}(2J+1)\begin{pmatrix} j_1 & j_2 & J \\ m_1 & m_2 & -M \end{pmatrix}\begin{pmatrix} j_1 & j_2 & J \\ m'_1 & m'_2 & -M \end{pmatrix} = \delta_{m_1,m'_1}\delta_{m_2,m'_2} \quad \text{(D.8)}$$

and

$$\sum_{m_1,m_2}\begin{pmatrix} j_1 & j_2 & J \\ m_1 & m_2 & -M \end{pmatrix}\begin{pmatrix} j_1 & j_2 & J' \\ m_1 & m_2 & -M' \end{pmatrix} = \frac{\delta_{J,J'}\delta_{M,M'}}{(2J+1)}. \quad \text{(D.9)}$$

When $j_1 = j_2 = j$, $m_1 = -m_2 = m$ and $M = 0$, we have

$$\sum_{m}(-1)^{j-m}\begin{pmatrix} j & j & J \\ m & -m & 0 \end{pmatrix} = \delta_{J,0}\sqrt{(2j+1)}. \quad \text{(D.10)}$$

D.3 Three Angular Momenta and 6-j Symbols

Given three angular momenta j_1, j_2 and j_3, there are three ways to couple them to give a total angular momentum of J:

$$\begin{aligned} &|(j_1 j_2)\, j_{12}, j_3; JM\rangle\,, \\ &|(j_1 j_3)\, j_{13}, j_2; JM\rangle\,, \\ &|j_1, (j_2 j_3)\, j_{23}; JM\rangle\,, \end{aligned} \tag{D.11}$$

where we couple the two angular momenta outside the parentheses to give J. Applying a procedure similar to that used to couple two angular momenta, we can write each of these in terms of the state vector $|j_1 m_1 j_2 m_2 j_3 m_3\rangle$. Recalling that this state vector is separable, we can insert a sum over states twice to couple the angular momenta pairwise. This gives for the first entry

$$\begin{aligned} &|(j_1 j_2)\, j_{12}, j_3; JM\rangle \\ &= \sum_{m_1, m_2, m_{12}, m_3} |j_1 m_1 j_2 m_2 j_3 m_3\rangle \, \langle j_1 m_1 j_2 m_2 |\, j_{12} m_{12}\rangle \, \langle j_{12} m_{12} j_3 m_3 | JM\rangle\,. \end{aligned} \tag{D.12}$$

In Eq. (D.12), we have two Clebsch–Gordon coefficients, $\langle j_1 m_1 j_2 m_2 |\, j_{12} m_{12}\rangle$ and $\langle j_{12} m_{12} j_3 m_3 | JM\rangle$, that can be evaluated with Eq. (D.5). A similar relationship holds for the other two coupling schemes. In general, we will be interested in the transformation between two of these coupling schemes. That is, we want to evaluate matrix elements of the type

$$\langle (j_1 j_2)\, j_{12}, j_3; JM | j_1, (j_2 j_3)\, j_{23}; JM\rangle\,,$$

that represents the projection of one coupling scheme onto another. This matrix element is just a number that can be evaluated in terms of 6-j symbols. The projection of the first scheme onto the third leads to

$$\begin{aligned} &\langle (j_1 j_2)\, j_{12}, j_3; J | j_1, (j_2 j_3)\, j_{23}; J\rangle \\ &= (-1)^{j_1 + j_2 + j_3 + J} \sqrt{(2 j_{12} + 1)\,(2 j_{23} + 1)} \begin{Bmatrix} j_1 & j_2 & j_{12} \\ j_3 & J & j_{23} \end{Bmatrix}, \end{aligned} \tag{D.13}$$

where we have dropped M from the left-hand side because the matrix element does not depend on M. The property of the 6-j symbols is that the following triples add as vectors: (j_1, j_2, j_{12}), (j_1, J, j_{23}), (j_2, j_3, j_{23}) and (j_{12}, j_3, J); otherwise, the 6-j symbol is identically zero. Numerical values for the 6-j symbols are tabulated in Appendix D of Cowan [2] while an analytical expression is given in Chapter 5 of Cowan [2] and Appendix 2 of [126]. It is also possible to evaluate the 6-j symbols in terms of a sum of the products of four 3-j symbols:

$$\begin{Bmatrix} j_1 & j_2 & j_{12} \\ j_3 & J & j_{23} \end{Bmatrix}$$
$$= (2J+1) \sum_{\substack{m_1,m_2,\\ m_{12},m_{23},M}} (-1)^{\kappa} \begin{pmatrix} j_1 & j_2 & j_{12} \\ m_1 & m_2 & m_{12} \end{pmatrix}$$
$$\times \begin{pmatrix} j_1 & J & j_{23} \\ -m_1 & M & -m_{23} \end{pmatrix} \begin{pmatrix} j_3 & j_2 & j_{23} \\ -m_3 & -m_2 & m_{23} \end{pmatrix}$$
$$\times \begin{pmatrix} j_3 & J & j_{12} \\ m_3 & -M & -m_{12} \end{pmatrix}, \tag{D.14}$$

where κ is the scalar sum $\kappa = j_1 + j_2 + j_{12} + j_3 + j_{23} + J$.

D.4
Four Angular Momenta and 9-j Symbols

Coupling four angular momenta is quite important because it allows us to transform between *LS* and *jj* coupling schemes. The general matrix element of concern has the form

$$\langle (j_1 j_2)\, j_{12}, (j_3 j_4)\, j_{34}; J | (j_1 j_3)\, j_{13}, (j_2 j_4)\, j_{24}; J \rangle \tag{D.15}$$
$$= \sqrt{(2j_{12}+1)\,(2j_{34}+1)\,(2j_{13}+1)\,(2j_{24}+1)} \begin{Bmatrix} j_1 & j_2 & j_{12} \\ j_3 & j_4 & j_{34} \\ j_{13} & j_{24} & J \end{Bmatrix}.$$

The 9-j symbol can be rewritten in terms of a sum over 6-j symbols for evaluation. Specifically,

$$\begin{Bmatrix} j_1 & j_2 & j_{12} \\ j_3 & j_4 & j_{34} \\ j_{13} & j_{24} & J \end{Bmatrix} \tag{D.16}$$
$$= \sum_k (-1)^{2k} (2k+1) \begin{Bmatrix} j_1 & j_3 & j_{13} \\ j_{24} & J & k \end{Bmatrix} \begin{Bmatrix} j_2 & j_4 & j_{24} \\ j_3 & k & j_{34} \end{Bmatrix} \begin{Bmatrix} j_{12} & j_{34} & J \\ k & j_1 & j_2 \end{Bmatrix}.$$

Appendix E
Tensor Algebra

In this appendix we will present the basic properties associated with spherical tensors. This is not intended to be a tutorial on the subject. The reader is directed to the references at the end of this appendix for a more detailed study.

E.1
Spherical Tensors

A three-dimensional Cartesian vector, $\vec{V}$, has components $\left(V_x, V_y, V_z\right)$, which can be written as V_i with $i = 1, 2, 3$, where $V_1 = V_x$, $V_2 = V_y$ and $V_3 = V_z$. The latter notation is the usual designation for a three-dimensional Cartesian tensor of rank 1. It is more convenient, when working with atoms, to use spherical coordinates (r, θ, ϕ) instead of x, y, z, where the conversion between the two systems of components is given by

$$x = r\sin\theta\cos\phi, \tag{E.1}$$

$$y = r\sin\theta\sin\phi, \tag{E.2}$$

$$z = r\cos\theta. \tag{E.3}$$

We will now encounter spherical vectors and higher-ranked spherical tensors. The usual notation for a spherical tensor of rank k is $T_{k,q}$. A rank-1 spherical tensor in three dimensions will have $k = 1$ and $q = -1, 0, 1$. Similar to the Cartesian tensor of rank 1, the spherical tensor of rank 1 is related to the components of a spherical vector. In spherical coordinates, $\vec{V}$ has components V_{+1}, V_0 and V_{-1}. The quantity $T_{k,q}$ is an irreducible tensor operator that is Hermitian if

$$\widehat{\mathbf{T}}^{\dagger}_{k,q} = (-1)^q \widehat{\mathbf{T}}_{k,-q}. \tag{E.4}$$

When $\widehat{\mathbf{T}}_{k,q}$ acts only on spatial coordinates (i.e., does not act on spins) then

$$\widehat{\mathbf{T}}_{k,q} = C_k Y_{kq}(\Omega), \tag{E.5}$$

where Y_{kq} are the spherical harmonics (see Appendix B.7) and C_k is a function that is independent of the angular coordinates. It is often convenient to

Light-Matter Interaction: Atoms and Molecules in External Fields and Nonlinear Optics.
W. T. Hill and C. H. Lee

ISBN: 978-3-527-40661-6

convert between Cartesian and spherical vectors. As an example, we will consider the spherical vector $V_q = f(r)Y_{1q}$, where we have taken C_1 to be purely a function of r and have dropped the index 1 from V. From Eq. (1.25), we write the components of V as of a spherical vector in the three spherical harmonics Y_{10} and $Y_{1,\pm1}$, from which it is clear that

$$V_{\pm1} = f(r)Y_{1,\pm1} = \mp f(r)\sqrt{\frac{3}{8\pi}}\sin\theta e^{\pm i\phi}, \tag{E.6}$$

$$V_0 = f(r)Y_{1,0} = f(r)\sqrt{\frac{3}{4\pi}}\cos\theta. \tag{E.7}$$

Expanding the exponential and recognizing $\sin\theta\cos\phi$, etc., as unit vectors in the x, etc., direction, we can write

$$V_{\pm1} = \mp\frac{1}{\sqrt{2}}\left(V_x \pm iV_y\right), \tag{E.8}$$

$$V_0 = V_z, \tag{E.9}$$

where the magnitude of V is $\sqrt{3/4\pi} f(r)$.

E.2
Commutation Relations

There are two commutation relations that are frequently encountered when working with spherical tensors:

$$\left[J_\pm, \widehat{T}_{k,q}\right] = \sqrt{(k \mp q)(k \pm q + 1)}\widehat{T}_{k,q\pm1}, \tag{E.10}$$

$$\left[J_z, \widehat{T}_{k,q}\right] = q\widehat{T}_{k,q}. \tag{E.11}$$

As an example, the following is how the orbital angular momentum $\widehat{L}_{-1}$ commutes with $L_\pm$ and L_z:

$$\left[L_\pm, \widehat{L}_{1,-1}\right] = \sqrt{2}\widehat{L}_{1,0}, \tag{E.12}$$

$$\left[L_z, \widehat{L}_{1,-1}\right] = -\widehat{L}_{1,-1}. \tag{E.13}$$

E.3
Reduced Matrix Elements

When an operator is represented as a spherical tensor, such as angular momentum ($\vec{\mathbf{L}}$) or an electric dipole ($e\vec{\mathbf{r}}$), we need to find matrix elements of the form $\langle \gamma jm | \widehat{T}_{k,q} | \gamma' j' m' \rangle$. The *Wigner–Eckart theorem*,

$$\langle \gamma jm | \widehat{T}_{k,q} | \gamma' j' m' \rangle = (-1)^{j-m} \begin{pmatrix} j & k & j' \\ -m & q & m' \end{pmatrix} \langle \gamma j \| \widehat{\mathbf{T}}_k \| \gamma' j' \rangle, \tag{E.14}$$

allows us to relate this matrix element to 3-j symbols and a reduced matrix element. Values for 3-j symbols are tabulated in several places[1] or easily calculated in programs such as MATHEMATICA®. The reduced or double-bar matrix element is equivalent to the *line strength* (see Section 4.3.4.1). Specifically, we relate the square of the reduced matrix element to the sum over the matrix elements,

$$|\langle \gamma j \| \widehat{\mathbf{T}}_k \| \gamma' j' \rangle|^2 = \sum_{q,m,m'} |\langle \gamma jm | \widehat{\mathbf{T}}_{k,q} | \gamma' j' m' \rangle|^2. \tag{E.15}$$

Clearly, the reduced matrix element is independent of angle (the orientation of the operator) and the magnetic quantum numbers. Consequently, neither the magnetic quantum number, m, nor q appear in the reduced matrix element. Its independence of q allows us to write

$$|\langle \gamma j \| \widehat{\mathbf{T}}_k \| \gamma' j' \rangle|^2 = (2k+1) \sum_{m,m'} |\langle \gamma jm | \widehat{\mathbf{T}}_{k,q} | \gamma' j' m' \rangle|^2, \tag{E.16}$$

with $2k+1$ being the number of possible orientations of the operator. For Hermitian operators (Eq. (E.5)), the reduced matrix elements obey

$$\langle \gamma j \| \widehat{\mathbf{T}}_{k,q} \| \gamma' j' \rangle = (-1)^{j-j'} \langle \gamma' j' \| \widehat{\mathbf{T}}_{k,q} \| \gamma j \rangle^*. \tag{E.17}$$

One evaluates the reduced matrix element by selecting an m state for which the left-hand side of Eq. (E.14) can be determined most conveniently. For example, the reduced matrix element for the spin operator, $\hat{T} = \vec{\mathbf{S}}$ with $k = 1$, for an electron in a state $|\gamma s m_s\rangle$, can be found as follows:

$$\langle \gamma s || \mathbf{S} || \gamma s \rangle = (-1)^{m_s - s} \frac{\langle \gamma s m_s | \mathbf{S} | \gamma s m_s \rangle}{\begin{pmatrix} s & k & s \\ -m_s & q & m_s \end{pmatrix}}. \tag{E.18}$$

The most convenient orientation of $\vec{\mathbf{S}}$ will be along the z-axis where $q = 0$ and $\vec{\mathbf{S}} \rightarrow \vec{\mathbf{S}}_z$. The numerator on the right-hand side of the expression will just be m_s. Evaluating the 3-j symbol leads to

$$\langle \gamma s || \mathbf{S} || \gamma s \rangle = \sqrt{s(s+1)(2s+1)} = \sqrt{\frac{3}{2}}. \tag{E.19}$$

1) See, for example, A. R. Edmonds, *Angular Momentum in Quantum Mechanics* [126].

The last step is made by setting $s = 1/2$. As promised, the reduced matrix element is independent of m_s. One can show that it is also independent of q by letting $q = \pm 1$ and showing that the right-hand side gives the same answer. Once we have a value for the reduced matrix element, we are able to evaluate specific matrix elements in terms of it. For example, the matrix element for $\mathbf{S}_\pm$, where $q = \pm 1$, is

$$\begin{aligned}\langle \gamma s m_s | \mathbf{S}_\pm | \gamma' s' m_s' \rangle &= \sqrt{(s \mp m_s)(s \pm m_s + 1)}\delta_{s,s'}\delta_{m_s',m_s\pm 1} \\ &= \sqrt{\left(\frac{1}{2} \mp m_s\right)\left(\frac{3}{2} \pm m_s\right)}\delta_{s,s'}\delta_{m_s',m_s\pm 1}.\end{aligned} \tag{E.20}$$

We recognize $\mathbf{S}_\pm$ as the raising and lowering operators.

E.4
Matrix Elements of Products of Operators

Consider an operator U_{k_2,q_2} that operates in a space spanned by the basis set $|\gamma J M\rangle$. If $|\psi\rangle$ is a function in this space, then we can write

$$\begin{aligned}|\psi\rangle &= U_{k_2,q_2} |\gamma' J' M'\rangle \\ &= \sum_{\gamma'',J'',M''} |\gamma'' J'' M''\rangle \langle \gamma'' J'' M''| U_{k_2,q_2} |\gamma' J' M'\rangle \\ &= \sum_{\gamma'',J'',M''} |\gamma'' J'' M''\rangle (-1)^{J''-M''} \begin{pmatrix} J'' & k_2 & J' \\ -M'' & q_2 & M' \end{pmatrix} \\ &\quad \times \langle \gamma'' J'' \| U_{k_2} \| \gamma' J' \rangle .\end{aligned} \tag{E.21}$$

Product of tensor operators. Often, we need to find matrix elements of tensor products of operators. If, for example, we have another operator, T_{k_1,q_1}, which also acts in this space, the matrix element of the product can be written as

$$\begin{aligned}&\langle \gamma J M | T_{k_1,q_1} U_{k_2,q_2} |\gamma' J' M'\rangle \\ &= \sum_{\gamma'',J'',M''} \langle \gamma J M | T_{k_1,q_1} |\gamma'' J'' M''\rangle (-1)^{J''-M''} \\ &\qquad \times \begin{pmatrix} J'' & k_2 & J' \\ -M'' & q_2 & M' \end{pmatrix} \langle \gamma'' J'' \| U_{k_2} \| \gamma' J' \rangle .\end{aligned}$$

Applying the Wigner–Eckart theorem a second time leads to

$$\begin{aligned}&\langle \gamma JM|\, T_{k_1,q_1} U_{k_2,q_2}\, |\gamma' J' M'\rangle \\ &= \sum_{\gamma'',J'',M''} (-1)^{J-M} \begin{pmatrix} J & k_1 & J'' \\ -M & q_1 & M'' \end{pmatrix} \langle \gamma J \|\, T_{k_1}\, \| \gamma'' J'' M''\rangle \\ &\qquad \times (-1)^{J''-M''} \begin{pmatrix} J'' & k_2 & J' \\ -M'' & q_2 & M' \end{pmatrix} \langle \gamma'' J''\, \| U_{k_2} \|\, \gamma' J'\rangle\,.\end{aligned}$$

Scalar product of two tensors. The preceding expression is an expression for a general product of two tensors when they both act in the same space. When the product is a scalar product, the expression can be simplified. Using the shorthand $\sum_q (-1)^q T_{k,-q} U_{k,q} \equiv T_k U_k \equiv \vec{T} \cdot \vec{U}$, the previous expression reduces to

$$\begin{aligned}\langle \gamma JM | T_k U_k | \gamma' J' M'\rangle &= \sum_{\gamma'' J'' M''} \langle \gamma J \| T_k \| \gamma'' J''\rangle \langle \gamma J \| U_k \| \gamma'' J''\rangle \\ &\quad \times (-1)^{J+J''} \sum_q (-1)^{q-M-M''} \begin{pmatrix} J & k & J'' \\ -M & -q & M'' \end{pmatrix} \\ &\quad \times \begin{pmatrix} J'' & k & J' \\ -M'' & q & M' \end{pmatrix} \\ &= \frac{\delta_{J,J'}\delta_{M,M'}}{2J+1} \sum_{\gamma'' J''} (-1)^{J''-J} \\ &\quad \times \langle \gamma J \|\, T_k\, \| \gamma'' J''\rangle \langle \gamma'' J'' \|\, U_k\, \| \gamma' J'\rangle\,. \qquad \text{(E.22)}\end{aligned}$$

The last step can be shown using the symmetry and summation properties of the 3-j symbols in Appendix D.2 and recognizing the fact that $2J - 2M$ is even.

Scalar products of tensors acting in different spaces. It is also common to need matrix elements of products when $T_{k,q}$ and $U_{k,q}$ commute and act in different spaces. Matrix elements of the scalar product between orbital and spin angular momenta ($\vec{\mathbf{L}} \cdot \vec{\mathbf{S}}$) in the state $|\gamma LSJM\rangle$ is an example. It can be shown that in this case the matrix element takes the form

$$\begin{aligned}&\langle (\gamma_1 j_1 \gamma_2 j_2)\, JM|\, T_k U_k \,\big|(\gamma_1' j_1' \gamma_2' j_2')\, J' M'\rangle \\ &= \delta_{j,j'}\delta_{m,m'} (-1)^{j_1'+j_2+J} \langle \gamma_1 j_1 \|\, T_k\, \| \gamma_1' j_1'\rangle \langle \gamma_2 j_2 \|\, U_k\, \| \gamma_2' j_2'\rangle \\ &\qquad \times \begin{Bmatrix} j_1 & j_2 & J \\ j_2' & j_1' & k \end{Bmatrix}. \qquad \text{(E.23)}\end{aligned}$$

Reduced matrix element of L in the state $|\gamma LSJM\rangle$. When a tensor operator acts in only part of the space, such as the orbital (**L**) or spin (**S**) angular momenta, we must evaluate reduced matrix elements of the form $\langle \gamma LSJ \,\|\mathbf{L}\|\, \gamma LSJ\rangle$. It is instructive to determine what such a reduced element means. To that end, we first look at the reduced matrix element of an operator $\hat{\mathbf{T}}_{k,q}$ in a state $|\gamma j_1 j_2 JM\rangle$. It is possible to show that

$$\langle \gamma j_1 j_2 J \| \mathbf{T}_k \| \gamma j_1' j_2' J' \rangle = (-1)^{j_1+j_2+J+k} \times \sqrt{(2J+1)(2J'+1)} \left\{ \begin{array}{ccc} j_1' & J' & j_2 \\ J & j_1 & k \end{array} \right\} \langle \gamma j_1 \| \mathbf{T}_k \| \gamma' j_1' \rangle .$$

The orbital angular momentum reduced matrix element then reduces to

$$\langle \gamma LSJ \| \mathbf{L} \| \gamma LSJ \rangle = (-1)^{L+S+J+1}(2J+1) \times \left\{ \begin{array}{ccc} L & J & S \\ J & L & 1 \end{array} \right\} \langle \gamma L \| \mathbf{L} \| \gamma L \rangle . \tag{E.24}$$

From the Wigner–Eckart theorem, it is possible to show that $\langle \gamma L \| \mathbf{L} \| \gamma L \rangle = \sqrt{L(L+1)(2L+1)}$. Evaluating the 6-j symbol leads to

$$\langle \gamma LSJ \| \mathbf{L} \| \gamma LSJ \rangle = \frac{J(J+1) + L(L+1) - S(S+1)}{2J(J+1)} \times \sqrt{J(J+1)(2J+1)} . \tag{E.25}$$

A semiclassical interpretation of Eq. (E.25) is possible if we define $\langle \cos(\vec{\mathbf{L}}, \vec{\mathbf{J}}) \rangle$, where $(\vec{\mathbf{L}}, \vec{\mathbf{J}})$ denotes the angle between $\vec{\mathbf{L}}$ and $\vec{\mathbf{J}}$. (A similar expression is possible for $\vec{\mathbf{S}}$ as well.) From the semiclassical vector model (Fig. E.1) coupling $\vec{\mathbf{L}}$, $\vec{\mathbf{S}}$ and $\vec{\mathbf{J}}$ and the law of cosines, it is possible to show that

$$\langle \cos(\vec{\mathbf{L}}, \vec{\mathbf{J}}) \rangle = \frac{J(J+1) + L(L+1) - S(S+1)}{2\sqrt{J(J+1)}\sqrt{L(L+1)}} , \tag{E.26}$$

$$\langle \cos(\vec{\mathbf{S}}, \vec{\mathbf{J}}) \rangle = \frac{J(J+1) + S(S+1) - L(L+1)}{2\sqrt{J(J+1)}\sqrt{S(S+1)}} . \tag{E.27}$$

Equation (E.26) allows us to rewrite Eq. (E.25) as

$$\langle \gamma LSJ \| \mathbf{L} \| \gamma LSJ \rangle = \sqrt{L(L+1)} \langle \cos(\vec{\mathbf{L}}, \vec{\mathbf{J}}) \rangle \frac{\langle \gamma J \| J \| \gamma J \rangle}{\sqrt{J(J+1)}} . \tag{E.28}$$

Thus, the reduced matrix element can be viewed as the average projection of **L** onto the average **J** axis.

One can write the left-hand side of Eq. (E.25) in a different way if we write $\vec{\mathbf{L}} \cdot \vec{\mathbf{J}}$ semiclassically as

$$\begin{aligned} \mathbf{S}^2 &= (\mathbf{J} - \mathbf{L})^2 \\ &= \mathbf{J}^2 + \mathbf{L}^2 - 2(\mathbf{L} \cdot \mathbf{J}) . \end{aligned} \tag{E.29}$$

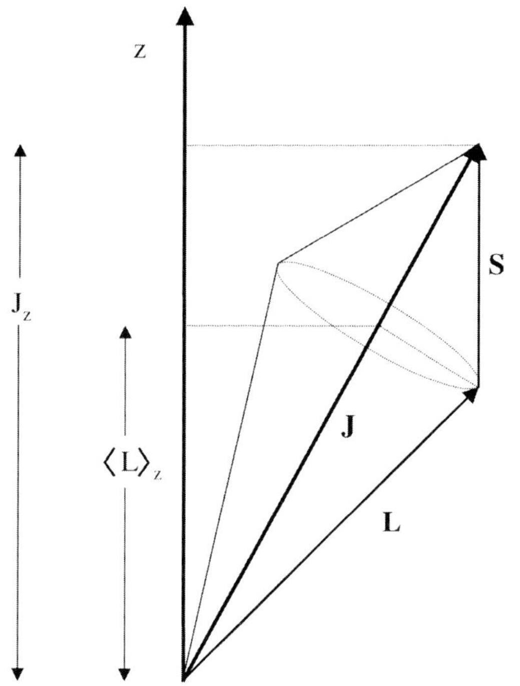

Fig. E.1 Semiclassical vector precession model.

Taking matrix elements with the state vector $|\gamma LSJM\rangle \equiv |\gamma JM\rangle$ of both sides of Eq. (E.29) and dividing by $\langle \gamma JM|\mathbf{J}^2|\gamma JM\rangle = J(J+1)$ leads to

$$\frac{\langle \gamma JM|\vec{\mathbf{L}}\cdot\vec{\mathbf{J}}|\gamma JM\rangle}{\langle \gamma JM|\mathbf{J}^2|\gamma JM\rangle} = \frac{J(J+1)+L(L+1)-S(S+1)}{2J(J+1)}. \tag{E.30}$$

The matrix elements of $\mathbf{L}^2$ and $\mathbf{S}^2$ were found with the aid of Eqs. (D.5), (D.6) and (D.8). Comparing Eqs. (E.25) and (E.30) allows us to write

$$\langle \gamma LSJ\|\mathbf{L}\|\gamma LSJ\rangle = \langle \gamma JM|\mathbf{L}\cdot\mathbf{J}|\gamma JM\rangle \frac{\sqrt{J(J+1)(2J+1)}}{J(J+1)}. \tag{E.31}$$

Finally, the matrix element of **L** is given by

$$\langle \gamma LSJM|\mathbf{L}|\gamma LSJM\rangle = M\frac{J(J+1)+L(L+1)-S(S+1)}{2J(J+1)}, \tag{E.32}$$

or

$$\langle \gamma LSJM|\mathbf{L}|\gamma LSJM\rangle = M\frac{\langle \gamma LSJ\|\mathbf{L}\|\gamma LSJ\rangle}{\langle \gamma J\|J\|\gamma J\rangle}. \tag{E.33}$$

Semiclassically, this corresponds to the projection of the average value of $\vec{\mathbf{L}}$, while it precesses about $\vec{\mathbf{J}}$, onto the z-axis as shown in Fig. E.1.

Tensor product of two tensors. Finally, we end by stating without proof the reduced matrix element of the tensor product of two operators acting in the same and different spaces.[2] If we let

$$X_{K,Q} = T_{k_1,q_1} \times U_{k_2,q_2}, \tag{E.34}$$

then the reduced matrix element of $P_{k,q}$ when both operators act in the same space will be

$$\langle \gamma' j' \| X_K \| \gamma j \rangle = \sqrt{2K+1}(-1)^{K+j+j'} \sum_{\gamma'',j''} \left\{ \begin{array}{ccc} k_1 & k_2 & K \\ j & j' & j'' \end{array} \right\} \times \langle \gamma' j' \| T_{k_1} \| \gamma'' j'' \rangle \langle \gamma'' j'' \| U_{k_2} \| \gamma j \rangle. \tag{E.35}$$

When the operators act in different spaces, we have

$$\langle \gamma' j_1' j_2' J' \| X_K \| \gamma j_1 j_2 J \rangle = \sqrt{(2J+1)(2J'+1)(2K+1)} \times \sum_{\gamma''} \langle \gamma' j_1' \| T_{k_1} \| \gamma'' j_1 \rangle \langle \gamma'' j_2' \| U_{k_2} \| \gamma j_2 \rangle \cdot \left\{ \begin{array}{ccc} j_1' & j_1 & k_1 \\ j_2' & j_2 & k_2 \\ J' & J & K \end{array} \right\}. \tag{E.36}$$

Further Reading

1 A. R. Edmonds, *Angular Momentum in Quantum Mechanics*, Princeton University Press, Princeton, NJ, 1974 [ISBN 0-691-07912-9].

2 R. D. Cowan, *The Theory of Atomic Structure and Spectra*, University of California Press, Berkeley, CA, 1981.

3 I. I. Sobelman, *Atomic Spectra and Radiative Transitions*, Springer Series in Chemical Physics. Springer, New York, NY, 1979.

4 F. Yang and J. H. Hamilton, *Modern Atomic and Nuclear Physics*, McGraw-Hill, New York, NY, 1996 [ISBN 0-07-025881-3].

2) The derivation is discussed in Ref. [126].

References

1 G. Arfken and H. Weber. *Mathematical Methods for Physicists*. Academic, New York, NY, 2000. ISBN 0-12-059825-6, Fifth edition.

2 R. D. Cowan. *The Theory of Atomic Structure and Spectra*. University of California Press, Berkeley, CA, 1981. ISBN 0-520-03821-5.

3 H. A. Bethe and E. E. Salpeter. *Quantum Mechanics of One- and Two-Electron Atoms*. Plenum, New York, NY, 1977. ISBN 0-306-20022-8, First paperback printing.

4 Albert Messiah. *Quantum Mechanics*. Dover Publications, 2000. ISBN: 0486409244.

5 K. Bockasten. *Phys. Rev. A*, 9:1087, 1974.

6 J. Weiner and P.-T. Ho. *Light–Matter Interaction, Volume 1: Fundamentals and Applications*. Wiley, Hoboken, NJ, 2003. ISBN 0-471-25377-4, First edition.

7 NIST. Basic atomic spectroscopic data, a. http://physics.nist.gov/PhysRefData/Handbook/periodictable.htm.

8 NIST. Physical reference data, b. http://physics.nist.gov/PhysRefData/contents.html.

9 NIST. Physics laboratory products and services, c. http://physics.nist.gov/products.html.

10 J. Javanainen, J. H. Eberly, and Q. C. Su. Numerical simulations of multiphoton ionization and above-threshold electron spectra. *Phys. Rev. A*, 38:3430, 1988.

11 Q. Su and J. H. Eberly. Model atom for multiphoton physics. *Phys. Rev. A*, 44: 5997, 1991.

12 S. Augst, D. Strickland, D. D. Meyerhofer, S. L. Chin, and J. H. Eberly. *Phys. Rev. Lett.*, 63:2212, 1989.

13 L. V. Keldysh. *Sov. Phys. JETP*, 20:1307, 1965.

14 A. M. Perelomov, V. S. Popov, and M. V. Terent'ev. *Sov. Phys. JETP*, 23:924, 1966.

15 M. V. Ammosov, N. B. Delone, and V. P. Krainov. *Sov. Phys. JETP*, 64:1191, 1986.

16 T. D. G. Walsh, F. A. Ilkov, J. E. Dechker, and S. L. Chin. The tunnel ionization of atoms, diatomic and triatomic molecules using intense 10.6 μm radiation. *J. Phys. B: At. Mol. Opt. Phys.*, 27:3767, 1994.

17 S. Augst, D. D. Meyerhofer, D. Strickland, and S. L. Chin. *J. Opt. Soc. Am. B*, 8:858, 1991.

18 F. A. Ilkov, J. E. Decker, and S. L. Chin. *J. Phys. B: At. Mol. Opt. Phys.*, 25:4005, 1992.

19 B. Walker, B. Sheehy, L. F. DiMauro, P. Agostini, K. J. Schafer, and K. C. Kulander. Precision measurement of strong field double ionization of helium. *Phys. Rev. Lett.*, 73:1227, 1994.

20 X. M. Tong, Z. X. Zhao, and D. Lin. Theory of molecular tunneling ionization. *Phys. Rev. A*, 66:033402, 2002.

21 E. Schrödinger. *Ann. Phys.*, 80:457, 1926.

22 P. S. Epstein. *Phys. Rev.*, 28:695, 1926.

23 I. Walker. *Z. Phys.*, 38:635, 1926.

24 D. R. Inglis and E. Teller. *Astrophys. J.*, 90: 439, 1939.

25 G. Wentzel. *Z. Phys.*, 38:518, 1926.

26 A. Khadjavi, A. Lurio, and W. Happer. Stark effect in the excited states of rb, cs, cd and hg. *Phys. Rev.*, 167:128, 1968.

Light-Matter Interaction: Atoms and Molecules in External Fields and Nonlinear Optics.
W. T. Hill and C. H. Lee

ISBN: 978-3-527-40661-6

27 I. I. Sobelman. *Atomic Spectra and Radiative Transitions*. Springer Series in Chemical Physics. Springer, Berlin, 1979. ISBN 0-387-09082-7.

28 N. Ramsey. *Nuclear Moments*. Wiley, New York, NY, 1953.

29 H. Kopfermann. *Nuclear Moments. English version prepared from the second German edition by E. E. Schneider*. Pure and Applied Physics, Volume 2. Academic, New York, NY, 1958.

30 Landolt-Börnstein. *Zahlenwert und Funktionen*. Springer, Berlin, 1952. Sixth edition, Volume I, Part 5.

31 P. A. M. Dirac. *Proc. R. Soc. Lond. Ser. A*, 112:661, 1926.

32 P. A. M. Dirac. *Proc. R. Soc. Lond. Ser. A*, 114:243, 1927.

33 U. Fano and J. W. Cooper. Spectral distribution of atomic oscillator strengths. *Rev. Mod. Phys.*, 40:441, 1968.

34 W. T. Hill III, J. Sugar, T. B. Lucatorto, and K. T. Cheng. Analysis of the $5p^6 \rightarrow 5p^5 nl\, j = 1)$ rydberg series in ba^{2+}. *Phys. Rev. A*, 36:1200, 1987.

35 W. T. Hill III and C. L. Cromer. Laser-driven ionization and photoabsorption spectroscopy of atomic ions. In W. M. Yen and M. D. Levenson, editors, *Lasers, Spectroscopy and New Ideas: A Tribute to Arthur L. Schawlow*, volume 54 of *Optical Sciences*, page 183, New York, NY, 1987. Springer.

36 U. Fano. *Phys. Rev. A*, 2:353, 1970.

37 K. T. Lu. *Phys. Rev. A*, 4:579, 1971.

38 C.-M. Lee and K. T. Lu. *Phys. Rev. A*, 8: 1241, 1973.

39 U. Fano. Effects of configuration interaction on intensities and phase shifts. *Phys. Rev.*, 124:1866, 1961.

40 F. H. Faisal. *Theory of Multiphoton Processes*. Plenum, New York, NY, 1987. ISBN 0-306-42317-0.

41 P. Agostini and G. Petite. Photoelectric effect under strong irradiation. *Contemp. Phys.*, 29:57, 1988.

42 P. Agostini, F. Fabre, G. Mainfray, G. Petite, and N. K. Rahman. Free–free transitions following six-photon ionization of xenon atoms. *Phys. Rev. Lett.*, 42:1127, 1979.

43 H. Lefevbre-Brion and R. W. Field. *Perturbations in the Spectra of Diatomic Molecules*. Academic, Orlando, FL, 1986. ISBN 0124426905.

44 G. Herzberg. *Molecular Spectra and Molecular Structure I: Spectra of Diatomic Molecules*. Van Nostrand Reinhold, New York, NY, 1950.

45 K. P. Huber and G. Herzberg. *Molecular Spectra and Molecular Structure: Constants of Diatomic Molecules*. Van Nostrand Reinhold, New York, NY, 1978.

46 NIST. Chemical data, d. http://webbook.nist.gov/.

47 J. L. Dunham. *Phys. Rev.*, 41:721, 1932.

48 T. E. Sharp. Potential-energy curves for molecular hydrogen and its ions. *At. Data*, 2:119, 1971.

49 I. Kovács. *Rotational Structure in the Spectra of Diatomic Molecules*. Am. Elsevier, New York, 1969.

50 R. Rydberg. *Z. Phys.*, 80:514, 1933.

51 O. Klein. *Z. Phys.*, 76:221, 1932.

52 A. L. G. Rees. *Proc. Phys. Soc. A, B*, 59: 998, 1947.

53 R. T. Birge and H. Sponer. *Phys. Rev.*, 28: 259, 1926.

54 R. J. LeRoy and R. B. Bernstein. *J. Chem. Phys.*, 52:3869, 1970.

55 W. C. Stwalley. *Chem. Phys. Lett.*, 6:241, 1970.

56 NIST J. T. Hougen. Nbs monograph 115: The calculation of rotational energy levels and rotational line intensities in diatomic molecules. http://physics.nist.gov/Pubs/Mono115/contents.html.

57 M. L. Ginter. Private communication.

58 E. Wigner and E. E. Witmer. *Z. Phys.*, 51: 859, 1928.

59 D. D. Konowalow, M. E. Rosenkrantz, and M. L. Olson. The molecular electronic structure of the lowest $^1\Sigma_g^+$, $^3\Sigma_u^+$, $^1\Sigma_u^+$, $^3\Sigma_g^+$, $^1\Pi_u$, $^1\Pi_g$, $^3\Pi_u$, $^3\Pi_g$ states of Na_2. *J. Chem. Phys.*, 72:2612, 1980.

60 Z. Vager, R. Naaman, and E. P. Kanter. *Science*, 244:426, 1989.

61 J. H. Posthumus, L. J. Frasinski, A. J. Giles, and K. Codling. *J. Phys. B: At. Mol. Opt. Phys.*, 28:L349, 1995.

62 T. Zuo and A. D. Bandrauk. *Phys. Rev. A*, 52:R2511, 1995.

63 T. Seideman, M. Yu. Ivanov, and P. B. Corkum. The role of electron localization in intense-field molecular ionization. *Phys. Rev. Lett.*, 75:2819, 1995.

64 R. S. Mulliken. Intensities of electronic transitions in molecular spectra II: charge-transfer spectra. *J. Chem. Phys.*, 7:20, 1939.

65 S. Chelkowski and A. D. Bandrauk. Two-step coulomb explosions of diatoms in intense laser fields. *J. Phys. B: At. Mol. Opt. Phys.*, 28:L723, 1995.

66 P. A. Franken, A. E. Hill, C. W. Peters, and G. Weinreich. *Phys. Rev. Lett*, 7:118, 1961.

67 R. W. Boyd. *Nonlinear Optics*. Academic, San Diego, CA, 1992.

68 J. A. Armstrong, N. Bloembergen, J. Ducuing, and P. S. Pershan. *Phys. Rev.*, 127:1918, 1962.

69 T. Kuri and K. I. Kitayama. *J. Lightwave Technol.*, 21:3167, 2003.

70 R. Krahenbuhl and W. K. Burns. *IEEE Trans. Microwave Theory Tech.*, 48:860, 2000.

71 L. Xu, X. C. Zhang, and D. H. Auston. *Appl. Phys. Lett.*, 61:1784, 1992.

72 Q. Wu and X. C. Zhang. *Appl. Phys. Lett.*, 67:3523, 1995.

73 J. A. Giordmaine and R. C. Miller. *Phys. Rev. Lett.*, 14:973, 1965.

74 J. E. Bjorkholm and A. Ashkin. *Phys. Rev. Lett.*, 32:129, 1974.

75 L. F. Mollenauer, R. H. Stolen, and J. P. Gorden. *Phys. Rev. Lett.*, 45:1095, 1980.

76 F. Zernike and J. E. Midwinter. In *Applied Nonlinear Optics*, New York, NY, 1973. Wiley.

77 E. P. Ipper and C. V. Shank. In S. L. Shapico, editor, *Ultrashort Light Pulses*, pages 83–122, Berlin and New York, 1977. Springer.

78 P. D. Maker, R. W. Terhune, M. Nisenoff, and C. M. Savage. *Phys. Rev. Lett.*, 8:21, 1962.

79 R. L. Byer and R. L. Herbst. In Y. R. Shen, editor, *Tunable Infrared Generation*, Berlin, 1977. Springer.

80 D. Magde and H. Mahr. *Phys. Rev. Lett*, 18:905, 1967.

81 J. M. Manley and H. E. Rowe. *Proc. IRE*, 47:2115, 1959.

82 J. Warner. *Appl. Phys. Lett.*, 12:222, 1968.

83 J. Shah, T. C. Damen, B. Deve, and D. Block. *Appl. Phys. Lett.*, 50:1307, 1987.

84 N. Bloembergen. *Nonlinear Optics*. World Scientific, Singapore, 4th edition, 1996.

85 Y. R. Shen. *The Principles of Nonlinear Optics*. Wiley, New York, NY, 1984.

86 R. H. Pantell and H. E. Puthoff. *Fundamentals of Quantum Electronics*. Wiley, New York, NY, 1969.

87 M. D. Levenson and N. Bloembergen. Observation of two-photon absorption without doppler broadening on the $3s$–$5s$ transition in sodium vapor. *Phys. Rev. Lett.*, 32:645, 1974.

88 F. Biraben, B. Cagnar, and G. Grynberg. *Phys. Rev. Lett.*, 32:643, 1974.

89 N. Bloembergen, H. Lotem, and Jr R. T. Lynch. *Indian J. Pure Appl. Phys.*, 16:151, 1978.

90 M. D. Levenson and S. S. Kano. *Introduction to Nonlinear Laser Spectroscopy*. Academic, San Diego, CA, 1988.

91 G. P. Agrawal. *Nonlinear Fiber Optics*. Academic, San Diego, CA, 1989.

92 N. Bloembergen and P. S. Pershan. *Phys. Rev.*, 128:606, 1962.

93 N. Bloembergen and C. H. Lee. *Phys. Rev. Lett.*, 19:835, 1967.

94 C. H. Lee and V. Bhanthumnavin. *Opt. Commun.*, 18:326, 1976.

95 H. J. Simon and C. H. Lee. *Opt. Lett.*, 13: 440, 1988.

96 V. Bhanthumnavin and C. H. Lee. *Phys. Rev. A*, 50:2579, 1994a.

97 D. L. Butler, G. L. Burdge, C. H. Lee, and H. J. Simon. *Opt. Lett.*, 17:1125, 1992.

98 C. Ohlhoff, G. Lupke, C. Meyer, and H. Kurz. *Phys. Rev. B*, 55:4596, 1997.

99 N. Bloembergen, H. J. Simon, and C. H. Lee. *Phys. Rev.*, 181:1261, 1969.

100 V. Bhanthumnavin and N. Ampole. *Microwave Opt. Technol. Lett.*, 3:239, 1990.

101 V. Bhanthumnavin and C. H. Lee. *J. Appl. Phys.*, 75:3294, 1994.

102 B. Dick, A. Gierulski, G. Marowski, and G. A. Reider. *Appl. Phys. B*, 38:107, 1985.

103 T. F. Heinz, H. W. K. Tom, and Y. R. Shen. *Phys. Rev. A*, 28:1883, 1983.

104 N. Bloembergen, R. K. Chang, S. S. Jha, and C. H. Lee. *Phys. Rev.*, 174:813, 1968.

105 J. I. Dadap, X. F. Hu, N. M. Russell, J. G. Ekendt, J. K. Lowell, and M. C. Downer. *IEEE J. Sel. Top. Quantum Electron.*, 1:1145, 1995.

106 J. I. Dadap, P. T. Wilson, M. H. Anderson, and M. C. Downer. *Opt. Lett.*, 22:1, 1997.

107 C. H. Lee, R. Chang, and N. Bloembergen. *Phys. Rev. Lett.*, 18:167, 1967.

108 J. I. Dadap, X. F. Hu, M. H. Anderson, M. C. Downer, J. K. Lowell, and O. A. Aktsipetrov. *Phys. Rev. B*, 53:R7607, 1996.

109 I. I. Smolyaninov, A. V. Zayats, and C. C. Davis. *Opt. Lett.*, 22:1592, 1997.

110 I. I. Smolyaninov, H. Y. Liang, C. H. Lee, and C. C. Davis. *J. Appl. Phys.*, 89:206, 2001.

111 J. I. Gersten and A. N. Tzan. *J. Chem. Phys.*, 73:3023, 1980.

112 X. C. Zhang, Y. Jin, and X. F. Ma. *Appl. Phys. Lett.*, 61:2764, 1992.

113 N. M. Froberg, B. B. Hu, X. C. Zhang, and D. H. Auston. *Appl. Phys. Lett.*, 59:3207, 1991.

114 Q. Wu, Ph. D thesis, Rensselaer Polytechnic Institute, 1997.

115 Q. Wu and X. C. Zhang. *Appl. Phys. Lett.*, 68:1604, 1996.

116 Q. Wu and X. C. Zhang. *Appl. Phys. Lett.*, 71:1285, 1997.

117 X. F. Ma and X. C. Zhang. *J. Opt. Soc. Am. B*, 10:1175, 1993.

118 R. S. Van Dyck, P. B. Schwinberg, and H. G. Dehmelt. *Phys. Rev. Lett.*, 59:26, 1987.

119 J. Schwinger. On quantum-electrodynamics and the magnetic moment of the electron. *Phys. Rev.*, 73:416, 1948.

120 V. W. Hughes and T. Kinoshita. *Rev. Mod. Phys.*, 71:S133, 1999.

121 J. Mathews and R. L. Walker. *Mathematical Methods of Physics*. Benjamin Cummings, San Francisco, CA, 1970. ISBN 0-8053-7002-1, Second edition.

122 P. A. M. Dirac. *The Principles of Quantum Mechanics*. Oxford University Press, Oxford, UK, 1981. ISBN 0-19-852011-5, Fourth edition.

123 E. U. Condon and H. Odabaşi. *Atomic Structure*. Cambridge University Press, New York, NY, 1980. ISBN 0-521-21893-8, Fourth edition.

124 R. Courant and D. Hilbert. *Methods of Mathematical Physics, Volume I*. Interscience, New York, NY, 1953. First edition.

125 P. R. Halmos. *Introduction to Hilbert Space and the Theory of Spectral Multiplicity*. Interscience, Chelsea, NY, 1957. ISBN 0828400822, Second edition.

126 A. R. Edmonds. *Angular Momentum in Quantum Mechanics*. Princeton University Press, Princeton, NJ, 1974. ISBN 0-691-07912-9.

Index

Light-Matter Interaction: Atoms and Molecules in External Fields and Nonlinear Optics.
W. T. Hill and C. H. Lee

ISBN: 978-3-527-40661-6

Related Titles

Weiner, J., Ho, P.-T.

Light-Matter Interaction

Volume 1: Fundamentals and Applications

256 pages
2003
Hardcover
ISBN 0-471-25377-4

Demtröder, W.

Molecular Physics

Theoretical Principles and Experimental Methods

484 pages with 288 figures and 43 tables
Softcover
ISBN 3-527-40566-6

Bachor, H.-A., Ralph, T. C.

A Guide to Experiments in Quantum Optics

434 pages with 195 figures
2004
Softcover
ISBN 3-527-40393-0

Cohen-Tannoudji, C., Diu, B., Laloe, F.

Quantum Mechanics

Volume 1

914 pages
1977
Softcover
ISBN 0-471-16433-X

Cohen-Tannoudji, C., Diu, B., Laloe, F.

Quantum Mechanics

Volume 2

640 pages
1977
Softcover
ISBN 0-471-16435-6